Solar Collectors: Theory and Applications

Solar Collectors: Theory and Applications

Edited by Taylor Simpson

SYRAWOOD
PUBLISHING HOUSE
New York

Published by Syrawood Publishing House,
750 Third Avenue, 9th Floor,
New York, NY 10017, USA
www.syrawoodpublishinghouse.com

Solar Collectors: Theory and Applications
Edited by Taylor Simpson

International Standard Book Number: 978-1-64740-435-2 (Hardback)

Cataloging-in-publication Data

Solar collectors : theory and applications / edited by Taylor Simpson.
p. cm.
Includes bibliographical references and index.
ISBN 978-1-64740-435-2
1. Solar collectors. 2. Solar energy. 3. Solar heating. I. Simpson, Taylor.
TJ812 .S65 2023
621.472--dc23

TABLE OF CONTENTS

PREFACE

In my initial years as a student, I used to run to the library at every possible instance to grab a book and learn something new. Books were my primary source of knowledge and I would not have come such a long way without all that I learnt from them. Thus, when I was approached to edit this book; I became understandably nostalgic. It was an absolute honor to be considered worthy of guiding the current generation as well as those to come. I put all my knowledge and hard work into making this book most beneficial for its readers.

A solar collector refers to a type of device that gathers and concentrates solar radiation from the sun. This device is used primarily for solar heating, which enables the heating of water for personal use. These devices are generally a flat box made up of three primary components, including a coolant carrying tube, a transparent cover, and an insulated rear plate. The coating and material type of the solar collector affects the absorption of solar energy. Solar collectors come in a variety of shapes and sizes including evacuated tube collectors, point focus collectors, flat plate collectors, and line focus collectors. This book is a compilation of chapters that discuss the most vital concepts and emerging trends in the study of solar collectors. Most of the topics covered herein provide information on the theory and applications of solar collectors. The book will serve as a valuable source of reference for graduate and postgraduate students.

I wish to thank my publisher for supporting me at every step. I would also like to thank all the authors who have contributed their researches in this book. I hope this book will be a valuable contribution to the progress of the field.

Editor

1

Comparison of Solar Collector Testing Methods—Theory and Practice

Paweł Obstawski *, Tomasz Bakoń and Dariusz Czekalski

Institute of Mechanical Engineering, Department of Fundamental Engineering and Energetic, Warsaw University of Life Sciences, 02-787 Warsaw, Poland; tomasz_bakon@sggw.edu.pl (T.B.); dariusz_czekalski@sggw.edu.pl (D.C.)

* Correspondence: pawel_obstawski@sggw.edu.pl

Abstract: One of the most important problems of operating solar heating systems involves variable efficiency depending on operating conditions. This problem is more pronounced in hybrid energy systems, where a solar installation cooperates with other segments based on conventional carriers of energy or renewable sources of energy. The operating cost of each segment of a hybrid system depends mainly on the resulting efficiency of solar installation. For over 40 years, the procedures of testing solar collectors have been undergoing development, testing, comparison and verification in order to create a procedure that would allow determining the thermal behavior of a solar collector without performing expensive and complicated experimental tests, usually based on the steady state condition. The proper determination of the static and dynamic properties of a solar collector is of key significance, as they constitute a basis for the design of a solar heating installation, as well as a control system. It is therefore important to conduct simulating and operating tests enabling the performance of a comparative analysis intended to indicate the degree to which the static and dynamic properties of a solar collector depend on the method used for their determination. The paper compares the static and dynamic properties of a flat solar collector determined by means of various methods. Based on the produced results, it has been concluded that the static and dynamic properties of a collector determined using various methods may differ from each other even by 50%. This means that it is possible to increase the efficiency of a solar heating installation via the use of an adaptive control algorithm, enabling real-time calculation of the values of characteristic parameters of solar installation, e.g., the time constant under operating conditions.

Keywords: construction of solar thermal collector; equivalent thermal network; thermo-electric analogy; parametric identification; efficiency characteristics; time constant

---

Highlights

- Comparison of solar collector testing methods.
- Static and dynamic properties of the collector determined by different methods may differ from each other by up to 50%.
- Increase in the efficiency of the solar installation possible with an adaptive control system.

1. Introduction

One of the main operational problems encountered in solar heating installations involves their considerably lower efficiency reached under operating conditions compared to that assumed at the engineering stage. Despite the use of various techniques based on artificial neural networks [1], genetic algorithms [2] and other stochastic techniques [3] to optimize the operation of the solar heating system, the problem has still not been satisfactorily resolved. The low efficiency problem occurs very often in hybrid power supply systems, in which a solar system cooperates with other heating and cooling plants in a hybrid energy system, e.g., Solar-Assisted Heat Pumps, Ground-Source Heat Pumps, gas and oil heating. Under operating conditions, the thermal behavior of each component of a system has noticeable influence on the thermal behavior of other components, which affects efficiency in particular segments, and, as a result, the efficiency of the whole system.

The operating cost of each segment of a hybrid system depends mainly on the resulting efficiency of a solar installation. In practice, the lower efficiency of a solar system translates into higher consumption of fossil fuel, which generates higher operating costs. Therefore, experimental work is still being carried out on new collector constructions [4] and the possibility of long-term energy accumulation using phase change materials [5], such as trigeneration systems [6]. It is possible to achieve optimization of the operation of a solar installation (solar system capacity increase) using smooth flow control of the working medium that should be adjusted to the existing operating conditions (solar radiation intensity, temperature of the medium, water temperature in the storage tank) [7]. It is possible to develop the control system of a solar segment in a hybrid installation based on the dynamic properties of the solar collector represented by the time constant determined using a step response [8]. For over 40 years, the procedures of testing solar collectors have been undergoing development, testing, comparison and verification in order to create a procedure that would allow determining the thermal behavior of a solar collector without performing expensive and complicated experimental tests, usually based on the steady state condition. The proper determination of the static and dynamic properties of a solar collector is of key significance, as they constitute a basis for the design of a solar heating installation, as well as a control system. In spite of the continuing development of the concepts of adjusting the operation of a solar heating installation based on the simplest proportional controller [9], proportional and on-off fluid flow control algorithm [10], proportional-derivative control algorithm [11], through continuous adjustment using a Proportional Integral Differential PID regulator [12,13] and ending with complicated fuzzy algorithms [14,15], learning algorithm [16], and artificial neural networks [17] improving the performance of thermal solar collector, the ensuring of required control quality still remains an issue. They probably arise from the fact that the static and dynamic parameters of solar collectors determined in laboratory tests can be different under operating conditions. Although the quality standard ISO 9459-4:2013 [18] was developed to test solar heating systems under operating conditions, it does not allow determining the thermal behavior of a solar collector. It is therefore important to conduct simulating and operating tests enabling the performance of a comparative analysis intended to indicate the degree to which the static and dynamic properties of a solar collector depend on the method used for their determination.

2. An Overview of Test Methods for Flat-Plate Solar Collectors

The basic model of a flat-plate solar collector on which the quality standards Ashrae [18], EN12975-1:2010 [19] and ISO9806:2017 [20] are based includes an equation developed by Hottel, Whilier and Bliss (Table 1), referred to as the HWB equation [21]. It is based on the energy balance of a collector relative to the average temperature of the medium flowing through it. The model combines parameters of the collector that are still considered to be characteristic, e.g., the effective heat capacity $(mc)_e$, the heat transfer coefficient F′ and the heat loss coefficient U_L.

$$(mc)_e \frac{dT_{fo}}{dt} = F'A_c[(\tau\alpha)_e I_T - U_L(T_{fm} - T_A)] + \dot{m}c_p(T_{fi} - T_{fo}) \quad (1)$$

The model and methodology enable the determination of the static and dynamic parameters of a solar collector under indoor and outdoor conditions. The static properties of a collector are represented by efficiency characteristics, which are regarded as the main quality parameter enabling the performance of a comparative analysis of flat-plate solar collectors from different manufacturers. The dynamic properties of a collector are represented by a step response. Due to the fact that the collector testing methodology presented in quality standards [18–20] had adopted highly rigorous conditions, often difficult to fulfill under external conditions, the methodology of determining solar collector parameters and the structure of the HWB equation have been modified by many researchers. Most often, the solar collector was treated by researchers as a homogeneous body, and the model was based on an assumption that the thermal output of the collector in a specific period of time is a sum of the components of the effective temperature increase of the working medium [22–24]. Some researchers divided the collector into segments following the direction of heat flow, preparing an energy balance for each segment [25]. Important introduced modifications also involved the methodology. Main modifications concerned the number of tests—one test [22,25–27], two tests [23, 24,28–30], performed during the collector testing experiment, the flow of agent through the collector during the test; constant flow [22–24,29,30]; variable flow [27,28]; as well as atmospheric conditions effective during the experiment—the value of solar radiation variable during the experiment [22,24–30], the value of the inlet temperature constant [22,24,27], variable [30] during the experiment. All described methods of testing solar collectors were based on a collector model in a differential equation form (Table 1). This form of solar collector models does not allow a full analysis of all parameters describing the dynamic behavior. A solution to this problem involves their conversion to a transfer function form using the Laplace transform. An analysis of the transient states of the collector using the Laplace transform was used both for the HWB equation [21], its numerous modifications [31–36], as well as collector models based on innovative concepts, e.g., piston flow [37,38]. Models belonging to this group are usually used to simulate the operation of existing solar collectors. They treat the solar collector as a homogeneous body and they are based on four parameters that are considered characteristic: thermal capacity, heat removal coefficient, thermal efficiency coefficient and heat loss coefficient.

Much wider possibilities in an analysis of the static and dynamic properties of a collector are provided by methods based on thermo-electric analogy (Table 1). This group of methods is advantageous in its ability to analyze the impact of construction parameters, e.g., geometry of the collector components and materials used in their production, on the static and dynamic properties of the collector already at the design stage. The thermo-electric analogy is based on the assumption that electrical current corresponds to the heat rate, a voltage drop in a reactive element corresponds to the temperature, and electric capacity corresponds to the heat capacity of the element. The method was first presented by Kamminga [39]. In the collector structure, Kamminga identified three homogeneous bodies: the transparent cover, the absorber and the working medium. Therefore, the collector model was developed based on three differential equations, describing the process of heat exchange between particular components of the collector. The model included heat capacity values of the glass cover, the absorber and the working medium, heat resistance between the glass cover and the environment, between the absorber and the transparent cover as well as between the working medium and the absorber. Similar to the HWB equation, the method developed by Kamminga has been modified by numerous researchers.

The modifications primarily involved the heat network construction of a collector model and the methodology of calculating the values of the individual coefficients of the model. The base model developed by Kamminga [39] was approximated by some researchers to two components [40,41], the glass cover being the most frequently omitted component. One of the advantages of the Equivalent Thermal Network ETN method involves the ability to perform highly detailed analyses regarding the construction of the collector, taking multiple factors into account. Also, for this reason, the ETN structure of the collector model can be very complex [42–47], and the simulation studies even considered the sky temperature [47]. Important modification also involved the methodology. Numerous simplifying assumptions were introduced into the ETN model of the collector. The heat capacity of the components was omitted, which resulted in only allowing an analysis of the steady states of the collector, more precisely an analysis of the impact of construction parameters on the efficiency of the collector [40,42,43]. The simplifying assumptions also involved heat exchange between the construction elements of the collector [41], with no consideration of the flow of the working medium [42]; the analyses were performed for a single collector channel only [43], not taking into account transverse heat flow in the collector; the analyses were justified only for a laminar flow of the agent [46]. The complex ETN models enable an analysis of the static and dynamic properties of the collector, while the dynamic properties of the collector could be analyzed both in terms of time and frequency [39].

A certain limitation of the presented methods involved the impossibility to relate the static and dynamic properties of the collector determined at the design stage or under test conditions to operating conditions. It should be noted that the presented methods enable the determination of characteristic parameters of the collector in an idle state (with no heat load), and under operating conditions, the collector bears the load of a reservoir. For this reason, methods have also been developed to enable the modeling of both the collector and the whole solar heating installation based on operating data. Artificial neural networks (ANN) were the most frequently used tool. The ANN enables a capacity analysis and prediction of the output of the working medium of both a solar heating installation [48–54] and a single solar collector [55–59].

An advantage of the ANN involves the possibility to predict selected operating parameters, e.g., the outlet temperature of the medium with forced presets, e.g., the variable intensity of solar radiation. It is also possible to determine the static and dynamic properties of a solar installation under operating conditions by means of the ANN; however, due to the form of the model, analytical design of the control system is considerably limited. Moreover, the main drawback of the ANN involves the necessity to use long strings of time series for network teaching. Methods based on parametric identification are free of that drawback. The basis for the methodology and the application of the parametric identification method used in solar installation diagnostics was developed by Obstawski [59] (Table 1). The developed method has been modified twice [60,61].

The author used the parametric identification method (PI) to determine the dynamic parameters of a solar installation under operating conditions. Based on the measurement of solar radiation and the inlet and outlet temperatures of fluid, it was possible to develop a solar segment model in the form of a differential equation. The solar collector segment was considered a double input and single output object. Therefore, it was possible to perform an impact analysis of solar radiation and inlet temperature variation in response to changes in outlet temperature and the dynamic behavior of a solar collector segment using the inverse Laplace transform. Aleksiejuk et al. [56] tried to develop the concept of solar installation diagnostics under operating conditions presented by Obstawski [55].

The authors tried to compare two types of models: the analog model in an ETN form proposed by Chochowski [40] and the digital model prepared based on the measured solar radiation and the inlet and outlet temperatures of fluid under operating conditions using the parametric identification method. The values of parameters of the analog model in an ETN form, which is a modification of the model, were calculated based on the construction parameters of the collector. Both types of models were of the third order, which made it possible to compare the structure of models. The values of parameters of the models were different. The step characteristics for the modeled solar collector segment calculated for both types of models were also different. This means that the dynamic parameters of a solar segment under operating conditions can be different from those obtained in a design process. It is a serious problem concerning the design of a solar heating installation, as well as an important problem involving the design of a control system. It is therefore important to perform a comparative analysis intended to indicate the degree to which the static and dynamic properties of a solar collector depend on the method used for their determination.

Table 1. Characteristic parameters for each method and conditions for which it is correct.

Item		Collector Model	Characteristic Parameters of the Collector	Testing Conditions	Method of the Equation Coefficient Estimation	Test Stand	Number of Homogeneous Bodies in the Collector Structure
1	HWB [21]	$(mc)_e \frac{dT_{fo}}{dt} = F'A_c[(\tau\alpha)_e I_T - U_L(T_{fm} - T_A)] + \dot{m}c_p(T_{fi} - T_{fo})$	$F'(\tau\alpha)$, $F'U_L$, $(mc)_e$	$\dot{m} = C$, $T_{fi} = C$, $I_T = C$	Method of least squares	Specified in the standards	1
2	Roger's method [22]	$q_u(j) = \sum_{n=1}^{N}[F_R(\tau\alpha)_e K_n I_j(n)] - F_R U_L \overline{(T_f - T_A)}$	$F_R(\tau\alpha)_e$, $F_R U_L$	$\dot{m} = C$, $T_{fi} = C$, $I_T \neq C$, $T_A \neq C$	Linear regression method	In accordance with the standard	1
3	Saunier's method [23]	$(M_a + M_e)c_p \frac{\partial T_{fm}}{\partial t} = \eta_0 A_p I_T - (a_1 + U_a)A_p(T_{fm} - T_A) - a_2 A_p(T_{fm} - T_A)^2 + (P_p + \sigma P_e)$	M_e, η_0, U_1, U_2	$P_e = C$, $=$ $q_u = 0$, $\dot{m} = \textit{large value}$	Linear regression method	Not compliant with the standard	1
4	Filter method [24]	$q_u(j) = \int_0^\infty F_R(\tau\alpha)_e y(t)\, I_T(t-\tau)dt - F_R U_L \overline{(T_f - T_A)}$	$F_R(\tau\alpha)_e$, $F_R U_L$	$\dot{m} = C$, $T_{fi} = C$, $I_T \neq C$, $T_A \neq C$	Method of least squares	In accordance with the standard	1
5	DSC method [25]	$\frac{(mc)_e}{N_c}\frac{dT_{fmp}}{dt} = \frac{F'}{N_c}\left[(\tau\alpha)_e I_e - U_L(T_{fmp} - T_A)\right] - \dot{m}c_p(T_{fmp} - T_{fmp-1})$	$F'(\tau\alpha)_e$, $F'U_L$, $(mc)_e$	$\dot{m} \neq C$, $I_T \neq C$, $T_A \neq C$	Parametric identification	In accordance with the standard	1
6	Excel's method [26]	$\overline{C}[T_{fm}(t_2) - T_{fm}(t_1)] = \int_{t_1}^{t_2}(P_p + \delta P_e + \eta_0 I_T)dt - (a_1 + U_a)A_p \int_{t_1}^{t_2}[T_{fm}(t) - T_A]dt - a_2 A_p \int_{t_1}^{t_2}[T_{fm}(t) - T_A]^2 dt$	η_0, U_1, U_2	$P_e = C =$ $q_u = 0$,	Statistic tool based on successive approximation method, linear regression method and numerical integration method	Not compliant with the standard	1
7	Perers' method [27]	$q_u = F'(\tau\alpha)_e K_{\tau\alpha b}(\Theta) I_b + F'(\tau\alpha)_e K_{\tau\alpha d}(\Theta) I_d - F'a_1\Delta T - F'a_2\Delta T^2 - F'U_3\Delta T_w - F'U_{sky}\Delta T_{sky} - (mc)_e\frac{dT_{fm}}{dt} - U_p\Delta T$	$F'(\tau\alpha)_e$, $K_{\tau\alpha b}$, $K_{\tau\alpha d}$, $F'a_1$, $F'a_2$, $F'U_3$, $F'U_{sky}$, $(mc)_e$	$\dot{m} \neq C$, $T_{fi} = C$, $I_T \neq C$	Method of least squares	In accordance with the standard	1
8	Wijeysundera's method [28]	$(mc)_e\frac{dT_s}{dt} = A_p F_R[(\tau\alpha)_e I_T(t) - U_L(T_{fi} - T_A)] - (AU)_t(T_s - T_A) - (AU)_i \in_i (T_s - T_A) - (AU)_0 \in_0 (T_{f0} - T_A)$	$F_R(\tau\alpha)_e$, $F_R U_L$, (mc),	$\dot{m} \neq C$, $I_T \neq C$, $T_A \neq C$	Using the Levenberg-Marquardt method	Not compliant with the standard	1
9	QDT method [29]	$q_u = F'(\tau\alpha)_e I_T - F'U_L(T_{fm} - T_A) - (mc)_e\frac{dT_{fm}}{dt}$	$F'(\tau\alpha)_e$, $F'U_L$, (mc)	$I_T = \textit{forced}$, $\dot{m} = C$, $T_{fi} = C$, $T_A \neq C$,	Linear regression	In accordance with the standard	1
10	NDM method [30]	$T_{fo}(\tau) = T_0 \exp\left(\frac{F'U_L}{(mc)_e}\tau_d\right) + \sum_{k=0}^{n-1}\left[\frac{F'(\tau\alpha)_e}{(mc)_e}I_T(\tau - k\Delta t) + \frac{F'(\tau\alpha)_e}{(mc)_e}T_A(\tau - k\Delta t)\right]\exp\left(\frac{F'U_L}{(mc)_e}k\Delta t\right)\Delta t$	$F_R(\tau\alpha)_e$, $F_R U_L$, (mc),	$\dot{m} = C$, $T_{fi} \neq C$, $I_T \neq C$, $T_A \neq C$	Linear regression	In accordance with the standard	1
11	Kong [31]	$\frac{d^2T_{fo}}{dt^2} + A\frac{dT_{fo}}{dt} + BT_{fo} = C\frac{dT_{fi}}{dt} + DT_{fi} + E_1 G_b + E_2 G_b + FT_a$ Where: A, B, C, D, E, F - binding equations of parameters of model	$F'(\tau\alpha)_e$, $(mc)_f$, $(mc)_{ab}$	$\dot{m} = C$	Iteration method	-	2

Table 1. *Cont.*

Item		Collector Model	Characteristic Parameters of the Collector	Testing Conditions	Method of the Equation Coefficient Estimation	Test Stand	Number of Homogeneous Bodies in the Collector Structure
12	Amrizal [37]	$(mc)_e \frac{(T_{ix}^{it} - T_{ix}^{it-1})}{\Delta T} =$ $A_P\left[F'(\tau\alpha)_{en}K_\theta(\theta)G^{it} - F'U_L\left(T_{ix}^{it} - T_{ix}^{it-1}\right) - \dot{m}c_p\left(T_{ix}^{it} - T_{ix}^{it-1}\right)\right]$	$F'(\tau\alpha)$, $F'U_L$, $(mc)_e$	$\dot{m} = C$	-	In accordance with the standard	1
13	Buzás [32]	$\overline{T}_{fo}(s) = \frac{T_z}{T_z s+1}T_{fo}(0) + \frac{T_z}{T_z s+1}\frac{A\eta_0}{C_f}I(s) +$ $\frac{T_z}{T_z s+1}\left(\frac{\dot{m}}{V} - \frac{U_L A}{2C_f}\right)\overline{T}_i(s) + \frac{T_z}{T_z s+1}\frac{U_L A}{C}\overline{T}_a(s)$	Tz, C_f,	$\dot{m} = C$	-	-	1
14	Deng [33]	$Q_u =$ $\dot{m} * c_p\left[\frac{1}{B}\frac{d^2T_{fo}}{dt^2} - \frac{A}{B}\frac{dT_{fo}}{dt} + \frac{C}{B}\frac{dT_{fi}}{dt} + \frac{E_1}{B}I_T + \frac{E_2}{B}I_D + \frac{F}{B}(T_{fi} - T_a)\right]$ Where: A, B, C, E, F - binding equations of parameters of model	$F_R(\tau\alpha)_e$, $(mc)_e$, U_L	$\dot{m}= C$	Weighted least squares	-	2
15	Deng [34]	$Q_{u,Dyn}(n+1) = f(n)[Q_u(n) - f(n)]\exp\left(-\frac{\Delta T}{T_z}\right)$	$F_R(\tau\alpha)_e$, $(mc)_e$, U_L	$\dot{m}= C$	Least squares linear regression	In accordance with the standard	1
16	Deng [35]	$Q_{u,Dyn}(n+1) - Q_{u,SST}(n) =$ $\left[Q_{u,Dyn}(n) - Q_{u,SST}(n)\right]\exp\left(-\frac{\Delta T}{T_z}\right)$	$F'(\tau\alpha)_e$, $F'U_L$, $F'(mc)_e$,	$\dot{m}= C$	Multi-linear regression	In accordance with the standard	1
17	Deng [36]	$\dot{m}c_p\left(T_{ix}^{it} - T_{ix}^{it-1}\right) =$ $A_p\left[F'(\tau\alpha)_e K_\theta(\theta)G_{gl}{}^{it} - F'U_L\left(T_{ix}^{it} - T_a^{it}\right)\right]$	$F'(\tau\alpha)_e$, $F'U_L$	$\dot{m} = C$	Nonlinear least square	-	1
18	Kaminga's method [39]	$C_g\frac{\partial T_g}{\partial t}(x,t) = \frac{T_a - T_g(x,t)}{R_{ag}} + \frac{T_p(x,t) - T_g(x,t)}{R_{pg}}$ $C_p\frac{\partial T_p}{\partial t}(x,t) = \frac{T_p(x,t) - T_g(x,t)}{R_{pg}} + \frac{T_f(x,t) - T_p(x,t)}{R_{pf}} + (\tau\alpha)_e I_T(t)$ $C_f\left\{\frac{T_f}{\partial t}(x,t) + u\frac{\partial T_f(x,t)}{\partial x}\right\} = \frac{T_p(x,t) - T_f(x,t)}{R_{pf}}$	C_g, C_p, C_f, R_{ag}, R_{pg}, R_{pf}	$T_a \neq C, E \neq C$	The algorithm for equation coefficient calculations was not provided	Simulation	3
19	Mintsa Do Ango [40]	$R_{rad} = \frac{1}{h_{rad} * S}$, $R_{conv} = \frac{1}{h_{conv} * S}$	R_{rad}, R_{conv}	$h_{rad} \neq C$, $h_{conv} \neq C$	The algorithm for equation coefficient calculations was provided	Simulation	2
20	Tsilingiris [42]	$I_T - U_L(T_P - A) - \frac{k_p}{b}(T_P - T_{pf}) = 0$ $\frac{k_p}{b}(T_P - T_{pf}) - h_f 1(T_{pf} - T_f) = 0$ $h_f 1(T_{pf} - T_f) - q_o - h_f 2(T_f - T_{fb}) = 0$ $h_{f2}(T_f - T_{bf}) - \frac{k_p}{b}(T_{bf} - T_b) = 0$ $\frac{k_p}{b}(T_{bf} - T_b) - U_b(T_b - T_a) = 0$	F', U_L	$h_f 1 \neq C$, $h_f 2 \neq C$, $k \neq C$	The algorithm for equation coefficient calculations was not provided	Simulation	3
21	Eisenmann [43]	$F' = \left[w\left(\frac{1}{(w-e)F+e} + \frac{U_L}{h_{gf}\pi D_i}\right)\right]^{-1}$	F', U_L	$w \neq C, g \neq C$, $D_i \neq C$	The algorithm for equation coefficient calculations was not provided	Simulation	3

Table 1. *Cont.*

Item		Collector Model	Characteristic Parameters of the Collector	Testing Conditions	Method of the Equation Coefficient Estimation	Test Stand	Number of Homogeneous Bodies in the Collector Structure
22	Henning [41]	$C_A \frac{\partial T_A}{\partial t} = (\tau\alpha)G - h_i(T_A - T_f) - U_L(T_A - T_{a,met})$ $C_f\left(\frac{\partial T_f}{\partial t} + \frac{\partial T_{Af}}{\partial z}\right) = h_i(T_A - T_f)$	$(\tau\alpha)_e$, U_L, C_A, C_f	$T_A \neq C$, $h_i \neq C$	The algorithm for equation coefficient calculations was not providedNetwork solution by means of iteration	Simulation	2
23	Chochowski [44]	$(mc)_i \frac{dT_i}{dt} + T_i \sum_{j=1}^{n+1} \Lambda_{ij} - \sum_{j=1}^{n+1} \Lambda_{ij} T_j = P_i + \Lambda_{i0} T_0$	η, $(mc)_e$	$\dot{m} \neq C$, $T_a \neq C$, $I \neq C$, $\Lambda \neq C$	The algorithm for equation coefficient calculations was provided	Simulation	3
24	Farkas [45]	$\Lambda T(\tau) + C\dot{T}(\tau) = Q(\tau)$	(mc), η	$\dot{T} \neq C$, $Q \neq C$	The algorithm for equation coefficient calculations was provided network solved using the Galerkin's method	Simulation	4
25	Cristofari [46]	$CC_i \frac{dT_i}{dt} = U(j)U_{ij}\left(T_j - T_i\right) + \Sigma\ (i)K_i$	(mc), K_i,	$\dot{m}$ = laminar, $T_a \neq C$, $I_T = \neq C$	The algorithm for equation coefficient calculations was provided network solved using the Runge Kutta Merson method	Simulation	6
26	Rodriguez-Hidalgo [47]	$\sum_i m_i c_{pi} \frac{dT_i}{dt} = Q_{up} - Q_{solar} + Q_D + Q_{useful}$	(mc), η	$\dot{m} \neq C$, $T_a \neq C$, $I_T \neq C$	The algorithm for equation coefficient calculations was provided	Simulation	7
27	Obstawski [59]	$G_1(s) = \frac{T_{fout}(s)}{I(s)} = \frac{b_2 s^2 + b_1 s + b_0}{s^3 + a_2 s^2 + a_1 s + a_0}$ $G_2(s) = \frac{T_{fout}(s)}{T_{fin}(s)} = \frac{b_2 s^2 + b_1 s + b_0}{s^3 + a_2 s^2 + a_1 s + a_0}$	T_z, η	$\dot{m} = C$, $T_a \neq C$, $I_T \neq C$	Parametric identification	Diagnostics of a collector heat loaded with a storage tank	1
28	Aleksiejuk [60]	$G_1 = \frac{T_{ab}(s)}{I_T(s)} \cdot \frac{T_{fo}(s)}{T_{ab}(s)}$, $G_2 = \frac{T_{fo}}{T_{in}}$	T_z, η	$\dot{m} = C$, $T_a \neq C$, $I_T \neq C$	Parametric identification	Diagnostics of a collector heat loaded with a storage tank	3
29	Obstawski [61]	$T_{fo}(s) = \frac{\frac{A_c \cdot F_R}{\dot{m}c_p}}{\left(\frac{(mc)_e}{\dot{m}c_p}\right)s+1} \cdot I_T(s) + \frac{\frac{(\dot{m}c_p - U_L K)}{\dot{m}c_p}}{\left(\frac{(mc)_e}{\dot{m}c_p}\right)s+1} \cdot T_{fi}(s)$	F_R, U_L, $(mc)_e$,	$\dot{m} = C$, $T_a \neq C$, $I \neq C$	Parametric identification	Diagnostics of a collector heat loaded with a storage tank	1

3. Methodology

In order to perform a comparative analysis, the static and dynamic properties of a solar collector will be determined using three methods: the steady state method according to the quality standard ISO 9806 under outdoor conditions [20], the ETN method developed by Chochowski [44] and the PI method developed by Obstawski [61]. The produced results shall be compared to the static and dynamic properties determined by the producer according to the quality standard ISO 9806 under indoor conditions and listed in the technical documentation of the collector.

The steady state method according to the quality standard ISO 9806 is based on the HWB Equation (1). The equation is the simplest relation combining instantaneous values of power delivered to the collector with the power convected by the working medium and the heat losses of the collector. The equation allows an analysis of both the steady state and transient states. When analyzing the steady state, it is possible to determine the efficiency characteristics of a collector. During elimination of the product of heat capacity and the time-derivative average temperature of the medium in the collector from the equation, the element of an increase in the internal temperature of the collector is eliminated. In such a case, it is possible to determine the efficiency of the collector, being a ratio of the effective power provided by the collector and the power of solar radiation on the collector aperture surface area Equation (2).

$$\eta = \frac{\dot{m}\, c_p (T_{fi} - T_{fo})}{A_c I_T} = \eta_0 - F'\, U_L T_m \tag{2}$$

When assuming the loss coefficient, non-linearly dependent on the temperature, Equation (2) is in the form of Equation (3), and the relation between the coefficients a_1 and a_2 and the collector heat loss coefficient U_L may be described using Equation (4).

$$\eta = \eta_0 - a_1 T_m - a_2 I_T (T_m)^2 \tag{3}$$

where $T_m = \frac{T_f - T_A}{I_T}$–reduced temperature

$$U_L = a_1 + a_2 (T_f - T_A) \tag{4}$$

This method, recognized as normative, is also referred to as the steady state method, since the determination of coefficients of Equation (1) or its modification Equations (2) and (3) requires the performance of an experiment for constant medium flow through the collector, fixed inlet temperature of the medium in the collector and the fixed value of solar irradiance. This method can be used under indoor and outdoor conditions. In accordance with the standard, the experiment should be performed for solar irradiance exceeding 700 W/m^2 on the collector aperture surface area (usually, this experiment is performed for solar irradiance of 1000 W/m^2). If the test is performed under outdoor conditions, it is required that the diffuse radiation share is lower than 30% in global irradiance, whereas the sun ray incidence angle on the absorber surface does not change by more than 20° during the experiment. The experiment shall be performed for at least four input temperature values, for which it is necessary to determine at least four independent measuring points, resulting in 16 points in total. The temperature shall be evenly distributed within the nominal operating range of the collector as much as possible, whereas the standard recommends that, as far as possible, one value of the medium inlet temperature should be selected so that the average value of the medium temperature in the collector would amount to ±3 K of ambient temperature. In such a case, it is possible to determine the collector zero-loss efficiency η_0. The coefficients a_1 and a_2 are determined statically using the method of least squares for a second-degree polynomial function. If the above conditions are fulfilled, this method allows the determination of efficiency characteristics under both indoor conditions, with the use of a solar radiation simulator, and under outdoor conditions.

The second method used in the determination of the static and dynamic properties of the collector will be the ETN method developed by Chochowski [44]. The flat-plate solar collector was presented in the form of a two-terminal-pair network, and three homogeneous bodies were identified in the collector structure: the transparent cover, the absorber and the working medium presented as active nodes (Figure 1). The method allows an analysis of both the steady and transient states of the collector. In addition, it is possible to simulate the operation of collector batteries that may be interconnected serially or in parallel. Using this method, a heat steam is the equivalent of electrical current, thermal resistances–resistors and temperature distribution–voltage reduction.

Node No. 1, in which energy is emitted in half of the thickness with a power of P1 absorbing part of the solar radiation falling on the plane, corresponds to the glass cover (average temperature of T1).

Node No. 2, in which, as a result of absorbing solar rays passing through the glass cover, thermal energy with a power of P2 is emitted on the surface, represents the absorber (average temperature of T2).

Node No. 3 represents the working medium flowing through the solar collector with average temperature of T3 determined from:

$$T_3 = \frac{1}{2}(T_{fi} + T_{fo}) \quad (5)$$

Thermal streams P flow between the network nodes and the loss center (environment) with thermal resistance R. Thermal resistances present in the diagram (Figure 1) have the following physical sense:

$R_{1,0}$—thermal resistance between the front cover and the environment [K/W]
$R_{1,2}$—thermal resistance between the front cover and the absorber [K/W]
$R_{2,0}$—thermal resistance between the absorber and the environment, calculated towards the bottom of the collector [K/W]
$R_{2,3}$—thermal resistance between the absorber and the working medium [K/W]
$(mc)_1$—thermal capacity of the glass cover [J/K]
$(mc)_2$—thermal capacity of the absorber [J/K]
$(mc)_3$—thermal capacity of the working medium [J/K]

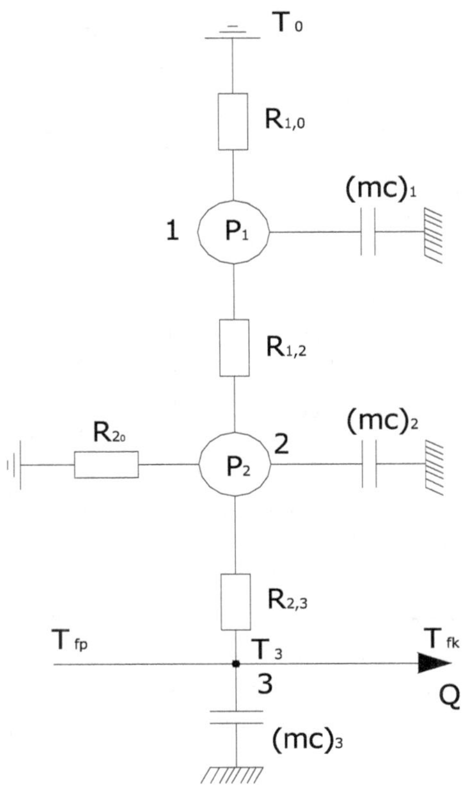

Figure 1. Thermal schema of the collector for a transient state.

The presented thermal network was solved in the form of a system of differential equations combining heat capacity values of specific homogeneous bodies and heat resistance between the particular components of the collector Equation (6).

$$(mc)_i \frac{dT_i}{dt} + T_i \sum_{j=1}^{n+1} \Lambda_{ij} - \sum_{j=1}^{n+1} \Lambda_{ij} T_j = P_i + \Lambda_{i0} T_0 \tag{6}$$

The solution of Equation (6) follows the solving system of linear equations, which takes on the matrix form of Equation (7), where the matrix c Equation (8) is a diagonal matrix of thermal capacities of particular components and the matrix Λ Equation (9) is a matrix of thermal conductivities of particular network nodes.

$$c \frac{d\,\overline{T}(t)}{dt} + \Lambda\,\overline{T}(t) = \overline{P}(t) \tag{7}$$

where:

$$c = \begin{matrix} (mc)_1 & 0 & 0 \\ 0 & (mc)_2 & 0 \\ 0 & 0 & (mc)_3 \end{matrix} \tag{8}$$

$$\Lambda = \begin{bmatrix} \Lambda_{11} & \Lambda_{12} & \Lambda_{13} \\ \Lambda_{21} & \Lambda_{22} & \Lambda_{23} \\ \Lambda_{31} & \Lambda_{32} & \Lambda_{33} \end{bmatrix} \tag{9}$$

By both sides multiplying by the inverse matrix c^{-1}, Equation (6) takes the following form Equation (10):

$$\frac{d\,\overline{T}(t)}{dt} = -\left(c^{-1} \times \Lambda\right) \times \overline{T} + c^{-1} \times \overline{P} \tag{10}$$

The matrix notation Equation (11) obtains as result of substituting matrices Equations (8) and (9) to matrix Equation (10) and can be after transformed into the form of Equation (12).

$$\frac{d}{dt}\begin{bmatrix} T_1 \\ T_2 \\ T_3 \end{bmatrix} = \begin{bmatrix} mc_1 & 0 & 0 \\ 0 & mc_2 & 0 \\ 0 & 0 & mc_3 \end{bmatrix} \times \begin{bmatrix} \Lambda_{11} & \Lambda_{12} & \Lambda_{13} \\ \Lambda_{21} & \Lambda_{22} & \Lambda_{23} \\ \Lambda_{31} & \Lambda_{32} & \Lambda_{33} \end{bmatrix} \times \begin{bmatrix} T_1 \\ T_2 \\ T_3 \end{bmatrix} + \begin{bmatrix} mc_1 & 0 & 0 \\ 0 & mc_2 & 0 \\ 0 & 0 & mc_3 \end{bmatrix} \times \begin{bmatrix} P_1 \\ P_2 \\ P_3 \end{bmatrix} \tag{11}$$

$$\frac{d}{dt}\begin{bmatrix} T_1 \\ T_2 \\ T_3 \end{bmatrix} = \begin{bmatrix} \frac{1}{mc_1} & 0 & 0 \\ 0 & \frac{1}{mc_2} & 0 \\ 0 & 0 & \frac{1}{mc_3} \end{bmatrix} \times \begin{bmatrix} \Lambda_{11} & \Lambda_{12} & \Lambda_{13} \\ \Lambda_{21} & \Lambda_{22} & \Lambda_{23} \\ \Lambda_{31} & \Lambda_{32} & \Lambda_{33} \end{bmatrix} \times \begin{bmatrix} T_1 \\ T_2 \\ T_3 \end{bmatrix} + \begin{bmatrix} \frac{1}{mc_1} & 0 & 0 \\ 0 & \frac{1}{mc_2} & 0 \\ 0 & 0 & \frac{1}{mc_3} \end{bmatrix} \times \begin{bmatrix} P_1 \\ P_2 \\ P_3 \end{bmatrix} \tag{12}$$

The matrix of the system and a matrix of the input Equation (13) is the solution to the matrix of state variables.

$$\frac{d}{dt}\begin{bmatrix} T_1 \\ T_2 \\ T_3 \end{bmatrix} = \begin{bmatrix} \frac{\Lambda_{11}}{mc_1} & \frac{\Lambda_{12}}{mc_1} & \frac{\Lambda_{13}}{mc_1} \\ \frac{\Lambda_{21}}{mc_2} & \frac{\Lambda_{22}}{mc_2} & \frac{\Lambda_{23}}{mc_2} \\ \frac{\Lambda_{31}}{mc_3} & \frac{\Lambda_{32}}{mc_3} & \frac{\Lambda_{33}}{mc_3} \end{bmatrix} \times \begin{bmatrix} T_1(t) \\ T_2(t) \\ T_3(t) \end{bmatrix} + \begin{bmatrix} \frac{P_1}{mc_1} \\ \frac{P_2}{mc_2} \\ \frac{P_3}{mc_3} \end{bmatrix} \times u(t) \tag{13}$$

The matrix of the system expresses the ratio of thermal conductivity of network nodes to their thermal capacity. The individual homogeneous bodies temperatures of a solar collector are variables of the condition, while the input column matrix is the ratio of the heat output of a node to its thermal capacity. The presented method allows not only an analysis of transient states of the temperature values of a particular component, but also the heat flow distribution between them. In addition, it is possible to simulate the operation of collector batteries that may be interconnected serially or in parallel. The document also provides an accurate algorithm enabling calculation of particular coefficient values of the differential equation.

The third method used in the determination of the static and dynamic properties of the collector will be one based on parametric identification developed by Obstawski [61]. The diagnostics method for a solar heating system allows the determination of static properties under operating conditions, i.e., the heat loss coefficient U_L, the collector efficiency coefficient F′, the heat removal factor F_R, and heat capacity values j $(mc)_e$, as well as dynamic properties: gain kp, the time constant T_c of the collector. The calculation method is based on two equivalent models: one based on a database using parametric analysis methods, and the other based on an energy balance combining the characteristic parameters of a collector (modification of the HWB equation). The model in the form of a differential Equation (12), obtained as a result of parametric identification, is converted by using discrete Laplace transform into discrete operator transmittance Equation (14), and then, by means of relation Equation (16), into continuous transmittance. Likewise, the analytical model Equation (13) can be converted into equivalent operator transmittance Equation (17) by using the Laplace transform Equation (15). Comparison of models adequately Equations (14)–(19) makes it possible to determine the characteristic parameters of the diagnosed system (Figure 2). The diagnostics method for a solar collector system enables the determination of its static and dynamic properties based on operating data, which makes it versatile.

$$A(z)y(t) = B(z)u(t-n_k) + e(t) \tag{14}$$

$$\frac{dT_{out}(t)}{dt} = \frac{I(t) \times A_C \times F_R - \dot{m}\, c_p \times (T_{out}(t) - T_{in}(t)) - U_{LK} \times (T_{in}(t) - T_{amb}(t))}{(mc)_{eK}} \tag{15}$$

$$G(s) = \frac{kp_1}{T_{C1}s+1} \times \frac{kp_2}{T_{C2}s+1} \times \ldots \times \frac{kp_n}{T_{Cn}s+1} \tag{16}$$

$$T_{out}(s) = \frac{\frac{A_c \cdot F_R}{\dot{m}\, c_p}}{\left(\frac{(mc)_{eK}}{\dot{m}\, c_p}\right)s+1} \times I(s) + \frac{\frac{(\dot{m}\, c_p - U_{LK})}{\dot{m}\, c_p}}{\left(\frac{(mc)_{eK}}{\dot{m}\, c_p}\right)s+1} \times T_{in}(s) + \frac{\frac{U_{LK}}{\dot{m}\, c_p}}{\left(\frac{(mc)_{eK}}{\dot{m}\, c_p}\right)s+1} \times T_{amb}(s) \tag{17}$$

$$T_{out}(s) = \frac{kp_1}{T_C(s)+1} \times I(s) + \frac{kp_2}{T_C(s)+1} \times T_{in}(s) \tag{18}$$

$$T_{out}(s) = \frac{\frac{A_c \cdot F_R}{\dot{m}\, c_p}}{\left(\frac{(mc)_{eK}}{\dot{m}\, c_p}\right)s+1} \times I(s) + \frac{\frac{(\dot{m}\, c_p - U_{LK})}{\dot{m}\, c_p}}{\left(\frac{(mc)_{eK}}{\dot{m}\, c_p}\right)s+1} \times T_{in}(s) \tag{19}$$

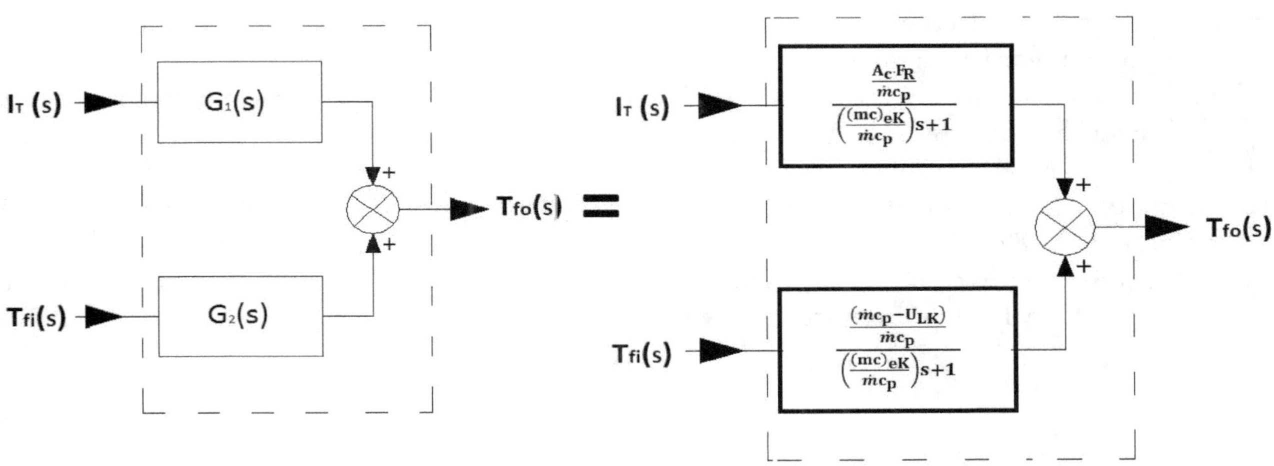

Figure 2. Block diagram of the solar segment model developed based on energy balance for a transient state.

The only requirement is the maintenance of constant medium flow rate in the system during the experiment. Thus, it is possible to determine the static and dynamic properties of the collector under operating conditions, and to specify the impact of solar irradiance variation and collector heat load on its properties.

4. Comparison of Characteristic Parameter Values Determined for a Collector Using Different Methods

A flat-plate solar collector is a non-fixed object operating in transient states [41]. In consequence, both the static and dynamic properties of the flat-plate solar collector are variable and depend on operating conditions and heat load [61]. In hybrid energy installations, cooperation between segments of the system has the most significant impact on the behavior of the solar collector. Thus, when designing the control system of a solar heating installation, it is very important to know how and within what limits the values of the characteristic parameters of the collector that describe its static and dynamic properties change, starting from the design process, through standard tests under indoor and outdoor conditions compliant with the ISO 9806:2017 standard [20], and heat load with a hot water storage tank under operating conditions. Making a mistake in the design process for controlling a solar installation in a hybrid energy system will decrease efficiency of the solar segment and can cause instability of the solar segment under certain operating conditions.

The presented groups of methods allow an analysis of the variability of characteristic parameter values. An example of an analysis will be presented for a flat-plate collector whose static properties, represented by efficiency characteristics, as well as dynamic properties—the time constant substitute—were determined under indoor conditions in accordance with the ISO 9806:2017 [20] standard (Table 2), and will be regarded as model. Therefore, the particular tests were performed in accordance with the standard for a volume flow of the working medium amounting to 0.02 dm^3/min per square meter of the tested collector aperture. The presented analyses of efficiency characteristics and step response enable the time constant substitute and the effective heat capacity of the tested collector to be determined using measurement data from two test stands. The first stand allows the determination of efficiency characteristics and time constant substitute of the collector under outdoor conditions in accordance with the ISO 9806:2017 [20] standard, and enables performance testing of heat loaded with a 100 dm^3 storage tank with one or two collectors connected in parallel (Figure 3). The operating modes of the installation can be changed by setting the position of electro valves. The measurement of fluid temperature and ambient temperature is performed by PT1000 sensors in a four-transverse system.

The sensors are made with class A accuracy. Sensors that measure fluid temperature are located in special capillaries soldered inside the pipes. Flow rate is measured by a turbine flow meter working in a current loop. Solar irradiance is measured by a class 1pyranometer. Fluid pressure is measured by a sensor with a 0–10 bar range and a 0–10 VDC voltage output. The hydraulic installation uses a circulation pump with the ability to continuously adjust flow rate, using a 0–10 VDC external voltage signal. All measurement components are integrated into one measurement system based on a Programmable Logic Controller (PLC), where the control algorithm of installation operation has been implemented. The PLC is connected using an Ethernet protocol to the Supervisory Control and Data Acquisition (SCADA) system with visualization and measured data acquisition (Figure 4). The measured physical values are recorded with a sampling period of 1 s.

Table 2. Characteristic parameters of the tested collector according to iso 9806 indoor conditions.

Normal Characteristic	Effective Heat Capacity [kJ/m²K]	Time Constant Substitute [s]	Collector Area [m²]	Aperture Area [m²]	Weight [kg]	Absorber Material	Glass Thickness [mm]	Insulation Thickness [mm]	Maximum Pressure [bar]	Volume [l]
$\eta = 0.79 - 4.17T_m - 0.01T_m^2$	6.73	80	2.0	1.76	43	Aluminum	4.0	Basalt 40	6	1.3

Figure 3. Test stand (system I).

The second test stand is a hybrid energy system based on the cooperation of two solar heating installations (flat and vacuum solar collectors) and a ground heat pump with a capacity of about 10.5 kW (Figure 5) [7]. The solar heating installation based on flat solar collectors is composed of four sections connected in parallel, consisting of five collectors each. The total area of aperture of the flat-plate collectors is 40 m^2. This section of the hybrid system is thermally loaded with a buffer tank about 1000 L in volume. A buffer tank about 300 L in volume provides thermal load for vacuum solar collectors with an aperture area of 6 m^2. The priority function of heating domestic hot water in this system is served by a ground heat pump. The nominal minimum temperature of domestic hot water in the buffer tank is 40 °C and it is the initial operating temperature value for flat and vacuum collectors. The lower priority source of energy for the heat pump involves a ground heat exchanger. An alternative source of energy includes a buffer tank about 2000 L in volume, which is connected to the flat-plate collector via a plate heat exchanger. Such a hydraulic construction of the hybrid installation makes it possible to accumulate energy surplus from the flat-plate collectors in the buffer tank and protect the solar installation against preheating. The energy accumulated in this buffer tank is used to regenerate the ground heat exchanger. The PT1000 class A temperature sensor in a four-transverse system has been located in a measurement capillary soldered inside the pipes. The flow rate in every segment of the system is measured by pulse flow meters with a resolution of 1 impulse per 1 L. An external voltage signal within a range of 0–10 VDC makes it possible to change the flow rate. Solar irradiance is measured by a class 1 pyranometer. The visualization of the system control unit and the acquisition of the measured data of test stand II (Figure 6) are similar to test stand I described above.

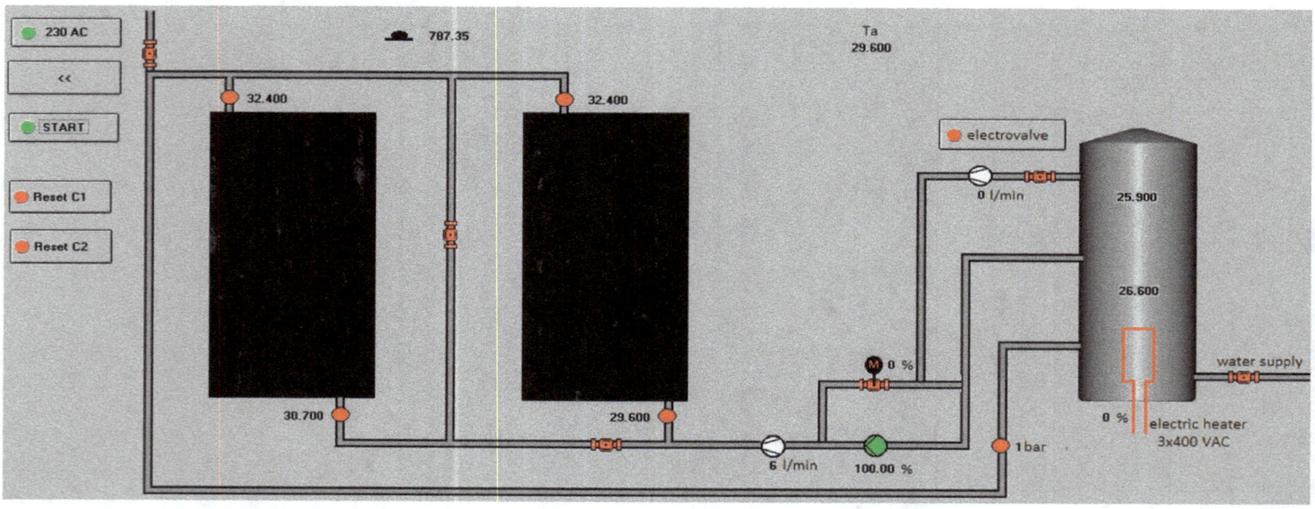

Figure 4. Visualization of work of test stand I in the Supervisory Control and Data Acquisition (SCADA) system.

Figure 5. Test stand (system II).

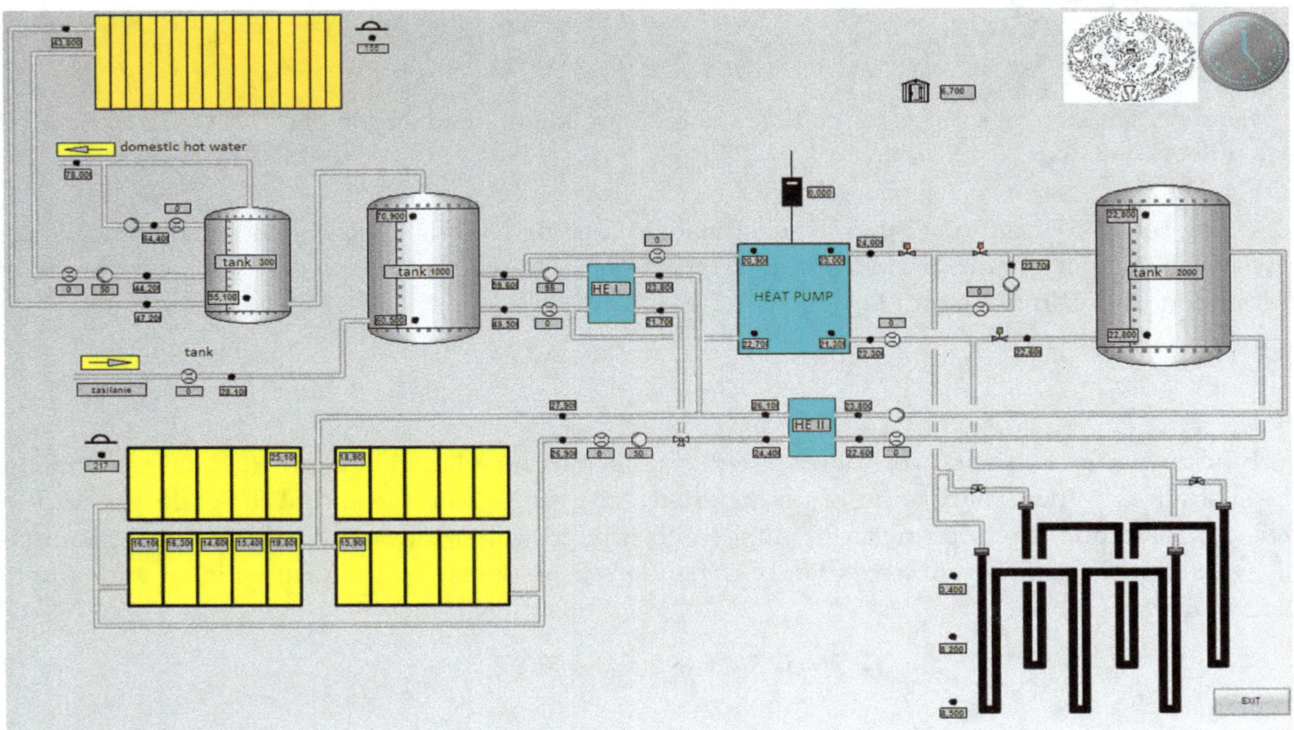

Figure 6. Visualization of work of test stand II in the SCADA system.

4.1. Analysis of Static and Dynamic Properties According to ISO 9806 under Outdoor Conditions

Figure 7 presents the step response of the tested collector determined based on an experiment performed according to ISO 9806:2017 [20] outdoor conditions in test stand I. The characteristics were determined for an inlet temperature of the working medium of 20.5 °C, an ambient temperature of 20 °C, solar irradiance of 806 W/m^2 ± 6 W/m^2 and a flow rate of 2.16 dm^3/min, as per the standard. In accordance with the methodology described in the ISO 9806:2017 standard [20], a time constant substitute of 171 s was read from the step response.

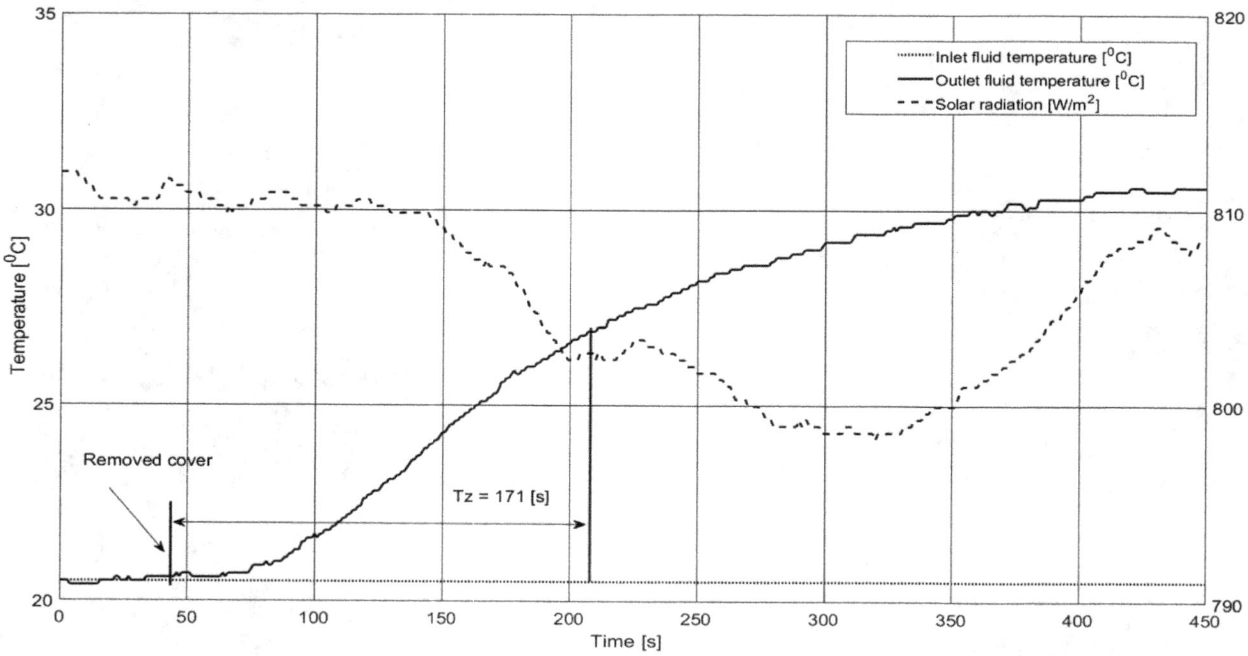

Figure 7. The step response of the collector determined according to the ISO 9806:17 standard under outdoor conditions.

Using relation Equation (20) [57], effective heat capacity of 14.3 kJ/m^{2}K was calculated. Figure 8 shows the efficiency characteristics of the tested collector developed based on data measured under outdoor conditions in test stand I.

$$(mc)_e = \frac{\dot{m} \times c_p \times T_z}{A_p} \tag{20}$$

Efficiency characteristics were determined described using relation Equation (21) using the method of least squares. When analyzing the characteristics, it should be noticed that the collector zero-loss efficiency is slightly lower than the efficiency determined under indoor conditions and amounts to 77%, whereas the coefficient of temperature-dependent heat losses is significantly higher and amounts to 54.8 W/m^{2}K^2 (Table 2).

$$\eta = 0.77 - 4.92\ T_m - 54.8{T_m}^2 \tag{21}$$

Using relations Equations (2) and (4), the values of F′ and U_L were calculated in relation to ambient temperature (Table 2). The value of F_R was calculated using relation Equation (22).

$$F_R = \frac{\dot{m}\ c_p}{A_p U_L}\left[1 - \exp\left(-\frac{A_p U_L F'}{\dot{m}\ c_p}\right)\right] \tag{22}$$

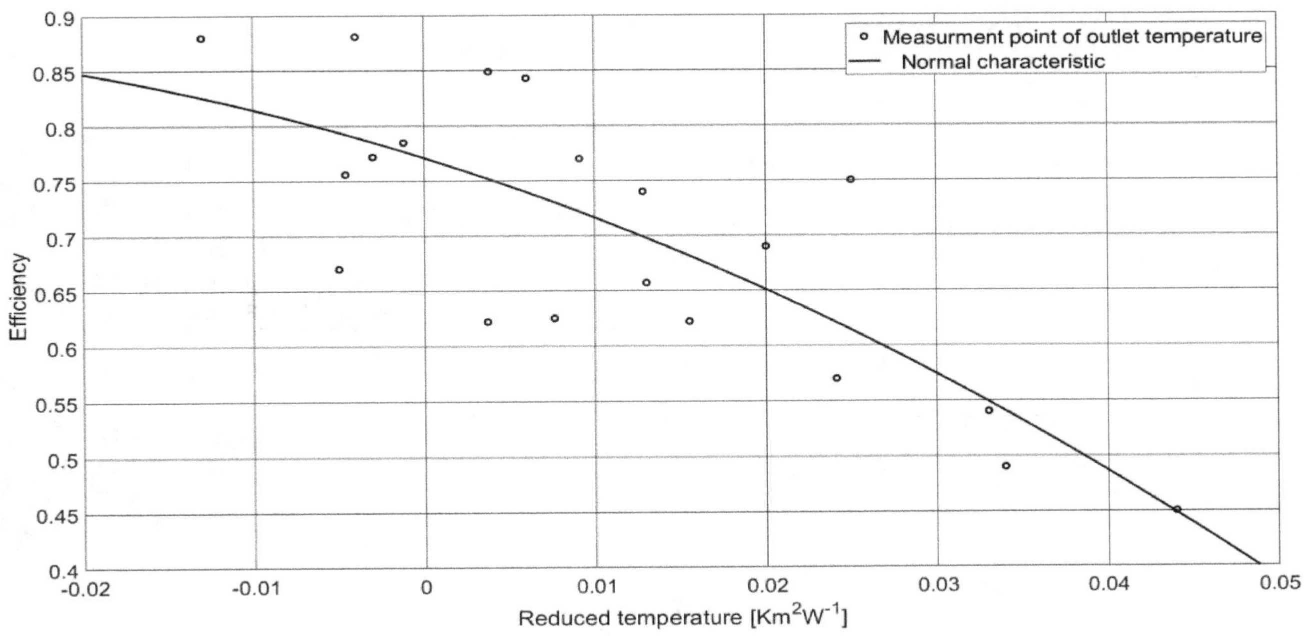

Figure 8. Collector efficiency characteristics determined according to the ISO 9806 standard under outdoor conditions.

4.2. *Analysis of Static and Dynamic Properties Using the Equivalent Thermal Network (ETN)*

The ETN method developed by Chochowski [44] was used to analyze the static and dynamic properties of the tested solar collector during the design process. The specific values of heat resistance were calculated based on the collector material data. The thermal network was implemented in Simulink (Figure 9). The simulation was performed in accordance with the standard for a flow rate of 2.16 dm^3/min. Figure 10 shows the step response of the tested collector, prepared for a solar irradiance value of 1000 W/m^2 and a working medium inlet temperature of 20 °C. In contrast to other methods, the ETN method allows an analysis of the impact of structural parameters on the energy flow and temperature distribution between specific homogeneous bodies of the collector. Therefore, it was possible to develop a step response for the glass cover, the absorber and the working medium.

When analyzing the step response family (Figure 10), it should be noticed that the absorber is characterized by the shortest time constant of 30 s, and the transparent glass cover is characterized by the longest time constant of 86 s. The working medium time constant is 77 s. When comparing efficiency characteristics determined using the thermal network method (Figure 11) and efficiency characteristics determined according to the standard under indoor and outdoor conditions, attention should be paid to the significantly reduced zero-loss efficiency value of 0.71 and the significantly reduced coefficient of temperature-dependent heat losses of 0.0001 W/m^2K^2 (Table 2). This coefficient value probably results from the linear character of the collector model. Using relations Equations (2), (4) and (22), the values of F′, F_R and U_L were calculated in relation to ambient temperature (Table 2).

$$\eta = 0.71 - 4.07T_m - 0.0001{T_m}^2 \quad (23)$$

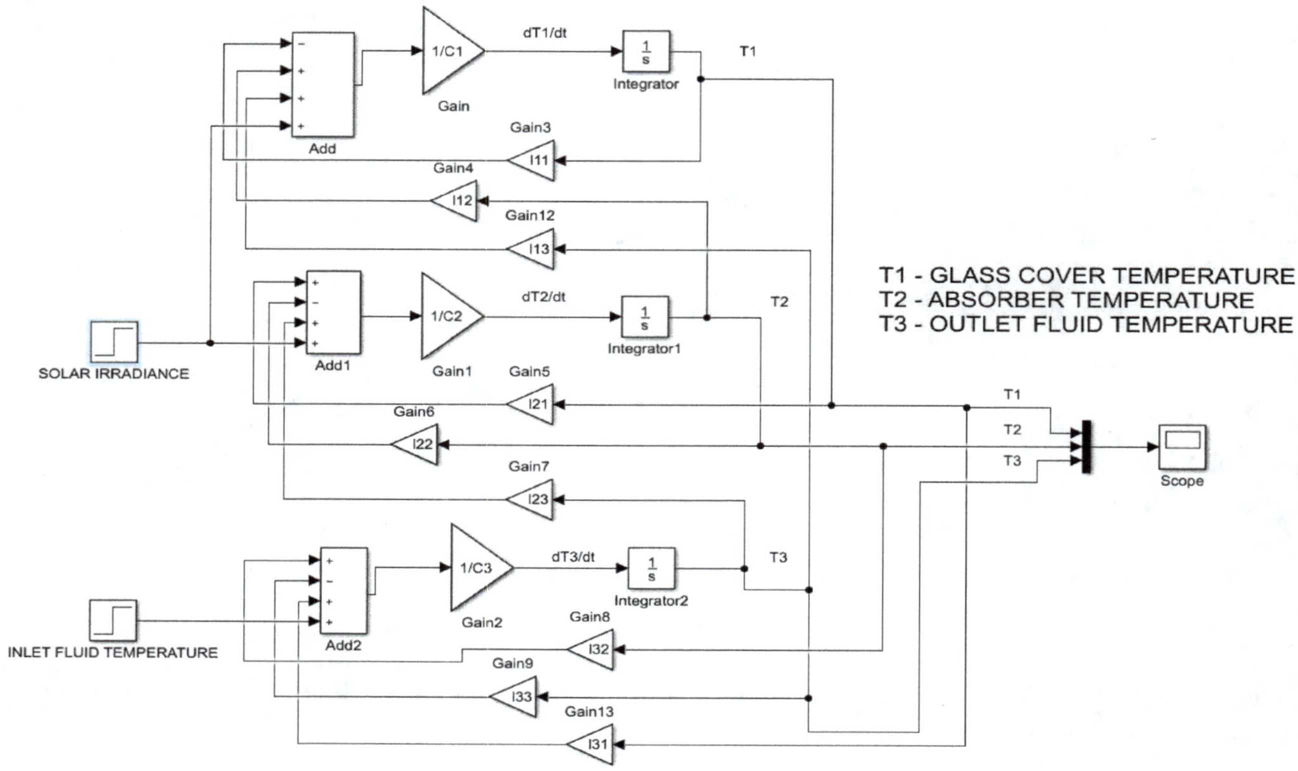

Figure 9. The equivalent thermal network implemented in Simulink.

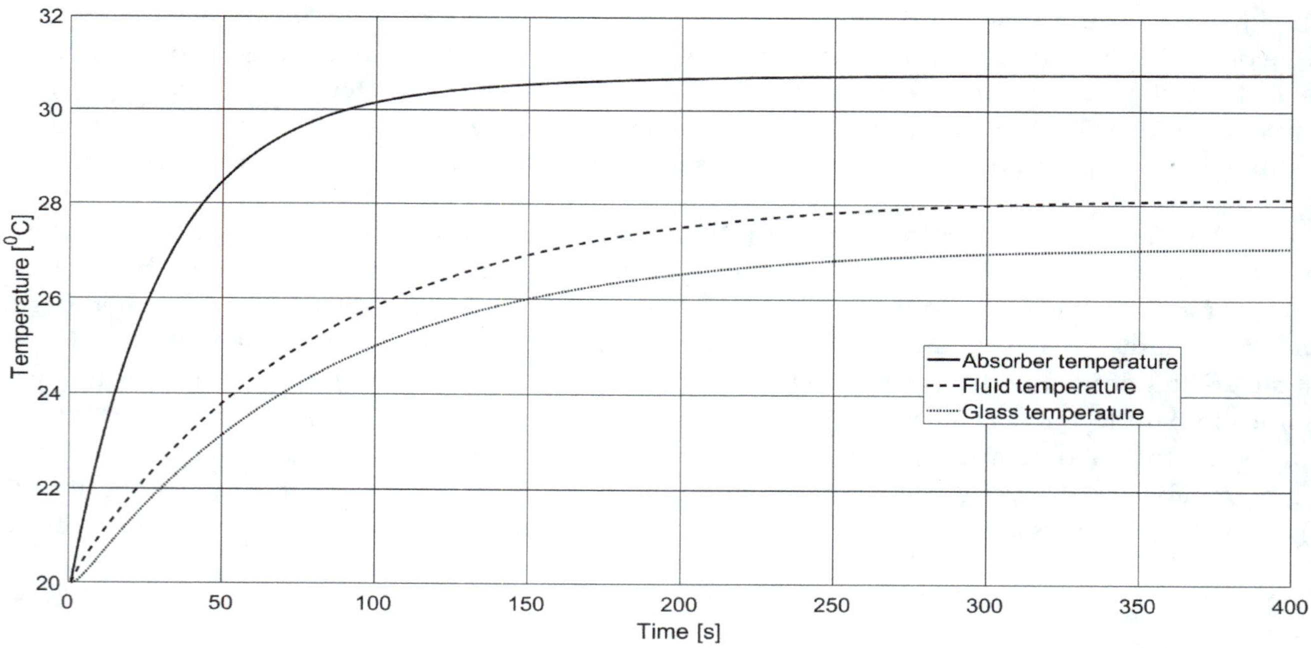

Figure 10. The step response of the collector determined using the ETN method according to Chochowski.

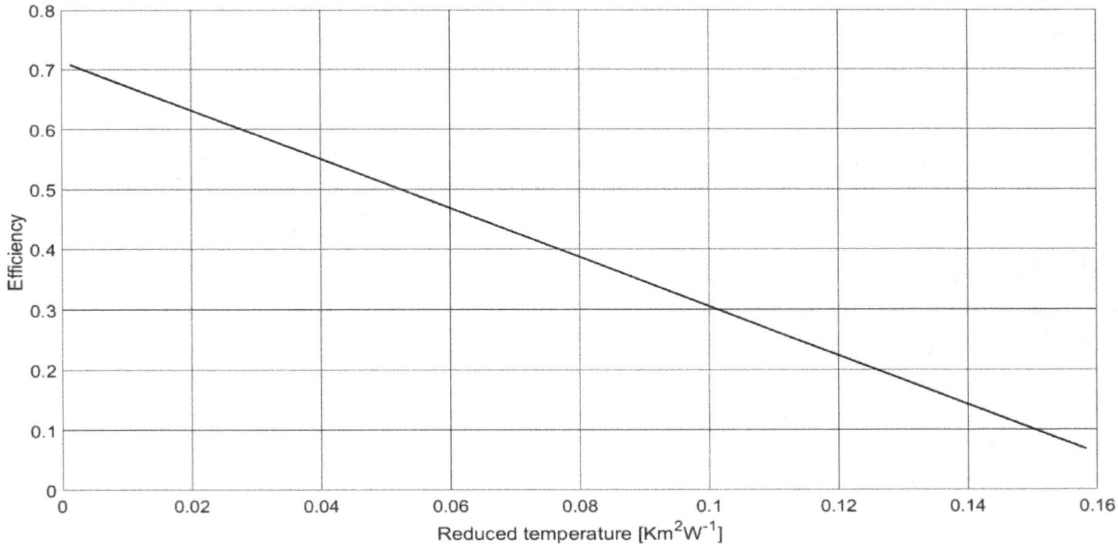

Figure 11. Efficiency characteristics of the collector determined using the ETN method.

4.3. *Analysis of Static and Dynamic Properties of the Collector Using the Parametric Identification Method for Heat Load with a Hot Water Storage Tank*

As already mentioned, the solar collector is a non-fixed object, whose characteristic parameters depend on the operating conditions. The determination of characteristic parameters under operating conditions is possible with the use of the method developed by Obstawski et al. [59] classified into the third group of methods, based on parametric identification. The method allows the determination of the static and dynamic properties of the collector based on the recorded time series of working medium temperature and solar irradiance. It is thus possible to compare the characteristic parameters of the collector assumed at the design stage to those determined during standard tests and under operating conditions. The analysis will be presented based on an example of the operation of two solar systems:

(a) two collectors loaded with a 100 dm^3 storage tank (test stand I)
(b) a battery of 20 collectors with a total absorber area of 40 m^2, loaded with a 1000 dm^3 storage tank (test stand II).

In the case of system I, the volume flow of the medium through a single collector is compliant with the ISO 9806 standard and amounts to 0.02 kg/m^2 per the absorber area. In the case of system II, the experiment was performed for a flow rate 50% lower than the standard value. The applied method enables the determination of collector parameters for pre-flow in accordance with the normative method.

4.3.1. Collector Parameters Determined for a 100 dm^3 Load (Test Stand I)

Figure 12 shows the values of solar irradiance and temperature of the medium during the experiment. It should be noticed that at the beginning of the experiment, variable weather conditions prevailed. Solar irradiance was within a range of 400–950 W/m^2. After 6000 s, solar conditions stabilized and the solar irradiance value reached 950 W/m^2. In the last phase of the test, variable weather conditions appeared again. For a constant volume flow rate of the working medium in the system, amounting to 4.0 dm^3/min, the working medium temperature increased from 30 to 78 °C. Based on the recorded measurement data using the parametric identification method, a model was created in the form of differential Equation (24), which was verified and then transformed, using Tustin's method, into a continuous one, and was approximated to an operator transform form described by the general Equation (25).

$$\begin{aligned} A(z) &= 1 - 0.857z^{-1} - 0.09157z^{-2} - 0.04576z^{-3} \\ B_1(z) &= 5.341e^{-5}z^{-1} \\ B_2(z) &= 0.0053889z^{-1} \end{aligned} \tag{24}$$

$$T_{fo}(s) = \frac{0.0094}{207.9s + 1} \times I_T(s) + \frac{0.949}{207.9s + 1} \times T_{fi}(s) \quad (25)$$

Having the gain coefficient $kp_1 = 0.0094$ and $kp_2 = 0.949$ for the initial temperature of the medium within a system operation range of 30–75 °C and solar irradiance of 970 W/m^2, the efficiency characteristics (Figure 13) was determined as described by Equation (26). Attention should be paid to the significantly reduced value of the zero-loss efficiency coefficient and the heat loss coefficient (Table 2). Figure 14 presents the step response of the tested collector. The heat loaded collector is characterized by a significantly higher time constant of 207.9 s, which means that the solar collector heat capacity under the load will be significantly higher than in a state with no load. The values of characteristic parameters are determined using relation Equation (19) and the model presented in Table 1 (item 30).

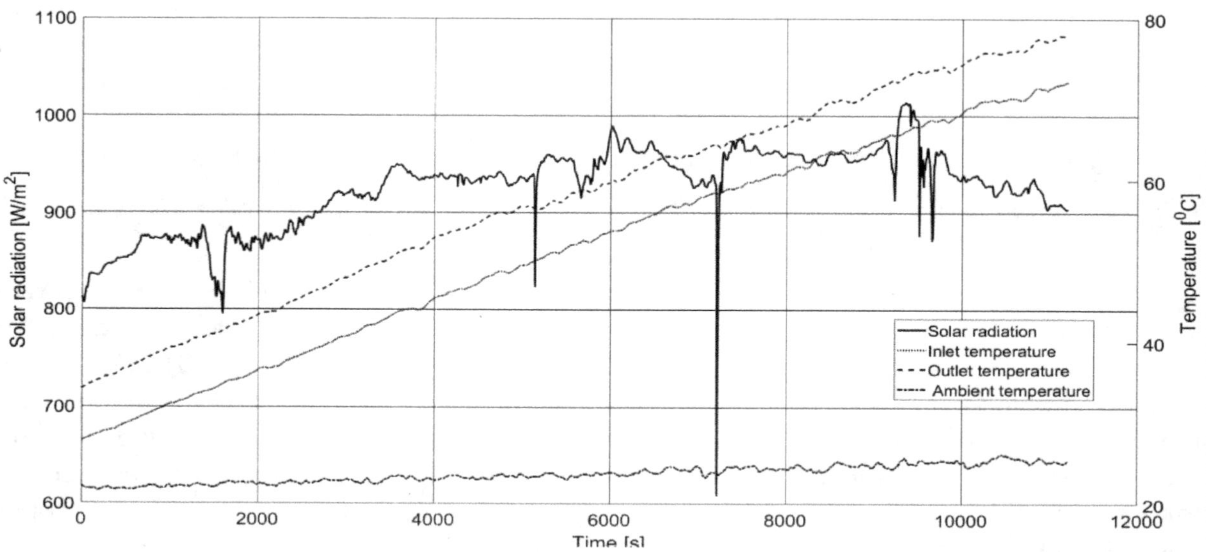

Figure 12. Solar irradiance and working medium temperature values in the system during the experiment.

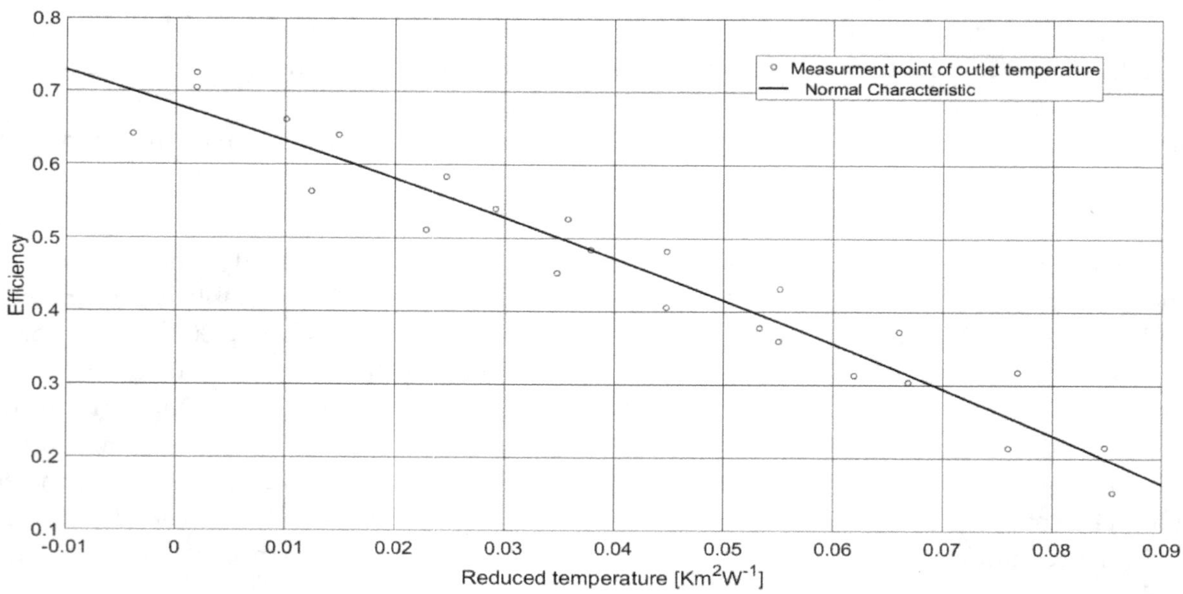

Figure 13. Efficiency characteristics of the collector heat loaded with a 100 dm^3 storage tank.

$$\eta = 0.68 - 4.8T_m - 10.0T_m{}^2 \quad (26)$$

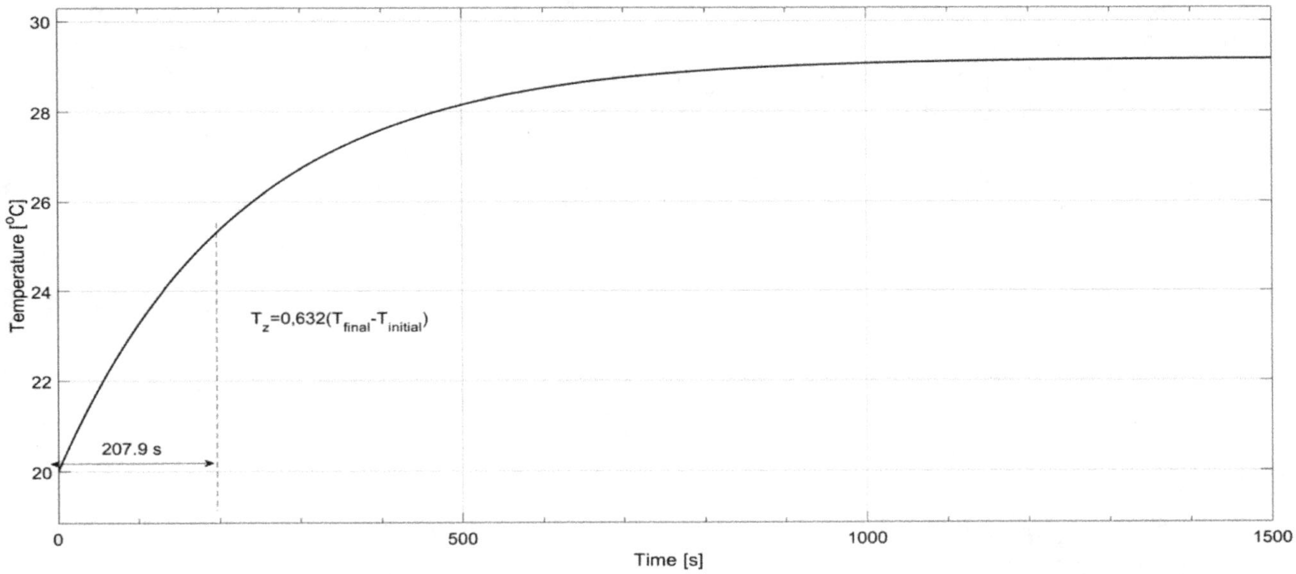

Figure 14. Step response of the collector heat loaded with a 100 dm^3 storage tank.

4.3.2. Collector Parameters Determined for a 1000 dm^3 Load (Test Stand II)

The method developed by Obstawski enables diagnostics of solar collectors working in large installations, such as the one presented in Figure 5. In practice, for such heating systems, the volume flow through the collector results from numerous factors, e.g., from the method of collector interconnections, and it is usually different from the standard flow rate. This method enables the values of characteristic parameters that were calculated for flow rate under operating conditions to be recalculated using the flow rate value according to the ISO 9806:2017 standard [6]. The experiment was performed for a constant volume flow rate in the system which equaled 0.41 dm^3/s. When assuming a uniform flow of medium through specific sections, it may be estimated that the flow of medium through one collector amounted to 0.0208 dm^3/s, thus being nearly 50% lower than during standard tests. Figure 15 shows the values of solar irradiance and the temperature of medium during the experiment. Due to the fact that the solar heating system is a segment of a hybrid power supply system based on a compressor heat pump, the automatic control system allows effective operation of the system once a temperature of 40 °C is exceeded in the hot water storage tank, which is the initial temperature of the working medium.

Based on the recorded measurement data using the parametric identification method, a model was created in the form of differential Equation (27), which was verified and then transformed, using Tustin's method, into a continuous one, and it was approximated to an operator transform form described by Equation (28).

$$\begin{aligned} A(z) &= 1 - 0.772\ z^{-1} - 0.2132\ z^{-2} + 0.07099\ z^{-3} \\ B1(z) &= 0.00264\ z^{-1} - 0.0006327\ z^{-2} \\ B2(z) &= 0.1621\ z^{-1} - 0.09136\ z^{-2} \end{aligned} \tag{27}$$

$$T_{fo}(s) = \frac{0.0154}{298.35s + 1} \times I_T(s) + \frac{0.947}{298.35s + 1} \times T_{fi}(s) \tag{28}$$

Having the values of the gain coefficient and the time constant substitute, the characteristic parameters of the collector, collector efficiency characteristics Equation (29) and the time constant were determined (Table 2).

$$\eta = 0.417 - 1.533\ T_m - 4.36\ {T_m}^2 \tag{29}$$

However, it should be noticed that the flow of medium through the collector in the modeled battery was two times lower than during standard tests. Knowing the value of the heat removal coefficient F_R, the coefficient of heat losses U_L, the battery aperture area A_C and the specific heat of the medium, by means of the equations (Table 1, item 30), it is possible to determine the efficiency characteristics of the collector batteries for the standard flow rate, which, for the analyzed collector battery, is 0.704 dm^3/s (Figure 16). By increasing the flow rate up to the value from the ISO 9806:2017 [6] test standard, the modeled collector can be described using the transfer function Equation (30).

$$T_{fo}(s) = \frac{0.0077}{149.16s + 1} \times I_T(s) + \frac{0.97}{149.16s + 1} \times T_{fi}(s) \quad (30)$$

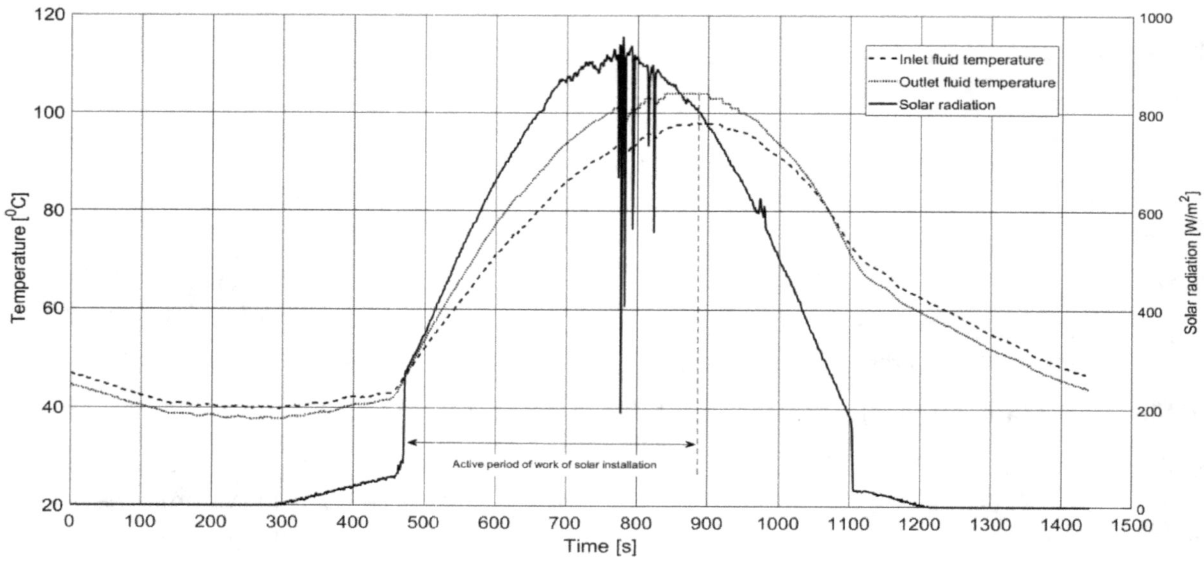

Figure 15. Solar irradiance and working medium temperature values in the system during the experiment (test stand II).

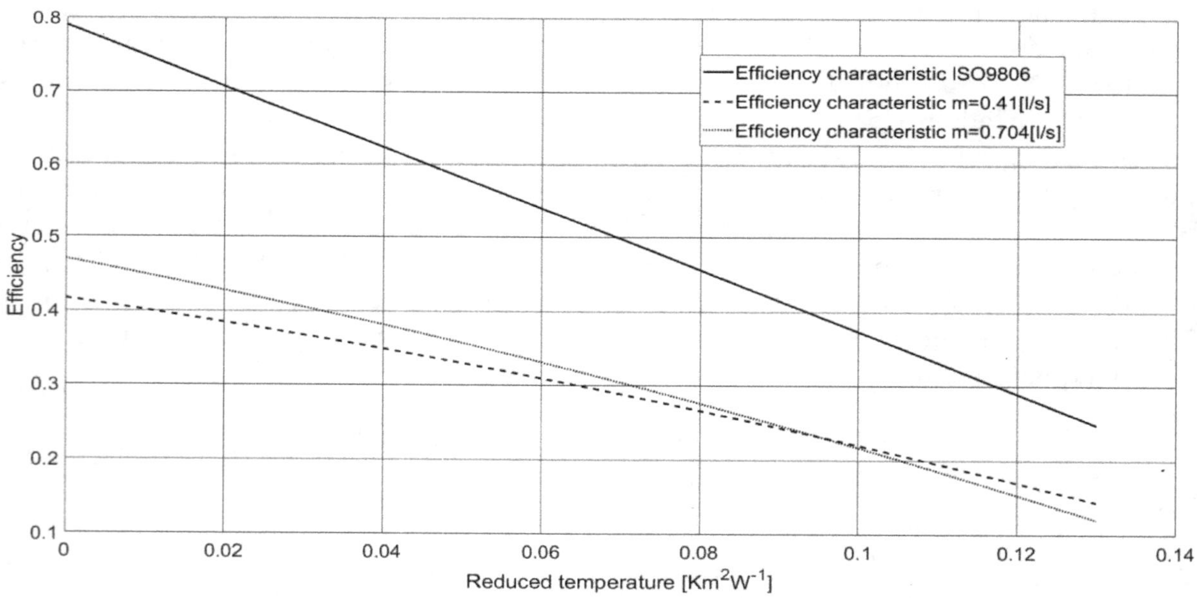

Figure 16. Efficiency characteristics of a battery of 20 collectors heat loaded with a 1000 dm^3 storage tank.

Figure 17 shows step characteristics of the modeled solar collector. The flow rate of fluid impacts the value of the time constant. An increase in the flow rate of fluid decreases the value of the time constant and increases the value of solar collector gain, which can be observed when analyzing collector model parameters after calibration (an increase in the flow rate of fluid) (Figure 16) (Table 2).

$$\eta = 0.47 - 2.001\ T_m - 5.289\ T_m^2 \quad (31)$$

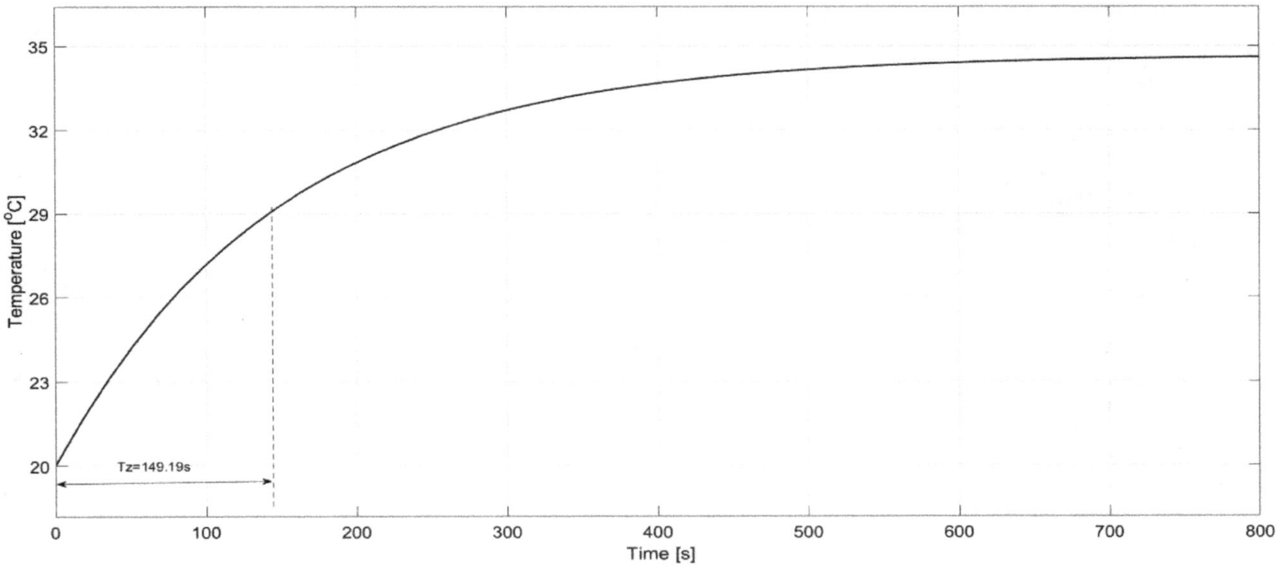

Figure 17. Step response of a battery of 20 collectors heat loaded with a 1000 dm^3 storage tank.

5. Analysis of the Results and Discussion

When analyzing the parameters of a flat solar collector that has been tested using different methods, it should be noticed that the values of specific parameters are different. The biggest differences in the values of solar collector efficiency occur when the collector is thermally loaded with a large buffer tank (test stand II). A difference in the value of collector efficiency between an indoor test performed in accordance with the ISO 9806 standard and the parametric identification method developed by Obstawski for the operating conditions reaches up to 40%. A difference in the value of collector efficiency between the indoor test of the ISO 9806 standard and other methods is smaller and reaches 4%. An almost identical situation occurs in the case of values of the F_R, F′ and U_L coefficients. The smallest difference in the value of the F_R, F′ and U_L coefficients is observed between the normative methods (the ISO 9806 standard) and the ETN method (developed by Chochowski). The difference ranges between 2% and 20%. When comparing values of the F_R, F′ and U_L coefficients to their values calculated under operating conditions, the difference is higher, nearly 50%, e.g., for the value of the U_L coefficient. When analyzing the value of the time constant calculated using various methods, significant similarities can be observed only for the ISO 9806 standard and the ETN method. In other methods, the difference in the value of the time constant reaches up to 270% (the ISO 9806 standard and test stand I) (Table 3).

The time constant value is a very important parameter when designing the control system of a solar installation. An error in the calculation of the time constant can cause a decrease in the efficiency of a solar installation, and in the case of hybrid energy systems it can increase the consumption of a conventional energy carrier. This error can also cause unstable operation of the solar installation. Differences in the values of particular coefficients calculated using different methods are probably the results of nonlinear static characteristics of solar heating installations. Each method is based on linear models in the form of differential equations. Each operating point change in a solar installation described by the parameters of work—e.g., the inlet and outlet fluid temperature, the value of solar irradiance, the ambient temperature—has influence on the value of the coefficients of a solar collector model. This can be observed when analyzing the results provided by the ISO 9806 standard, both indoor and obtained in the ETN methods.

Table 3. Comparison of the characteristic parameters of collectors based on different methods.

Method	Normal Characteristics	Effective Heat Capacity [kJ/m^{2}K]	F_R [-]	F' [-]	U_L [W/m^{2}K]	Time Constant T_z (s)	Flow through the Collector dm^3/min	Working Medium	Solar Irradiance W/m^2
ISO 9806 indoor	$\eta = 0.79 - 4.17T_m - 0.01{T_m}^2$	6.73	0.94	0.95	4.17	80	2.07	water	1000
ISO 9806 outdoor	$\eta = 0.77 - 4.92T_m - 54.8{T_m}^2$	14.3	0.8	0.81	4.92	171	2.16	water	806 ± 5W
ETN Chochowski	$\eta = 0.71 - 4.07T_m - 0.0001{T_m}^2$	6.71	0.75	0.76	4.07	77	2.16	water	1000
PI Obstawski installation I	$\eta = 0.68 - 4.8T_m - 10.0{T_m}^2$	14.6	0.66	0.67	3.59	207.9	2,07	glycol	1000
PI Obstawski installation II	$\eta = 0.417 - 1.533T_m - 4.36\,{T_m}^2$	10.47	0.54	0.55	2.07	298.35	1.24	glycol	1000
PI Obstawski installation II (after calibration)	$\eta = 0.47 - 2.001\,T_m - 5.289\,{T_m}^2$	10.47	0.54	0.55	2.07	149.19	2.07	glycol	1000

Calculations of the characteristic coefficients of a solar collector in both methods, e.g., the time constant, the effective heat capacity and heat loss, are convergent. This is a result of very stable and steady conditions during the experiments. In the case of the ISO 9806 standard indoor, experiments are performed under indoor conditions using a simulator of solar irradiance. In the case of ETN, the parameters of ETN are calculated based on the values of the coefficients of each solar collector component material, and the values of characteristic parameters of the collector are the result of computer simulations. In the case of other methods, the parameters vary. Even in the case of the outdoor method according to the ISO 9806 standard, the parameters change insignificantly (they are quasi-stabile) during experiments. As a result, designing the control systems of solar installations requires an adaptive control algorithm enabling calculation of the values of operating parameters in a solar heating system in real time. The biggest limitation of the presented and tested methods except the ISO 9806:2017 standard involves the fact of omitting the delay time, which is very important when designing a control system, because the response of a controlled object, such as a solar collector, can be in the same time as the command.

6. Conclusions

Based on the performed analyses, it shall be concluded that the static and dynamic properties of a flat-plate solar collector are variable and depend on prevailing operating conditions. When considering the parameters of a collector determined during standard tests (ISO 9806 indoor) as reference values, it shall be noticed that the most similar results concerning efficiency were produced using the ISO 9806 outdoor and the ETN methods. In the case of a collector heat loaded with a hot water storage tank, the efficiency of the collector is lower than the reference value. Similar conclusions may be formulated when analyzing the values of other parameters characterizing the steady state of the collector–F′, F_R and U_L. The heat capacity value of the collector and the resulting time constant substitute depend not only on structural parameters (similar results of the ISO 9806 standard indoor and the ETN method), but also on other factors, for instance related to the system structure and the collector heat load, the temperature of fluid, i.e., parameters defining the operating point of a solar heating installation that vary under operating conditions. Optimizing the operation of a solar heating installation requires developing an adaptive control algorithm enabling real-time calculations of characteristic parameters of the solar installation, e.g., the time constant, the delay time under operating conditions.

Author Contributions: Conceptualization, P.O.; literature review, P.O., T.B., D.C., original draft preparation, P.O; writing—review and editing, P.O., T.B.; funding acquisition, P.O., D.C., T.B. All authors have read and agreed to the published version of the manuscript.

Nomenclature

$(\mathbf{mc})_e$	Effective heat capacity of the collector per unit area [J/m^{2}K]
$(\tau\alpha)_e$	Effective transmittance-absorptance product [-]
$K_{\tau\alpha b}$	Incidence angle modifier for direct radiation [-]
η_0	Efficiency of the collector [-]
$\dot{m}$	Mass flow rate [kg/s]
a_1, a_2	First and second order heat loss coefficient [W/m^{2}K]
A(z), B(z), C(z), D(z)	Polynomials
A_C	Gross area of the collector [m^2]
A_p	Area of the absorber plate [m^2]
b	Plate thickness [m]
C	Heat capacity [J/K]
CC	Thermal capacity [W/K]
c_p	Specific heat of fluid [J/kgK]
D	Diameter [m]
e	Width of contact zone [m]
F	Fin efficiency
F′	Collector efficiency factor [-]

F_R	Collector heat removal factor [-]
G	Transfer function
H	Conductance [W/K]
h	Convective heat transfer coefficient [W/m^{2}K]
I_T	Solar radiation [W/m^2]
I_D	Diffuse solar irradiance of an inclined surface [W/m^2]
K	Incidence angle modifier [-]
kp	Gain
m	Mass [kg]
M_a	Effective mass of the auxiliary unit [kg]
M_e	Effective mass of the collector [kg]
n	Number of points in a data sample [-]
P	Thermal power [W]
P_e	Electric heater power [W]
P_p	Pump power [W]
Q	Heat power [W]
q_u	Rate of useful energy collected per unit area [W/m^2]
R	Thermal resistance [K/W]
s	Laplace operator
t, τ	Time [s]
T_A	Ambient temperature [K]
T_{fi}	Fluid inlet temperature [K]
T_{fm}	Mean fluid temperature [K]
T_{fo}	Fluid outlet temperature [K]
T_m	Reduced temperature [m^{2}K/W]
T_s	Storage tank temperature [K]
T_{sky}	Effective sky temperature [K]
T_z	Time constant [s]
U_L	Heat loss coefficient of the collector [W/m^{2}K]
U_p	Piping heat loss coefficient per unit collector area [W/m^{2}K]
U_{sky}	Heat loss coefficient dependent on wind speed [J/m^{3}K]
V	Heat transfer fluid volume in the collector (m^3)
w	Fin width [m]
y	Impulse response function [-]
z	Discrete Laplace operator
Λ, k	Thermal conductivity [W/K]

Index

a	ambient
ab	absorber
b	back plate
conv	convection
d	downwards
f	fluid
g	glass
gl	global
h	heater
i, j	current parameters (numbering of nodes)
p	plate
rad	
solar	solar irradiance
up	upwards
useful	instantaneous useful power
w	wall

References

1. Diez, F.J.; Navas-Gracia, L.M.; Martinez-Rodriguez, A.; Correa-Guimaraes, A.; Chico-Santamarta, L. Moddeling of flat-plate solar collector using artificial neural network for different working fluid (water) flow rates. *Sol. Energy* **2019**, *188*, 1320–1331. [CrossRef]
2. Banos, R.; Manzano-Agugliaro, F.; Montoyab, F.G.; Gil, C.; Alcayde, A.; Gómez, J. Optimization methods applied to renewable and sustainable energy: A review. *Renew. Sustain. Energy Rev.* **2011**, *15*, 1753–1766. [CrossRef]
3. Sharma, N.; Varun, S. Stochastic techniques used for optimization in solar systems: A review. *Renew. Sustain. Energy Rev.* **2012**, *16*, 1399–1411. [CrossRef]
4. Fan, M.; Zheng, W.; You, S.; Zhang, H.; Jiang, Y.; Wu, Z. Comparision of different dynamic thermal performance prediction models for the flat-plate solar collector with a new V-corrugated absorber. *Sol. Energy* **2020**, *204*, 406–418. [CrossRef]
5. Carmona, M.; Palacio, M. Thermal modeling of a flat plate solar collector with laten heat storage validate with experimental data in outdoor conditions. *Sol. Energy* **2019**, *177*, 620–633. [CrossRef]
6. Marrasso, E.; Roselli, C.; Sasso, M.; Tarioello, F. Analysis of a hybrid solar-assisted trigeneration system. *Energies* **2016**, *9*, 705. [CrossRef]
7. Chochowski, A.; Czekalski, D.; Obstawski, P. Monitoring of renewable energy sources hybrid system operation. *Electr. Rev.* **2009**, *8*, 92–95. (In Polish)
8. Hou, H.J.; Wang, Z.F.; Wang, R.Z.; Wang, P.M. A new method for the measurement of solar collector time constant. *Renew. Energy* **2005**, *30*, 855–865. [CrossRef]
9. Saltiel, C.; Sokolov, M. Optimal control of a multicomponent solar collector system. *Sol. Energy* **1985**, *34*, 463–473. [CrossRef]
10. Araújo, A.; Silva, R.; Pereira, V. Solar thermal modeling for rapid estimation of auxiliary energy requirements in domestic hot water production: On-off versus proportional flow rate control. *Sol. Energy* **2019**, *177*, 68–79. [CrossRef]
11. Morteza, A.; Ardehali, M.; Yae, K.H.; Smith, T.F. Development of proportional-sum-derivative control methodology. *Sol. Energy* **1996**, *57*, 251–260.
12. Winn, C.B. *Controls in Solar Energy Systems, an Annual Review of Research and Development*; Springer: London, UK, 1982; Volume 1, pp. 209–240.
13. Löf, G. *Active Solar Systems*; MIT Press: Cambridge, MA, USA, 1993.
14. Badescu, V. Optimal control of flow in solar collector systems with fully mixed water storage tanks. *Energy Convers. Manag.* **2008**, *49*, 169–184. [CrossRef]
15. Budea, S.; Bădescu, V. Improving the Performance of Systems with Solar Water Collectors used in Domestic Hot Water Production. *Energy Procedia* **2017**, *112*, 398–403. [CrossRef]
16. Bettoni, D.; Soppelsa, A.; Fedrizzi, R.; Del Toro Matamoros, R.M. Analysis and Adaptation of Q-Learning Algorithm to Expert Controls of a Solar Domestic Hot Water System. *Appl. Syst. Innov.* **2019**, *2*, 15. [CrossRef]
17. *ISO 9459-4:2013—Solar Heating—Domestic Water Heating Systems—Part 4: System Performance Characterization by Means of Component Tests and Computer Simulation*; European Committee for Standardization: Brussels, Belgium, 2013.
18. *ASHRE Standard 93-77: Methods of Testing to Determine the Thermal Performances of Solar Collectors*; American Society of Heating, Refrigerating and Air-Conditioning Engineers: Washington, DC, USA, 1997.
19. *EN 12975-1+A1:2010—Thermal Solar Systems and Components—Solar Collectors—Part 1: General Requirements*; European Committee for Standardization: Brussels, Belgium, 2010.
20. *EN ISO 9806:2017—Solar Energy—Solar Thermal Collectors—Test Methods*; European Committee for Standardization: Brussels, Belgium, 2017.
21. Duffie, J.A.; Beckman, W.A. *Solar Engineering of Thermal Processes*; John Wiley & Sons: Hoboken, NJ, USA, 1991.
22. Emery, M.; Rogers, B.A. On a solar collector thermal performance test method for use in variable conditions. *Sol. Energy* **1984**, *33*, 117–123. [CrossRef]
23. Saunier, G.Y.; Chungpaibulpatana, S. A New inexpensive dynamic method of testing to determine solar thermal performance. In *Proceedings ISES Solar World Congress, Australia*; Szokolay, S.V., Ed.; Pergamon Press: New York, NY, USA, 1983; Volume 2, pp. 910–916.

24. Wang, X.A.; Xu, Y.F.; Meng, X.Y. A filter method for transient testing of collector performance. *Sol. Energy* **1987**, *38*, 125–134. [CrossRef]
25. Muschaweck, J.; Spirkl, W. Dynamic solar collector performance testing. *Sol. Energy Mater. Sol. Cells* **1993**, *30*, 95–105. [CrossRef]
26. Chungpaibulpatana, S.; Exell, R.H.B. The effects of using a one-node heat capacitance model for determining solar collector performance parameters by transient test methods. *Sol. Wind Technol.* **1988**, *5*, 411–421. [CrossRef]
27. Perers, B. Dynamic method for solar collector array testing and evaluation with standard database and simulation programs. *Sol. Energy* **1993**, *50*, 517–526. [CrossRef]
28. Wijeysundera, N.E.; Hawlade, M.N.A.; Foong, K.Y. Estimation of collector performance parameters from daily system test. *Sol. Energy* **1996**, *118*, 30–36. [CrossRef]
29. Amer, E.H.; Nayak, J.K.; Sharma, G.K. A transient test methods for flat-plate collectors. *Energy Convers. Manag.* **1998**, *39*, 549–558. [CrossRef]
30. Amer, E.H.; Nayak, J.K.; Sharma, G.K. A new dynamic method for testing of flat-plate solar collectors. *Energy Convers. Manag.* **1999**, *40*, 803–823. [CrossRef]
31. Kong, W.; Wang, Z.; Li, X.; Li, X.; Xiao, N. Theoretical analysis and experimental verification of a new dynamic test method for solar collectors. *Sol. Energy* **2012**, *86*, 398–406. [CrossRef]
32. Buzás, J.; Kicsiny, R. Transfer functions of solar collectors for dynamical analysis and control design. *Renew. Energy* **2014**, *68*, 146–155. [CrossRef]
33. Deng, J.; Yang, X.; Wang, P. Study on the second-order transfer function models for dynamic tests of flat-plate solar collectors Part I: A proposed new model and a fitting methodology. *Sol. Energy* **2015**, *114*, 418–426. [CrossRef]
34. Deng, J.; Yang, X.; Wang, P. Study on the second-order transfer function models for dynamic tests of flat-plate solar collectors Part II: Experimental validation. *Sol. Energy* **2017**, *141*, 334–346. [CrossRef]
35. Deng, J.; Xu, Y.; Yang, X. A dynamic thermal performance model for flat-plate solar collectors based on the thermal inertia correction of the steady-state test method. *Renew. Energy* **2015**, *76*, 679–686. [CrossRef]
36. Deng, J.; Ma, R.; Yuan, G.; Chang, C.; Yang, X. Dynamic thermal performance prediction model for the flat-plate solar collectors based on the two-node lumped heat capacitance method. *Sol. Energy* **2016**, *135*, 769–779. [CrossRef]
37. Amrizal, N.; Chemisana, D.; Rosell, J.I.; Barrau, J. A dynamic model based on the piston flow concept for the thermal characterization of solar collectors. *Appl. Energy* **2012**, *94*, 244–250. [CrossRef]
38. Deng, J.; Yang, M.; Ma, R.; Zhu, X.; Fan, J.; Yuan, G.; Wang, Z. Validation of a simple dynamic thermal performance characterization model based on the piston flow concept for flat-plate solar collectors. *Sol. Energy* **2016**, *139*, 171–178. [CrossRef]
39. Kamminga, W. Experiences of a solar collector test method using Fourier transfer functions. *Heat Mass Transf.* **1985**, *28*, 1393–1404. [CrossRef]
40. Mintsa Do Ango, A.C.; Medale, M.; Abid, C. Optimization of the design of a polymer flat plate solar collector. *Sol. Energy* **2013** *87*, 64–75. [CrossRef]
41. Henning, H.M.; Sasse, M. A collector hardware simulator: Theoretical analysis and experimental result. *Sol. Energy* **1995**, *55*, 39–48. [CrossRef]
42. Tsilingiris, P.T. Heat transfer analysis of low thermal conductivity solar energy absorbers. *Appl. Therm. Eng.* **2000**, *20*, 1297–1314. [CrossRef]
43. Eisenmann, W.; Vajen, K.; Ackermann, H. On the correlations between collector efficiency factor and material content of parallel flow flat-plate solar collectors. *Sol. Energy* **2004**, *76*, 381–387. [CrossRef]
44. Chochowski, A. *Thermal States Analysis of a Flat-plate Solar Collector*; SGGW Publishers: Warsaw, Poland, 1991. (In Polish)
45. Farkas, I. Identification of flat plate solar collectors used in agriculture. In *Proceedings of the 9th IFAC-IFORS Symposium*; Bányász, C.S., Keviczky, L., Eds.; Elsevier: Budapest, Hungary, 1991; Volume 1, pp. 475–479.
46. Cristofari, C.; Notton, P.; Louche, A. Modelling and performance of a copolymer solar water heating collector. *Sol. Energy* **2002**, *72*, 99–112. [CrossRef]
47. Rodriguez-Hidalgo, M.C.; Rodriguez-Aaumente, P.A.; Lecuona, A.; Gutierrez-Ureta, G.L.; Ventas, R. Flat plate thermal solar collector efficiency: Transient behavior under working conditions. Part I: Model description and experimental validation. *Appl. Therm. Eng.* **2011**, *31*, 2394–2404. [CrossRef]

48. Kalogirou, S.; Panteliou, S.; Dentsoras, A. Modelling of solar domestic water heating systems using artificial neural networks. *Sol. Energy* **1999**, *65*, 335–342. [CrossRef]
49. Kalogirou, S.; Panteliou, S.; Dentsoras, A. Artificial neural networks used for the performance prediction of a thermosyphon solar water heater. *Renew. Energy* **1999**, *18*, 87–99. [CrossRef]
50. Kalogirou, S.; Panteliou, S. Thermosyphon solar domestic water heating systems long-term performance prediction using artificial neural networks. *Sol. Energy* **2000**, *69*, 163–174. [CrossRef]
51. Kalogirou, S. Long-term performance prediction of forced circulation solar domestic water heating systems using artificial neural networks. *Appl. Energy* **2000**, *66*, 63–74. [CrossRef]
52. Li, H.; Liu, Z.; Liu, K.; Zhang, Z. Predictive power of machine learning for optimizing solar water heater performance: The potential application of high-throughput screening. *Int. J. Photoenergy* 2017. [CrossRef]
53. Li, H.; Liu, Z. Performance prediction and optimization of solar water heater via a knowledge-based machine learning method. In *Handbook of Research on Power and Energy System Optimization*; IGI Global: Hershey, PA, USA, 2018.
54. Ghritlahre, H.K.; Prasad, R.K. Application of ANN technique to predict the performance of solar collector systems—A review. *Renew. Sustain. Energy Rev.* **2018**, *84*, 75–88. [CrossRef]
55. Farkas, I.; Geczy-Vig, P. Neural network modeling of flat-plate solar collectors. *Comput. Electron. Agric.* **2003**, *40*, 87–102. [CrossRef]
56. Kalogirou, S.A. Prediction of flat-plate collector performance parameters using artificial neural networks. *Sol. Energy* **2006**, *80*, 248–259. [CrossRef]
57. Sözen, A.; Melnik, T.; Unvar, S. Determination of efficiency of flat-plate solar collectors using neural network approach. *Expert Syst. Appl.* **2008**, *35*, 1533–1539. [CrossRef]
58. Baccoli, R.; Carlini, U.; Mariotti, S.; Innamorati, R.; Solinas, E.; Mura, P. Graybox and adaptive dynamic neural network identification models to inter the steady state efficiency of solar thermal collectors starting from the transient condition. *Sol. Energy* **2010**, *84*, 1027–1046.
59. Obstawski, P. Modeling the Dynamics of Work of a Flat-Plate Solar Collector. Ph.D. Thesis, Warsaw University of Life Sciences, Warsaw, Poland, 2007. (In Polish)
60. Aleksiejuk, J.; Chochowski, A.; Reshetiuk, V. Analog model of dynamics of a flat-plate solar collector. *Sol. Energy* **2018**, *160*, 103–116. [CrossRef]
61. Obstawski, P.; Bakoń, T.; Czekalski, D. Diagnostic method of solar thermal system based on the short time on-line measurements. *Appl. Therm. Eng.* **2019**, *148*, 420–429. [CrossRef]

2

Study of Dynamics of Heat Transfer in the Flat-Plate Solar Collector

Joanna Aleksiejuk-Gawron * and Andrzej Chochowski

Department of Fundamentals in Engineering and Energy, Institute of Mechanical Engineering, Warsaw University of Life Sciences, 02-787 Warsaw, Poland; andrzej_chochowski@sggw.edu.pl
* Correspondence: joanna_aleksiejuk@sggw.edu.pl

Abstract: Flat plate solar collector has been presented as an example of a heat-exchanger with two input signals, solar radiation intensity and temperature of working medium on the input, and one output signal, the temperature of a working medium on the output. The dynamics of heat exchange were analyzed for two models of a solar collector—an analog one using a thermoelectric analogy, and a digital one—determined experimentally in on-line mode using the parametric identification method. The characteristics of both models were compared in terms of their step and frequency response for selected construction and operational parameters. Tests of step responses determined for the analog model indicate that the dynamics of heat exchange in the solar collector depending on two input signals is varied. For step-forcing of input signals of the analog model, in both cases, a stable steady state is achieved, but while the first of the signals is inertial, the second one is oscillatory. The phenomenon of temperature oscillation at the collector outlet suggests the need to introduce a new physical quantity in the thermoelectric analogy-thermal inductance. Such an assessment of the dynamics of the solar collector can be useful for proper designing (construction parameters simulation) and diagnostics (operational parameters simulation) of the device.

Keywords: heat transfer; heat exchanger; solar collector; step response; thermoelectric analogy

---

1. Introduction

The main tasks of solar systems are effectively collection, processing, and accumulation of solar energy. It is connected with the need to recognize the thermal processes occurring in the solar collector, as well as to define the impact of the device construction and/or operational parameters on these processes. This task is connected not only with proper design of the components of such a system but also with automation of its functioning, especially if it is one of the segments of hybrid energy system (HES) [1,2]. The solution of control and regulation problems requires identification of dynamic properties of all cooperating component appliances, even at the stage of designing them best. Summarizing, proper recognition of heat transfer phenomena occurring in the solar collector in relation to construction and operational device parameters is directly related to the identification of the impact of these parameters on the dynamics of device operation. Therefore, this publication presents the dynamics of heating up of solar collector's working medium.

In the available studies, mathematical and physical models of solar collector are theoretical [3,4], based on energy balance differential equations. There are also digital models created on the basis of data from long-term measurements (monitoring system) [5,6].

In the papers [7,8], the dynamic properties of collector have been specified by the determination of heating characteristics of working medium with the assumption that solar radiation is the only input excitation signal and temperature at the collector's output is the output signal. The characteristics can be specified theoretically or experimentally. The theoretical characteristics is a solution of a differential equation describing heat transfer in the non-steady state [9,10]. Introducing certain simplifying assumptions is its disadvantage, but the possibility of conducting simulation research is its advantage. The step responses determined experimentally requires the measurements of temperature changes over time on the real object [11]. Time consumption and cost consumption are disadvantages of such a solution. Other elements of a solar heating installation are sometimes included in a steering block diagram, e.g., a hot water storage tank [12].

In automation, a solar collector is most often treated as a heating object with the inertial nature [13]. By analyzing the device frequency response, the recognition of dynamic properties in more detail is possible. Analysis of frequency responses is not widely common in solar collector modeling. More common is describing of solar radiation conditions using frequency characteristics for dynamic modeling of solar collector and heat exchanger [14–16]. A solar collector certainly can be treated as a heat exchanger.

The scientific literature describes the analysis of the impact of changes in solar collector parameters on device efficiency [17]. Additionally, the analyses on the influence of changes in the construction parameters on the collector efficiency and the working medium temperature rise not only in relation to flat-plate, vacuum tube but also in relation to the air collectors, parabolic trough collectors. Even the collectors made of plastic and concrete are performed as a result of simulation tests [18–28].

In addition to the construction parameters, there are analyses of operating parameters, such as: solar collector tilt angle [18,29], the volumetric flow rate of working medium [30] or working medium mass flow rate [28,31], and even the influence of weather conditions on device's efficiency [32]. In addition, the influence of the wind speed and flow rate of the working medium on the final outlet temperature has been studied for years [33]. The analyses concern the basic solar heating systems for water heating [34,35] and more complex hybrid [36] or industrial ones [37].

The application of analog model proposed by the Aleksiejuk et al. in [6] gives the possibility to simulate changes in solar collector construction and operational parameters without need for time-consuming and cost-consuming experiments. This publication presents the impact of changes in construction and operational parameters on the dynamics of heating up of solar collectors working medium.

In the process of designing or diagnostics of solar heating systems the collector plays a major role. It is important to choose such a level of detail in the model in order to be able to distinguish the influence of particular construction and operating parameters of a solar collector on the dynamics of the device. So far, the proposed models are first order models and treat the solar collector as a homogeneous body and therefore they are first-order models [11,13]. The article presents the application of the solar collector analog model, which distinguishes three homogeneous bodies: glass cover, absorber plate, and working medium. Hence, the analog model presented in the article is a third-order model. Two models in the form of electric four-terminal networks have been developed for the analysis of collector operation dynamics. They have one input signal (solar radiation intensity and, alternatively, the temperature of working medium on the collector input) and one output signal (temperature of the working medium on the collector output). In automation, such models are referred to as SISO (single input single output).

2. Materials and Methods

Both the analog and digital model were presented for the same device—flat plate solar collector. The flat plate solar collector is a part of an HES. Tested system is used to prepare hot water for the hotel and utility building. In addition to flat plate solar collectors, the HES includes vacuum tube solar collectors and heat pump, for which the lower energy source is a vertical ground exchanger (Figure 1). The main hot utility water tank has a capacity of 1000 dm^3, and a supplementary one 300 dm^3. The solar installation supplies thermal energy to a 2000 dm^3 tank, which, in turn, serves as energy storage for the heat pump. The HES is fully monitored, and the results of measurements of all the physical quantities (temperature, intensity of solar radiation, flow rate of working media, etc.) are registered daily with a resolution of one minute [38].

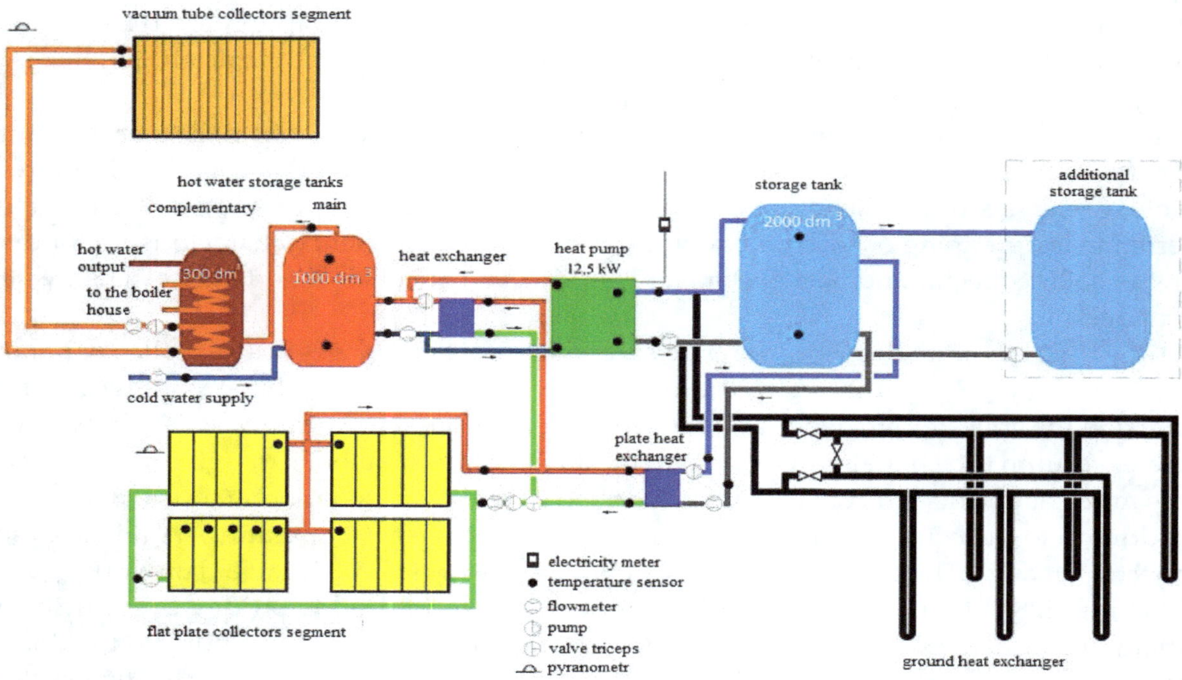

Figure 1. Hybrid energy system. Reproduced with permission from [6]

The authors have used the method of equivalent thermal network (ETN) based on thermoelectric analogy for building the analog model of a flat plate solar collector [6]. The analysis of heat exchange in a solar collector was carried out by solving an electrical circuit (Figure A1). Two collector models were developed, transforming its basic electrical circuit into two four-terminal networks, in which one input and one output signal (SISO models) were distinguished. The input signals are the solar radiation intensity $P(t)$ and the input temperature of the working medium $T_{fin}(t)$ and the output signal is the output temperature of the working medium $T_{fout}(t)$. This allows the influence of the input signals on the collector dynamics to be analyzed separately.

The assumption adopted in the construction of the electric circuit diagram of the collector is to divide the collector into three homogeneous bodies whose physical properties are concentrated in the nodes of the circuit, respectively: 1—glass cover, 2—absorber plate, and 3—working medium. Nodes 1, 2, and 3 are connected with each other and the environment by thermal resistance. Heat streams flow between the nodes and the environment. Each node has been assigned a corresponding heat capacity $C_i = (mc)_i$. The transient heat exchange is described in Equation (1):

$$C\frac{d\overline{T}(t)}{dt} + \Lambda\overline{T}(t) = \overline{P}(t) \tag{1}$$

where:

$$C = \begin{bmatrix} (mc)_1 & 0 & 0 \\ 0 & (mc)_2 & 0 \\ 0 & 0 & (mc)_3 \end{bmatrix}; \overline{T} = \begin{bmatrix} T_1 \\ T_2 \\ T_3 \end{bmatrix}; \overline{P} = \begin{bmatrix} P_1 + \frac{T_o}{R_{1,0}} \\ P_2 + \frac{T_o}{R_{20}} \\ \frac{T_{fin}}{R_f} \end{bmatrix} \tag{2}$$

$$\Lambda = \begin{bmatrix} \Lambda_{1,1} & \Lambda_{1,2} & \Lambda_{1,3} \\ \Lambda_{2,1} & \Lambda_{2,2} & \Lambda_{2,3} \\ \Lambda_{3,1} & \Lambda_{3,2} & \Lambda_{3,3} \end{bmatrix} = \begin{bmatrix} \frac{1}{R_{1,2}} + \frac{1}{R_{1,0}} & -\frac{1}{R_{1,2}} & 0 \\ -\frac{1}{R_{1,2}} & \frac{1}{R_{1,2}} + \frac{1}{R_{2,3}} + \frac{1}{R_{2o}} & -\frac{1}{R_{2,3}} \\ 0 & -\frac{1}{R_{2,3}} & \frac{1}{R_{2,3}} + \frac{1}{R_f} \end{bmatrix} \tag{3}$$

The basic circuit diagram on the basis of which the four-terminal networks were developed and solved is shown in Appendix A in Figure A1 and described in detail in [6]. Dynamic properties of the modeled collector are represented by the operator transmissions of two electric four-terminal networks. It was determined using Laplace transformation for a differential equation describing the dependence of the output signal on the input signal for a given four-terminal network. The collector can be presented as two separate SISO models represented by Equations (4) and (5):

$$G1(s) = \frac{T_{fout}(s)}{P(s)} = \frac{b_1' s + b_0'}{s^3 + a_2' s^2 + a_1' s + a_0'} \tag{4}$$

and

$$G2(s) = \frac{T_{fout}(s)}{T_{fin}(s)} = \frac{b_2'' s^2 + b_1'' s + b_0''}{s^3 + a_2'' s^2 + a_1'' s + a_0''} \tag{5}$$

transfer functions describing the relationships between the particular input (*in*)/output (*out*) signals: *P(s)*—transform of solar radiation intensity, $T_f(s)$—transform of working medium temperature. The coefficients of both the transfer functions depend on resistances and heat capacities of mentioned above flat plate solar collector (Tables A1–A4. The polynomial degree of the denominator in both transfer functions $G1(s)$ and $G2(s)$ is of the third degree. In this publication, the analog model has been applied for simulation the changes in construction and operational parameters of a flat plate solar collector. It was a theoretical simulation. For the method verification, analog model was compared to the results achieved for the digital model described in detail in [6].

A digital model was developed on the basis of a database of the real measurement results of the hybrid energy system. The method of parametric identification has been used for building the digital model; it models the process of the output signal on the basis of measurement results registered in online mode. It also allows for determining the dynamics of the modeled object by determining transfer function ($G1'(s)$, $G2'(s)$ according to the analog model) and step responses on its basis. The applied IT tool was MATLAB package System Identification Toolbox (SIT). ARX [n_a,n_b,n_k], the most frequently recommended one for modeling, has been selected from among the ones proposed for applying in the package of parametric models. The best matching with the process of the input signal has been the selection criterion of a parametric model, which corresponds to the minimization of the value of root mean square error (RMS). In this publication the accuracy of digital model has been increased. The process of selecting the appropriate digital model is described below. Selected model is the one with the highest degree of fit to the actual measurements.

Due to fact that the polynomial degree of the denominator in both transfer functions $G1'(s)$, $G2'(s)$, is of the third degree (the same as for the analog model), attention has been paid to whether we achieve minimization for the RMS error for the digital model with $n_a = 3$.

The choice of n_k parameter connected with the measurements of the analyzed phenomenon is very important. The correct increment of temperature of the working medium $\Delta T_f = T_{fout} - T_{fin}$, being the difference between temperature measured on the output of the collector T_{fout} and on the input T_{fin}, requires taking into account certain delay time in reading the measurement results of both temperatures [39–41]. It is connected with the time of working medium flowing through the collector. This time depends on the length of the flow canal and the yield of the working medium (working medium flow rate). The difference of temperature values on the input and output of the collector cannot be from the same moment of time. The more slowly the working medium flows, the higher temperature it heats to, but the delay of reading temperature on the output is higher. And the other way around, if the length of the stream of working medium is higher, the delay time is shorter. This means that the temperature measurement $T_{fin}(t_1 - n\Delta t)$ ought to correspond to the temperature measurement result $T_{fout}(t_1)$ at the moment t_1, where Δt is a cycle of measurements (a period connected with the frequency of measurements), and n is a number of retrograde measurement steps that must be taken into account for the accuracy of specifying ΔT_f. Similarly, one ought to consider entering the delay time upon reading the influence of the intensity of solar radiation. Therefore, the following doubts appear:

- Whether the delay time ought to be the same for both input signals, e.g., one measurement step or maybe a different one?

It seems that the process of heating elements of the collector under the influence of solar radiation intensity lasts longer than the flow of the working medium. The influence of n_k on the degree of matching the process of the modeled and real output signal (over 90%) and the convergence of step responses of both models (the analog and digital ones) have been checked before the final choice of a parametric model.

Matlab, a simulation program, determines a model's matching without the analysis of the physics of the phenomenon (black-box modeling). The choice of appropriate models depends on its creator (in the assumption of an expert in the field). The following models have been chosen:

- ARX [3,2,3] for transfer function $G1'(s)$ due to the best matching of step responses curve (Figure 2a) as well as due to the high 90.28% degree of model matching; the delay time of 120 s has been applied (2 measurement steps "backwards"),
- ARX [3,2,2] for transfer function $G2'(s)$ due to the best matching of step responses curve (Figure 2b), as well as due to the highest 92.73% degree of model matching; the delay time of 60 s has been applied (1 measurement step "backwards") taking into account the calculated time of the working medium flow through the collector from inlet to its outlet.

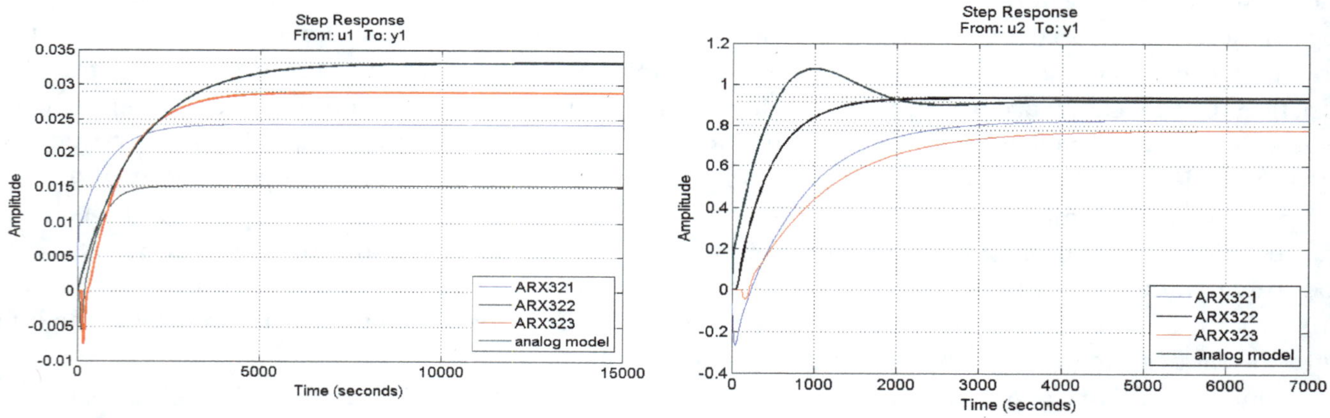

(a) (b)

Figure 2. Step responses for digital and analog model: (**a**) segment G1 and (**b**) segment G2.

The curve of the step responses shown in Figure 2a,b are the justification of decisions made. The model ARX [3,2,3] and the step responses determined from the real measurements matches best the step responses of the analog model in the case of transfer function $G1(s)$. The delay time of 120 s (2 steps) has been introduced here. The parametric model of ARX [3,2,2] with the delay of 1 measurement step (60 s), this is the time of the working medium flow through the collector, is the best solution for transfer function $G2(s)$, respectively. The nature of dynamics of the collector is inertial for the transfer function $G1(s)$ of the analog model; and it is a nonminimal phase one for the digital model. According to two separate SISO models in the analog model, there were also two digital models chosen for comparison.

Contrary to the analog model, it is not possible to divide an influence of the input signal on the end temperature in the digital model, as it is created on the real measurements results taken in the online mode. Regardless of the fact if we determine transfer function $G1'(s) = T_{fout}(s)/P(s)$ *or* $G2'(s) =$

$T_{fout}(s)/T_{fin}(s)$, the other of input signals affects a measurement result. Therefore, step responses of the digital model can differ from the ones presented for the analog model.

The digital model, obtained on the basis of real measurements, does not allow us to conduct simulation tests. For this reason, the next chapter presents simulations of changes in construction and operational parameters on the basis of an analog model.

3. Results

Table 1 contains the list of basic values of construction and operational parameters according to investigated solar collector. The diagrams of step response curves show the changes in parameters in relative values related to the parameters of the real collector. The scope of changes in the parameters retains a logical sense. The simulation was performed using the Matlab software.

Table 1. Study object construction and operational parameters: basic values.

Construction Parameters	
Thickness of glass σ_s (mm)	4
Thickness of absorber σ_a (mm)	2
Diameter of flow channels d (mm)	12
Operational Parameters	
Working medium flow rate Q (dm^3/min)	1.25
Ambient temperature T_a (°C)	22.5

In the case of influence of solar radiation intensity, frequency responses for the particular simulation variants were listed in Figures 3–6 below. For both input signals cases, the step responses for the particular simulation variants were listed in Figures 3–9 below. On each of the figure, in the legend, changes in the parameters are marked with a down arrow (below the basic value) or up arrow (above the basic value) with a value corresponding to its change i.e., 0.7 with the down arrow is equal to 0.7 times the basic value. The response for the digital model is indicated by the bold yellow line in each of graphs (added to compare, because the analog model is a theoretical one and digital is an experimental one as the result of real measurements). It is the point of reference for the comparison. The differences in the courses of the step and frequency responses caused by the simulation of the construction and operational parameters are the results of changes in the values of resistances (through the change in the heat exchange factors) and heat capacities. These quantities affect in turn the coefficient values of the polynomials of numerator and denominator of the transfer functions, being the quintessence of the collector dynamics.

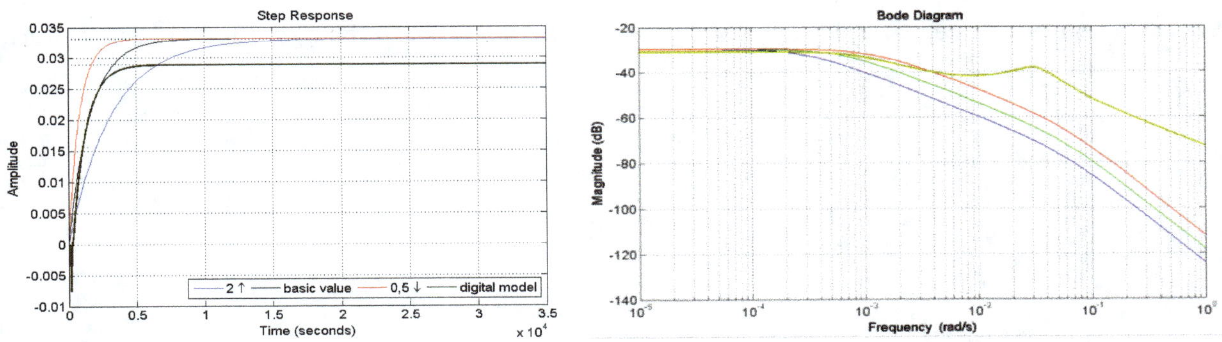

(a) (b)

Figure 3. Segment G1 step (**a**) and frequency responses (**b**) for simulation of changes in thickness of absorber.

absorber.

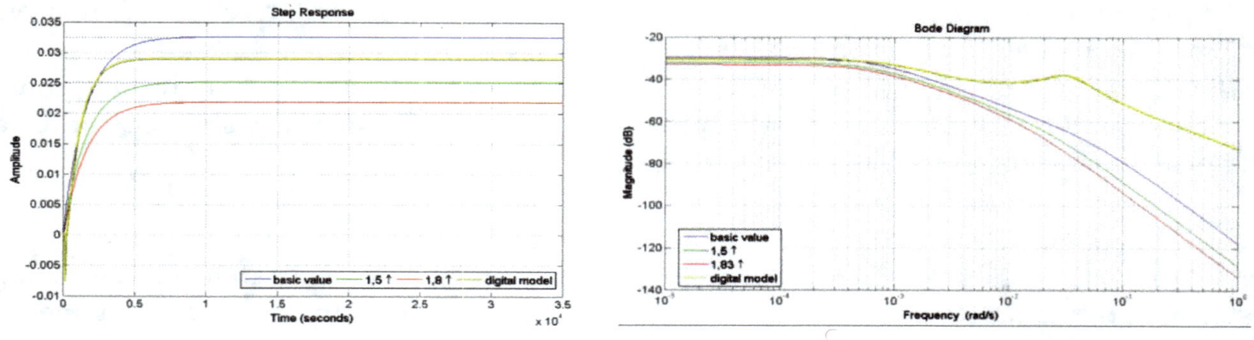

(**a**) (**b**)

Figure 4. Segment G1 step (**a**) and frequency responses (**b**) for simulation of changes in diameter of flow channels.

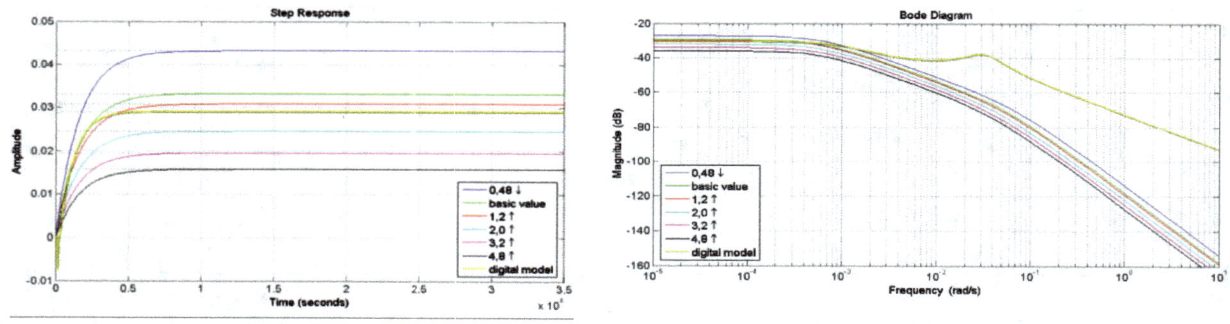

(**a**) (**b**)

Figure 5. Segment G1 step (**a**) and frequency responses (**b**) for simulation of changes in working medium flow rate.

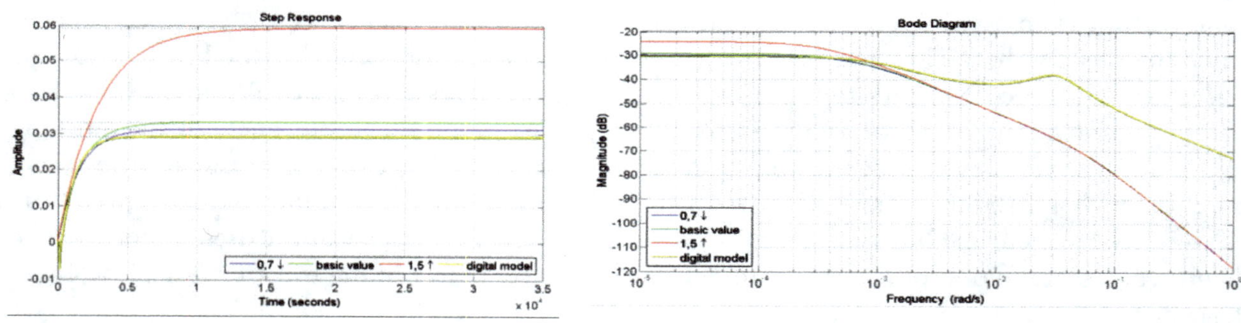

(**a**) (**b**)

Figure 6. Segment G1 step (**a**) and frequency responses (**b**) for simulation of changes in ambient temperature.

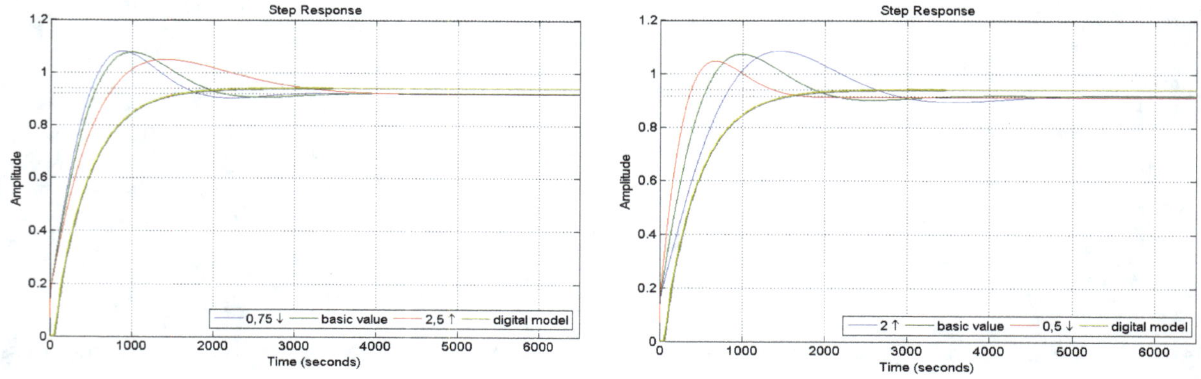

(**a**) (**b**)

Figure 7. Segment G2 step responses for simulation of changes in thickness of glass (**a**) and thickness of absorber (**b**).

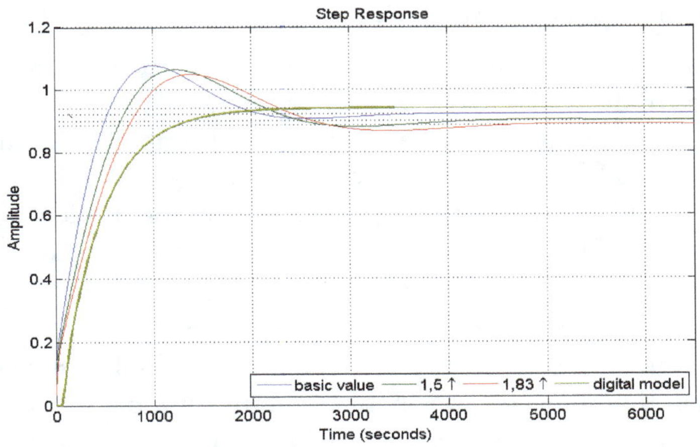

Figure 8. Segment G2 step responses for simulation of changes in diameter of flow channel.

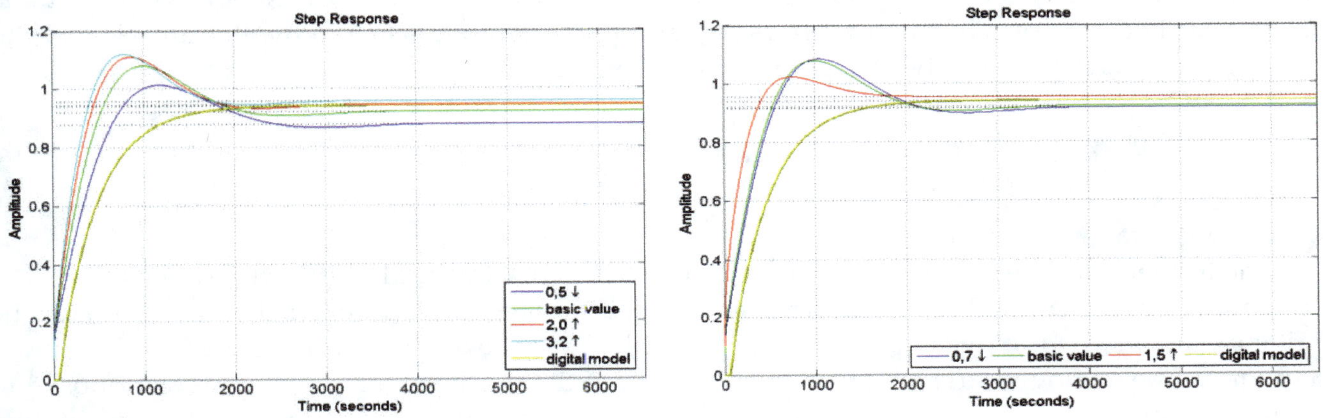

(**a**) (**b**)

Figure 9. Segment G2 step responses for simulation of changes in working medium flow rate (**a**) and ambient temperature (**b**).

3.1. Influence of Solar Radiation Intensity

The construction parameters, for which the simulation studies were performed, include thickness of absorber and diameter of flow channels. There were not any significant changes in simulation of the thickness of glass cover. The operational parameters, for which the simulation studies were performed, include working medium flow rate and ambient temperature. Additional calculations for the digital model were made only for operational parameters for three working medium flow values that have taken place. For these flow values, three days for which weather conditions (including daily distribution of solar radiation intensity and a similar total daily dose of solar radiation energy) were similar, were selected from the measurement database. Courses of solar radiation intensity were similar to each other, as well as daily radiation sum. Selection of days with a similar structure (large fluctuations) of temporary changes in solar radiation intensity was made on purpose. For slow changes in solar radiation intensity, the collector responds in a correct way, where the reaction is entirely the result of the effect. The cut-off frequency value indicates how the device suppresses solar radiation intensity fluctuations. With frequency of changes greater than the cut-off frequency, the collector only reacts to a certain mean value of radiation. It does not react to temporary high and low values.

Solar radiation intensity is a variable function of the fundamental component (optimal insolation) and higher harmonic components associated with cloudiness [42]. An analogous change in expenditure (which is generally constant) is not taken into account.

3.2. *Influence of Temperature of Working Medium on the Input*

The construction parameters, for which the simulation studies were performed, include thickness of glass cover, thickness of absorber, and diameter of flow channels. The operational parameters, for which the simulation studies were performed, include working medium flow rate and ambient temperature.

In the dynamics of heat exchange in the solar collector, for which the step response is determined, interesting phenomena of oscillatory changes appear in the analog model.

4. Discussion

From the step and frequency responses presented in Figures 3–9 it results that the second input, on the side of the signal—the temperature at the collector's input—has the stronger influence on the collector dynamics. For the digital model, this is the nonminimum phase character, i.e., the short-term slight temperature reduction appears first, then the inertial increase of the working medium temperature takes place. For analog model, this phenomenon is more complicated. The step response is of an inertial nature but with initial quickly damping oscillations of small value. The phenomenon is not clear and requires further studies. The other authors pay also attention to this fact in their studies [16,43–46]. The emerging oscillations on the step responses are related to the G2 segment where

$$G2(s) = \frac{T_{fout}(s)}{T_{fin}(s)} \tag{6}$$

which is a typical heat exchanger.

The influence of the intensity solar radiation on the working medium temperature rise for both the analog and digital models is not surprising. It is characterized by the inertia. However, for the digital model, inertia is bigger than for analog model. This probably results from the model creation technique. The analog model analyses the influence of the input signals separately, as opposed to the digital model, which cannot separate the influence of the input signals, because it is determined experimentally in the on-line mode.

In case of changes in the thickness of glass cover, the higher influence on the dynamics from the second segment (transfer function $G2(s)$) is observed. The time constant increases with the increase of the thickness of cover, the temperature oscillations are smaller. In case of the influence of the solar radiation intensity, the insignificant increase of the steady-state level for the thickest glass is observed. However, practically, the thickness of cover does not influence on the signal amplification.

In case of changes in the thickness of absorber, the influence on the dynamics, both on the side of the solar radiation intensity signal (G1) and on the side of the input temperature (G2) is observed. In case of the first segment, the character of the object is not changed, it is inertial. However, the increase of the time constant with the increase of the thickness of absorber is observed. The thicker absorber heats up slower, which is logical. When analyzing the second segment, it is observed (just like for the glass cover) that the time constant increases with the increase of the thickness of absorber and there are slower oscillations. The thinner the absorber, the higher the amplification of the output signal (outlet temperature of working medium). The absorber's heat energy is acquired better by the working medium because the transfer losses are smaller.

In case of changes in the diameter of flow channels, the flow rate of the working medium is changed. The influence on the dynamics, both on the side of the solar radiation intensity and on the side of the input temperature, is also observed here. In case of the first segment, the inertial character of the object is not changed. The time constant increases with the increase of the diameter of flow channels. In case of the second segment, the increase of the time constant with the increase of the diameter of flow channels is observed as well. The collector achieves the lower level of the steady state for the larger diameters because the flow rate is lower, and the heat transfer coefficient achieves smaller values.

From among operational parameters, the working medium flow rate has the greatest influence on dynamics of the collector operation, which is obviously known in literature. The influence of ambient temperature within the most common range of 15–25 °C is insignificant. If the temperature increases to the values corresponding to the medium temperature at the inlet of the collector, the heat losses decreases and the increase of the temperature level of steady state is evident.

Tests of step responses determined for the analog model indicate that the dynamics of heat exchange in the solar collector depending on two input signals: solar radiation intensity and inlet temperature of the working medium, is varied. In the analog (theoretical) model, it is possible to run simulations with only one input signal, if no changes are assumed for the other. There is no such possibility for the digital model, since it is determined in online mode, for actual parameters, where both solar radiation intensity and inlet temperature change. For step-forcing of input signals of the analog model, in both cases a stable steady state is achieved, but while the first signal interacts inertially on the output signal, the influence of the second one is oscillatory. Oscillations are quickly damped, and their parameters can be affected by such factors as the diameter of the flow channel (construction parameter) or the value of the flow of the working medium. Oscillations in the step response of all devices usually negatively affect work and hinder regulation and adjustment of work parameters. By designing the collector and selecting the appropriate construction parameters it is possible to reduce the occurrence of oscillations during its later operation.

The frequency characteristics make it possible to define the frequency of the input signal changes to which the collector proportionally transfers the changes to the output. Additionally, at what frequencies it suppresses these changes, reacting to certain average values over the time interval.

For step response in the digital model, these oscillations are not apparent, as the first signal (intensity of solar radiation) has a strong inertial effect. All the included charts present comparative characteristics for both models.

The theoretical analog model was built based on thermoelectric analogy. In electrical circuits, in transient states, voltage oscillations occur when there is a connected capacitive and inductive element (capacitor and induction coil). Then, the potential energy accumulated in the capacity C is mutually exchanged into the kinetic energy of the inductance L. Oscillations in the circuit suppressed by resistance R appear. So, if thermoelectric analogy is assumed to be acceptable in modeling thermal phenomena, then the potential energy accumulated in the thermal capacities of the structural components of devices can be converted into kinetic energy of emerging temperature field around these devices; hence the oscillatory nature of the step response determining the impact of the inlet temperature on the outlet temperature of the working medium (similarly as in electrical circuits). Therefore, it seems necessary to consider a solar collector electric diagram that takes this phenomenon into account. It seems necessary to define and introduce to thermoelectric analogy a new physical quantity-thermal inductance.

5. Conclusions

Having the analog model of the solar collector, it is possible to determine the influence of changes in the construction and operational parameters on the dynamics of the appliance operation, thus, to determine design guidelines for solar heating systems. It is not necessary to make prototypes of the appliances or to change the design at the later stage. In addition, it becomes possible to diagnose the solar collector as part of a solar heating system. Using the analog model, it is possible to determine whether a change in the construction and operating parameters has an impact on the time constant, gain factor, or delay time of the analyzed device model.

From among the construction parameters selected for the analysis, the following parameters have the greatest influence on the dynamics of the object: the thickness of absorber and the diameter of flow channels. The thickness of glass cover has definitely a lower influence on the dynamics of the object. When analyzing the diameter of flow channels, the changes in the amplification factor and the slight changes in the time constant values were observed. These studies should be extended by additional analyses in aspects of collector diagnostics. In case of the step responses for the second segment, the oscillating character of the phenomenon was observed. This issue should be developed in further studies. It seems that in modeling heating devices, using a thermoelectric analogy, it is necessary to introduce thermal inductance, which would explain the oscillatory nature of the G2 segment's influence on the output temperature in the solar collector. Other works have also highlighted this phenomenon [16,45].

In addition to the construction parameters, the operational parameters of the fluid solar collector operation are also an important aspect. While simulating the changes in the operational parameters the scale of their impact on the solar collector operation was shown.

By appropriate selection of the collector's construction parameters, it is possible to shape the expected course of step responses. It seems possible to obtain an inertial course (attenuation of oscillation) for $G2(s)$ transfer function for a given range of changes in operating parameters.

Author Contributions: Conceptualization and methodology, A.C. and J.A.-G.; software, J.A.-G.; validation, A.C. and J.A.-G.; formal analysis, investigation and data curation, A.C. and J.A.-G.; writing—original draft preparation and writing—review and editing, A.C. and J.A.-G.; project administration, A.C.; funding acquisition, A.C. Both authors have read and agreed to the published version of the manuscript.

Appendix A

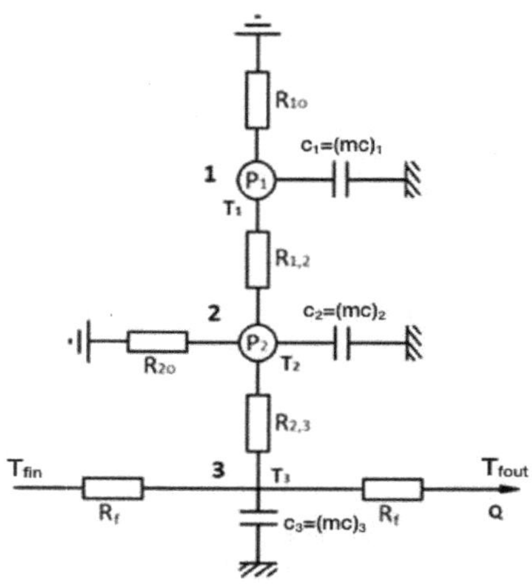

Figure A1. Electrical circuit diagram of an analog model of a flat-plate solar collector. Reproduced with permission from [6],

Table A1. Description of resistances and heat capacities of a modeled flat-plate solar collector.

Symbol and Description	Unit
R_{10}—resistance to glass conductivity and heat transfer from the outer surface of the glass cover to the ambient environment	K/W
R_{12}—resistance between glass cover and absorber	K/W
R_{23}—resistance being the sum of resistance to conductivity through the absorber and to transfer to the working medium	K/W
R_{20}—resistance being the total resistance of conductivity through the insulation surface, air gap and heat transfer to the ambient environment from the bottom surface of the collector	K/W
R_f—thermal resistance of the working medium	K/W
$R_{\lambda s}$—thermal resistance through the glass	K/W
$R_{\alpha s}$—thermal resistance from the outer surface of the glass cover	K/W
R_{as}—thermal resistance between the absorber and the inner surface of the glass cover	K/W
$R_{\lambda a}$—thermal resistance through the absorber	K/W
R_{α}—thermal resistance through the working medium	K/W
$R_{\lambda p}$—thermal resistance through the total air gap	K/W

Table A1. *Cont.*

Symbol and Description	Unit
$R_{\lambda iz}$—thermal resistance through the insulation	K/W
$R_{\lambda d}$—thermal resistance through the collector bottom casing	K/W
$R_{\alpha d}$—thermal resistance from the external bottom casing to the environment	K/W
α_{as}—heat transfer coefficient between the absorber and the inner surface of the glass cover	W/(m^2·K)
α_s—heat transfer coefficient from the external surface of the glass cover	W/(m^2·K)
α_d—heat transfer coefficient for the bottom casing of the collector	W/(m^2·K)
α—heat transfer coefficient of the working medium	W/(m^2·K)
λs—thermal conductivity of glass	W/(m·K)
λ—thermal conductivity of working medium	W/(m·K)
λa—thermal conductivity of absorber material	W/(m·K)
λp—thermal conductivity of air	W/(m·K)
λiz—thermal conductivity of insulation material	W/(m·K)
λd—thermal conductivity of bottom casing material	W/(m·K)
σs—glass thickness	m
σa—absorber thickness	m
σiz—insulation thickness	m
σd—bottom casing thickness	m
$\Sigma\sigma_p$—total air gap between the absorber and the insulation and the insulation and the bottom casing of the collector	m
S_s—glass cover surface	m^2
S_{as}—arithmetic mean of S_a and S_s	m^2
S_k—total internal surface area of the flow channels of the absorber	m^2
S_a—absorber surface	m^2
S_d—bottom casing surface	m^2
c_p—specific heat of the working medium	J/(m^3·K)
Q—working medium flow	m^3/s
mc_1—heat capacity of the glass cover	J/K
mc_2—heat capacity of the absorber	J/K
mc_3—heat capacity of the working medium	J/K

Table A2. List of equations used for thermal resistances and heat capacities calculation.

Equations	Number
$R_{1,o} = R_{\lambda s} + R_{\alpha s}$	(A1)
$R_{\lambda s} = \frac{0{,}5\delta_s}{\lambda_s S_s}$	(A2)
$R_{\alpha s} = \frac{1}{\alpha_s S_s}$	(A3)
$R_{1,2} = R_{\lambda s} + R_{as}$	(A4)
$R_{as} = \frac{1}{\alpha_{as} S_{as}}$	(A5)
$R_{2,3} = R_{\lambda a} + R_{\alpha}$	(A6)
$R_{\lambda a} = \frac{\delta_a}{\lambda_a S_k}$	(A7)
$R_{\alpha} = \frac{1}{\alpha\ S_k}$	(A8)
$R_{2,o} = R_{\lambda p} + R_{\lambda iz} + R_{\lambda d} + R_{\alpha d}$	(A9)
$R_{\lambda p} = \frac{\Sigma\ \delta_p}{\lambda_p S_a}$	(A10)
$R_{\lambda iz} = \frac{\delta_{iz}}{\lambda_{iz} S_a}$	(A11)
$R_{\lambda d} = \frac{\delta_d}{\lambda_d S_d}$	(A12)
$R_{\alpha d} = \frac{1}{\alpha_d S_d}$	(A13)
$R_f = \frac{1}{2\ c_p Q}$	(A14)
$(mc)_1 = c_s \cdot \rho_s \cdot S_s \cdot \sigma_s$	(A15)
$(mc)_2 = c_a \cdot \rho_a \cdot S_a \cdot \sigma_a$	(A16)
$(mc)_3 = c_p \cdot V_l$	(A17)

Table A3. Transfer function coefficients of the first segment $G1(s) = \frac{b'_1 s+b'_0}{s^3+a'_2 s^2+a'_1 s+a'_0}$ where: p_g—glass transmittance, a_a—absorber absorptivity, a_g—glass absorptivity.

Coefficient	Description
b′$_1$	$\frac{p_g \cdot a_a \cdot \frac{1}{S_a} \cdot R_{12} \cdot R_{10} \cdot mc_1}{R_{12} \cdot R_{10} \cdot R_{23} \cdot mc_1 \cdot mc_2 \cdot mc_3}$
b′$_0$	$\frac{\left(a_g \cdot \frac{1}{S_g} + p_g \cdot a_a \cdot \frac{1}{S_a}\right) \cdot R_{10} + p_g \cdot a_a \cdot \frac{1}{S_a} \cdot R_{12}}{R_{12} \cdot R_{10} \cdot R_{23} \cdot mc_1 \cdot mc_2 \cdot mc_3}$
a′$_2$	$\frac{R_{10} \cdot R_{23} \cdot mc_1 \cdot mc_3 + R_{12} \cdot R_{23} \cdot mc_2 \cdot mc_3 + \left(\frac{R_{10} \cdot R_{12}}{R_{20}}\right) \cdot R_{23} \cdot mc_1 \cdot mc_3 + R_{10} \cdot R_{23} \cdot mc_2 \cdot mc_3 + R_{10} \cdot R_{12} \cdot mc_1 \cdot mc_2 \cdot \left(\frac{R_{23}}{R_f}\right) + R_{10} \cdot R_{12} \cdot mc_1 \cdot mc_2}{R_{12} \cdot R_{10} \cdot R_{23} \cdot mc_1 \cdot mc_2 \cdot mc_3}$
a′$_1$	$\frac{R_{23} * \left(\frac{R_{12}}{R_{20}}\right) \cdot mc_3 + R_{23} \cdot \left(\frac{R_{10}}{R_{20}}\right) * mc_3 + R_{23} \cdot mc_3 + R_{10} \cdot mc_1 \cdot \left(\frac{R_{23}}{R_f}\right) + R_{12} \cdot mc_2 \cdot \left(\frac{R_{23}}{R_f}\right) + \left(\frac{R_{10} * R_{12}}{R_{20}}\right) \cdot \left(\frac{R_{23}}{R_f}\right) \cdot mc_1 + R_{10} \cdot mc_2 \cdot \left(\frac{R_{23}}{R_f}\right) + R_{10} \cdot mc_1 + R_{12} \cdot mc_2 + \left(\frac{R_{10} * R_{12}}{R_{20}}\right) \cdot mc_1 + R_{10} \cdot mc_2}{R_{12} \cdot R_{10} \cdot R_{23} \cdot mc_1 \cdot mc_2 \cdot mc_3}$
a′$_0$	$\frac{\left(\frac{R_{12}}{R_{20}}\right) \cdot \left(\frac{R_{23}}{R_f}\right) + \left(\frac{R_{10}}{R_{20}}\right) \cdot \left(\frac{R_{23}}{R_f}\right) + \frac{R_{23}}{R_f} + \frac{R_{12}}{R_{20}} + \frac{R_{10}}{R_{20}} + 1}{R_{12} \cdot R_{10} \cdot R_{23} \cdot mc_1 \cdot mc_2 \cdot mc_3}$

Table A4. Transfer function coefficients of the second segment $G2(s) = \frac{b''_2 s^2+b''_1 s+b''_0}{s^3+a''_2 s^2+a''_1 s+a''_0}$.

Coefficient	Description
b″	$\frac{R_{23} \cdot R_{12} \cdot R_{10} \cdot R_{20} \cdot mc_1 \cdot mc_2}{R_f \cdot R_{12} \cdot R_{10} \cdot R_{23} \cdot R_{20} \cdot mc_1 \cdot mc_2 \cdot mc_3}$
b″$_1$	$\frac{R_{12} \cdot R_{10} \cdot R_{20} \cdot mc_1 + R_{23} \cdot R_{10} \cdot R_{20} \cdot mc_2 + R_{23} \cdot R_{12} \cdot R_{10} \cdot mc_1 + R_{23} \cdot R_{12} \cdot R_{20} \cdot mc_2 + R_{23} \cdot R_{20} \cdot R_{10} \cdot mc_1}{R_f \cdot R_{12} \cdot R_{10} \cdot R_{23} \cdot R_{20} \cdot mc_1 \cdot mc_2 \cdot mc_3}$
b″$_0$	$\frac{R_{10} \cdot R_{20} + R_{12} \cdot R_{20} + R_{23} \cdot R_{10} + R_{23} \cdot R_{12} \cdot R_{23} \cdot R_{20}}{R_f \cdot R_{12} \cdot R_{10} \cdot R_{23} \cdot R_{20} \cdot mc_1 \cdot mc_2 \cdot mc_3}$
a″$_2$	$\frac{R_{23} \cdot R_{10} \cdot R_{20} \cdot mc_1 \cdot mc_2 + R_f \cdot R_{12} \cdot R_{10} \cdot R_{20} \cdot mc_1 \cdot mc_3 + R_f \cdot R_{23} \cdot R_{10} \cdot R_{20} \cdot mc_2 \cdot mc_3 + R_f \cdot R_{23} \cdot R_{12} \cdot R_{10} \cdot mc_1 \cdot mc_3 + R_f \cdot R_{23} \cdot R_{12} \cdot R_{20} \cdot mc_2 \cdot mc_3 + R_f \cdot R_{23} \cdot R_{20} \cdot R_{10} \cdot mc_1 \cdot mc_3 + R_f \cdot R_{12} \cdot R_{10} \cdot R_{20} \cdot mc_1 \cdot mc_2}{R_f \cdot R_{12} \cdot R_{10} \cdot R_{23} \cdot R_{20} \cdot mc_1 \cdot mc_2 \cdot mc_3}$
a″$_1$	$\frac{R_{12} \cdot R_{10} \cdot R_{20} \cdot mc_1 + R_{23} \cdot R_{10} \cdot R_{20} \cdot mc_2 + R_{23} \cdot R_{12} \cdot R_{10} \cdot mc_1 + R_{23} \cdot R_{12} \cdot R_{20} \cdot mc_2 + R_{23} \cdot R_{20} \cdot R_{10} \cdot mc_1 + R_f \cdot R_{10} \cdot R_{20} \cdot mc_3 + R_f \cdot R_{12} \cdot R_{20} \cdot mc_3 + R_f \cdot R_{23} \cdot R_{10} \cdot mc_3 + R_f \cdot R_{23} \cdot R_{12} \cdot mc_3 + R_f \cdot R_{23} \cdot R_{20} \cdot mc_3 + R_f \cdot R_{10} \cdot R_{20} \cdot mc_2 + R_f \cdot R_{12} \cdot R_{10} \cdot mc_1 + R_f \cdot R_{12} \cdot R_{20} \cdot mc_2 + R_f \cdot R_{20} \cdot R_{10} \cdot mc_1}{R_f \cdot R_{12} \cdot R_{10} \cdot R_{23} \cdot R_{20} \cdot mc_1 \cdot mc_2 \cdot mc_3}$
a″$_0$	$\frac{R_{10} \cdot R_{20} + R_{12} \cdot R_{20} + R_{23} \cdot R_{10} + R_{23} \cdot R_{12} + R_{23} \cdot R_{20} + R_f \cdot R_{12} + R_f \cdot R_{20} + R_f \cdot R_{10}}{R_f \cdot R_{12} \cdot R_{10} \cdot R_{23} \cdot R_{20} \cdot mc_1 \cdot mc_2 \cdot mc_3}$

References

1. Marrasso, E.; Roselli, C.; Sasso, M.; Tarioello, F. Analysis of a hybrid solar-assisted trigeneration system. *Energies* **2016**, *9*, 705. [CrossRef]
2. Zhang, C.; Sun, J.; Lubell, M.; Qiu, L.; Kang, K. Design and simulation of a novel hybrid solar-biomass energy supply system in northwest China. *J. Clean. Prod.* **2019**, *233*, 1221–1239. [CrossRef]
3. Khoukhi, M.; Maruyama, S. Theoretical approach of a flat-plate solar collector taking into account the absorption and emission within glass cover layer. *Sol. Energy* **2006**, *80*, 787–794. [CrossRef]
4. Beckman, W.; Duffie, J. *Solar Engineering of Thermal Processes*; J. Wiley and Sons: New York, NY, USA, 1974.
5. Buzas, J.; Kicsiny, R. Transfer functions of solar collectors for dynamical analysis and control design. *Renew. Energy* **2014**, *68*, 146–155. [CrossRef]
6. Aleksiejuk, J.; Chochowski, A.; Reshetiuk, V. Analog model of dynamics of a flat-plate solar collector. *Sol. Energy* **2018**, *160*, 103–116. [CrossRef]
7. Pierson, P.; Padet, J. Time constant for solar collectors. *Sol. Energy* **1990**, *44*, 109–114. [CrossRef]
8. Prapas, D.; Norton, B.; Milonidis, E.; Probert, S. Response function for solar-energy collectors. *Sol. Energy* **1988**, *40*, 371–383. [CrossRef]
9. Klein, S.A.; Duffie, J.A.; Beckmann, W.A. Transient considerations of flat plate solar collectors. *J. Eng. Power (Trans. ASME)* **1974**, *96A*, 109–113. [CrossRef]
10. Siebers, D.L.; Viskanta, R. Some aspects of the transient response of a flat-plate solar energy collector. *Energy Convers.* **1978**, *18*, 35–139. [CrossRef]
11. Obstawski, P.; Bakon, T.; Czekalski, D. Modification of the solar heating system diagnostic method under operating conditions. *Appl. Therm. Eng.* **2020**, *165*, 114474. [CrossRef]
12. Munroe, M.M. A method of determining the time constant of the flat-plate solar collector. *Energy Convers. Manag.* **1981**, *21*, 185–189. [CrossRef]
13. Obstawski, P.; Bakoń, T.; Czekalski, D. Diagnostic method of solar thermal system based on the short time on-line measurements. *Appl. Therm. Eng.* **2019**, *148*, 420–429. [CrossRef]
14. Guarracino, I.; Freeman, J.; Ramos, A.; Kalogirou, S.; Ekins-Daukes, N.; Markides, C. Systematic testing of hybrid PV-thermal (PVT) solar collectors in steady- state and dynamic outdoor conditions. *Appl. Energy* **2019**, *240*, 1014–1030. [CrossRef]
15. Lalot, S.; Desmet, B. The harmonic response of counter-flow heat exchangers—Analytical T approach and comparison with experiments. *Int. J. Therm. Sci.* **2019**, *135*, 163–172. [CrossRef]
16. Piotrowska, E.; Skowroński, P. Analysis of temperature oscillations parameters of heat exchanging systems. *Int. J. Heat Mass Transf.* **2018**, *127*, 651–662. [CrossRef]
17. Obstawski, P.; Bakoń, T.; Czekalski, D. Comparison of Solar Collector Testing Methods—Theory and Practice. *Processes* **2020**, *8*, 1340. [CrossRef]
18. Zhang, D.; Tao, H.; Wang, M.; Sun, Z.; Jiang, C. Numerical simulation investigation on thermal performance of heat pipe flat-plate solar collector. *Appl. Therm. Eng.* **2017**, *118*, 113–126. [CrossRef]
19. Andemeskel, A.; Suriwong, T.; Wamae, W. Effects of aluminium in thickness coated with solar paint on the thermal performance of evacuated tube collector. *Energy Procedia* **2017**, *138*, 429–434. [CrossRef]
20. Ferahta, F.Z.; Bougoul, S.; Ababsa, D.; Abid, C. Numerical Study of the Convection in the Air Gap of a Solar Collector. *Energy Procedia* **2011**, *6*, 176–184. [CrossRef]
21. Pandey, M.K.; Chaurasiya, R. A review on analysis and development of solar flat plate collector. *Renew. Sustain. Energy Rev.* **2017**, *67*, 641–650. [CrossRef]
22. Martinez-Rodriguez, G.; Fuentes-Silva, A.; Picon-Nunez, M. Solar thermal networks operating with evacuated-tube collectors. *Energy* **2018**, *146*, 26–33. [CrossRef]
23. Demou, A.D.; Grigoriadis, D.G.E. 1D model for the energy yield calculation of natural convection solar air collectors. *Renew. Energy* **2018**, *119*, 649–661. [CrossRef]
24. Guo, S.; Liu, D.; Chu, Y.; Chen, X.; Xu, C.; Liu, Q.; Guo, T. Dynamic behavior and transfer function of collector field in once- through DSG solar trough power plants. *Energy* **2017**, *121*, 513–523. [CrossRef]
25. O'Keeffe, G.; Mitchell, S.; Myers, T.; Cregan, V. Modelling the efficiency of a low-profile nanofluid-based direct absorption parabolic trough solar collector. *Int. J. Heat Mass Transf.* **2018**, *126*, 613–624. [CrossRef]
26. Zou, B.; Yao, Y.; Jiang, Y.; Yang, H. A new algorithm for obtaining the critical tube diameter and intercept factor of parabolic trough solar collectors. *Energy* **2018**, *150*, 451–467. [CrossRef]

27. Dorfling, C.; Hornung, C.K.; Hallmark, B.; Beaumont, R.J.J.; Fovargue, K.; Mackley, M.R. The experimental response and modelling of a solar heat collector fabricated from plastic microcapillary films. *Sol. Energy Mater. Sol. Cells* **2010**, *94*, 1207–1221. [CrossRef]
28. O'Hegarty, R.; Kinnane, O.; McCormack, S. Parametric analysis of concrete solar collectors. *Energy Procedia* **2016**, *91*, 954–962. [CrossRef]
29. Zou, Q.; Li, Z.; Wu, H. Modal analysis of trough solar collector. *Sol. Energy* **2017**, *141*, 81–90. [CrossRef]
30. Gertzos, K.P.; Caouris, Y.G. Optimal arrangement of structural and functional parts in a flat plate integrated collector storage solar water heater (ICSSWH). *Exp. Therm. Fluid Sci.* **2008**, *32*, 1105–1117. [CrossRef]
31. Badescu, V. Optimal control of flow in solar collectors for maximum exergy extraction. *Int. J. Heat Mass Transf.* **2007**, *50*, 4311–4322. [CrossRef]
32. Khin Chaw Sint, N.; Choudhury, I.A.; Masjuki, H.H.; Aoyama, H. Theoretical analysis to determine the efficiency of a CuO-water nanofluid based-flat plate solar collector for domestic solar water heating system in Myanmar. *Sol. Energy* **2017**, *155*, 608–619. [CrossRef]
33. Saito, A.; Utaka, Y.; Tsuchio, T.; Katayama, K. Transient response of flat plate solar collector for periodic solar collector intensity variation. *Sol. Energy* **1984**, *32*, 17–23. [CrossRef]
34. Huang, B.J.; Ton, W.Z.; Wu, C.C.; Ko, H.W.; Chamg, H.S.; Yen, R.H.; Wang, J.C. Maximum-power-point tracking control of solar heating system. *Sol. Energy* **2012**, *86*, 3278–3287. [CrossRef]
35. Azad, E. Theoretical and experimental investigation of heat pipe solar collector. *Exp. Therm. Fluid Sci.* **2008**, *32*, 1666–1672. [CrossRef]
36. Tian, Z.; Perers, B.; Furbo, S.; Fan, J. Thermo-economic optimization of a hybrid solar district heating plant with flat plate collectors and parabolic trough collectors in series. *Energy Convers. Manag.* **2018**, *165*, 92–101. [CrossRef]
37. Picon-Nunez, M.; Martinez-Rodriguez, G.; Fuentes-Silva, A.L. Design of solar collector networks for industrial applications. *Appl. Therm. Eng.* **2014**, *70*, 1238–1245. [CrossRef]
38. Chochowski, A.; Czekalski, D.; Obstawski, P. Monitoring of renewable energy sources hybrid system operations [in Polish]. *Prz Elektrotech* **2009**, *85*, 92–95.
39. Ampuno, G.; Roca, L.; Gil, J.; Berenguel, M.; Normey-Rico, J. Apparent delay analysis for a flat-plate solar field model designed for control T purposes. *Sol. Energy* **2019**, *177*, 241–254. [CrossRef]
40. Huang, B.; Wang, S. Identification of solar collector dynamics using physical model-based approach. *J. Dyn. Syst. Meas. Control* **1994**, *116*, 755–763. [CrossRef]
41. De Ron, A. Dynamic modelling and verification of a flat-plate solar collector. *Sol. Energy* **1979**, *24*, 117–128. [CrossRef]
42. Czekalski, D.; Chochowski, A.; Obstawski, P. Parametrization of daily solar irradiance variation. *Renew. Sustain. Energy Rev.* **2012**, *16*, 2461–2467. [CrossRef]
43. Weedy, B. The analogy between thermal and electrical quantities. *Electron. Power Syst. Res.* **1988**, *15*, 197–201. [CrossRef]
44. Abdel-Hamid, B. Modelling non-Fourier heat conduction with periodic thermal oscillation using the finite integral transform. *Appl. Math. Model.* **1999**, *23*, 899–914. [CrossRef]
45. Chochowski, A.; Piotrowska, E. Representation of transient heat transfer as the equivalent thermal network (ETN). *Int. J. Heat Mass Transf.* **2013**, *63*, 113–119.
46. Piotrowska, E.; Chochowski, A. Application of parametric identification methods for analysis of heat exchanger dynamics. *Int. J. Heat Mass Transf.* **2012**, *55*, 7109–7118. [CrossRef]

3

Low-Cost Solar Electricity using Stationary Solar Fields; Technology Potential and Practical Implementation Challenges to be Overcome: Outcomes from H2020 MOSAIC Project

Cristóbal Villasante [1,*], Saioa Herrero [1], Marcelino Sánchez [2], Iñigo Pagola [2], Adrian Peña [2], David Olasolo [1] and Ana Bernardos [2]

[1] Fundacion Tekniker, Polo Tecnológico de Eibar, Calle Iñaki Goenaga 5, 20600 Gipuzkoa, Spain; sherrero@tekniker.es (S.H.); david.olasolo@tekniker.es (D.O.)

[2] CENER (National Renewable Energy Centre of Spain), Ciudad de la Innovación 7, 31621 Navarra, Spain; msanchez@cener.com (M.S.); ipagola@cener.com (I.P.); apena@cener.com (A.P.); abernardos@cener.com (A.B.)

* Correspondence: cristobal.villasante@tekniker.es.

Abstract: At any time of the day, a spherical mirror reflects the rays coming from the sun along a line that points to the sun through the center of the sphere. This makes it possible to build concentrated solar power(CSP) plants with fixed solar fields and mobile receivers; that is, solar fields can be significantly cheaper and simpler, but challenging tracking systems for the mobile receiver need to be implemented. The cost-cutting possibilities for this technology have been under-researched. This article describes the MOSAIC concept, which aims to achieve low-cost solar energy by boosting the benefits of spherical reflectors while addressing their challenges. This new concept proposes to build large modular plants from semi-Fresnel solar bowls. One of these modules has been designed and is under construction in Spain. This article reports the main lessons learned during the design phase, describes the advantages and challenges of the concept, details the proposed routes to overcome them, and identifies the steps needed to develop a fully competitive industrial solution.

Keywords: CSP; solar; spherical concentrator; Fresnel; stationary reflector; tracking absorber

---

1. Introduction

Concentrated solar technology has the capacity to produce dispatchable green energy (both heat and electricity), which is a major advantage over other renewable energies [1,2]. In recent years, its production costs have been substantially reduced, but there is a pressing need to reduce them further in order to expand its market penetration. Reducing costs and maintaining or increasing energy conversion rates is the only way to reach this goal, but achieving both simultaneously is challenging [3].

Higher efficiency requires higher operating temperatures, which also increase the storage capacity of the systems and, therefore, their profitability [4,5]. However, achieving higher temperatures requires higher concentration ratios, which, in turn, requires a larger investment to pay for more precise tracking systems, more sophisticated materials, etc. Most of the costs of high-concentration solar technologies come from the solar field and especially from the sophisticated tracking systems needed [6].

The MOSAIC project [7] proposes a new plant configuration that provides high concentration ratios, even though it proposes a fixed solar field with great potential for reducing its costs. Furthermore, it proposes to group several of these concentration units in a modular plant, which would provide additional advantages.

CSP plant power blocks and storage systems are more cost-effective the larger they are; therefore, the plants tend to be large in size (100 MW or larger). For today's central receiver plants, this means huge towers and thousands of heliostats located at great distances (>1 km). This has cost implications, as it requires very demanding tracking and canting accuracies and very rigid structures. Moreover, it also involves significant atmospheric attenuation, which imposes limits on the size of the plant. However, it should be borne in mind that the aim is to achieve high temperatures, and therefore, high concentration ratios, which does not necessarily imply large concentrators since the concentration ratio is a dimensionless parameter. In contrast, the MOSAIC configuration achieves high powers (large collection areas) in a modular way, linking a number of modules, each with low atmospheric attenuation and moderate accuracy requirements.

2. The Current State of Research

Unlike parabolic mirrors, spherical concentrators (or solar bowls) can be fixed using a mobile receiver [8]. Hence, these systems are also known as SRTA (stationary reflector/tracking absorber). This is because a spherical mirror always focuses the solar radiation along a line passing through the sphere center and pointing to the sun (see Figure 1) irrespective of the position of the sun relative to the mirror. This concentrated flux is not uniform and grows from the surface of the mirror to the midpoint of the bowl's radius.

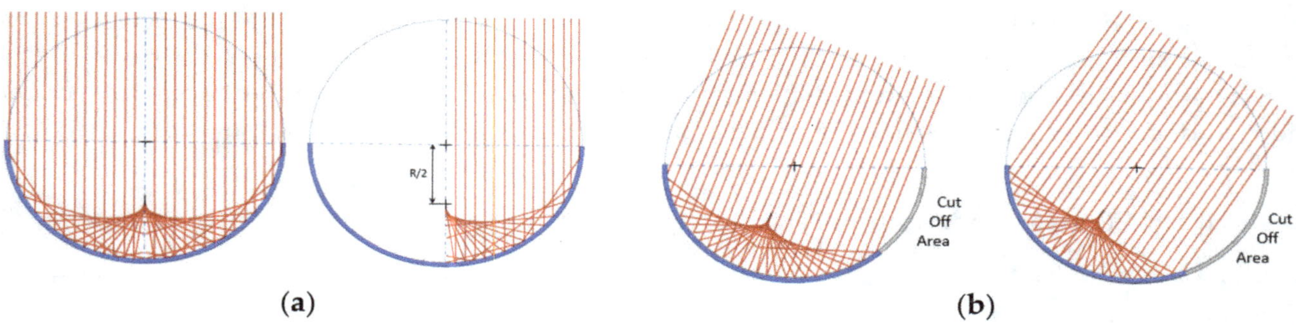

Figure 1. Spherical concentrators characteristics: (**a**) case of rays perpendicular to the aperture; (**b**) cases of rays not perpendicular to the aperture and its effect on reducing the effective aperture area.

Despite its well-established [9] cost reduction potential, the concept has been poorly studied. A small number of prototypes have been built at a relevant scale, and limited results have been published. According to Goodman [10], Adams [11] already recorded a stationary spherical reflector made of 40 flat glass facets built in Middle Colaba, Bombay, India, as early as 1878. Other researchers [12] refer to the 1928 Berland patent [13] as the birth of the technology.

Optical characteristics of the SRTA, their implications on the minimum receiver size, and the axial variation of the concentration ratio were studied in detail by Steward and Kreith [8]. At the same time, Kreider [14] analyzed its thermal performance and identified key design parameters.

Fixed reflectors make it possible to build very large mirrors, and to even integrate them in a building roof (see examples in Figure 2). However, this paper will focus on stand-alone configurations best suited for large-scale electricity production.

(a) (b) (c)

Figure 2. Stationary reflector/tracking absorber (SRTAs) incorporated in building roofs: (**a**) Steward residence with 9.14 m diameter bowl at Boulder, Colorado. The design began in 1968 [10]; (**b**) Solar bowl integrated into the roof of the community kitchen in Auroville Universal Township, Tamil Nadu, India [15]; (**c**) An industrial building concept with the roof constructed of SRTA modules as proposed by Cohen et al. [12].

In the 1970s, solar bowl programs were launched in the USA, France, and Israel and new developments started in India in the 1980s and Europe in the 1990s:

- USA: The United States Department of Energy (DOE) DOE launched the Crosbyton solar power project (CSPP), and the analog design verification system (ADVS) began tests in 1980 in Crosbyton (Texas). The reflector of the ADVS was a spherical bowl with 19.8 m aperture diameter, 60° semi-rim angle, and was tilted 15° [16]. Its receiver was 5.7 m long made of Inconel alloy 617, producing superheated steam at 538 °C. The system was intended to act as a testbed for the design of a 5 MW hybrid solar/fossil fuel power plant. Such a plant would have had 10 bowls, each with 60 m aperture diameter, but after the test period, the ADVS was dismantled.
- France: CNRS (Centre National de la Recherche Scientifique) developed the Pericles project to study SRTAs for the vicinity of the equator or tropical areas. A test bowl was built in Marseilles, France, and later rebuilt in Recife, Brazil. The system was operational in 1980. It had a 10 m aperture diameter, 120° rim angle, and no inclination. It was known as mini-Pericles since 30–40 m diameter bowls were studied within the project. The bowl included an additional mobile element (rotary visor) to reduce shadows and improve efficiency in the early morning and late in the day [17]. Gilotherm TH oil was used as Heat Transfer Fluid (HTF) up to 330 °C.
- Israel: Several SRTA versions were built in the 1970s in Haifa, Israel, at the Technion–Israel Institute of Technology. The largest prototype had a 10 m diameter dish that produced steam at 300 °C, but no public report has been found; it was operational in 1979. Design, construction, and testing of a smaller version (2.52 m diameter) operated with PAZTHERM 22™ are described in [18]. Other papers describe successive versions in the range of 2.4 m to 8.7 m aperture diameter [12]. Ref. [19] shows a video of a recent SRTA test bench.
- India: The first dish, with a 3.5 m aperture diameter, 120 rim-angle, inclined to 12° was developed in Auroville, Tamil Nadu, India, was already reported by Harper in 1982 [20]. It used small flat glasses glued to concrete. In the 1990s, a much larger solar bowl was integrated into Auroville's community kitchen [15]. This new project involved not only Harper but also other technology enthusiasts who had been involved in previous projects in France or the USA, such as Goodman, Authier, or Debilly. They considered that solar bowl technology was relatively simple and low-cost and that it had the potential to use labor more intensively than capital, making it suitable for developing countries. In fact, the system had been supplying steam for cooking since 2001. The solar kitchen included a spherical bowl with 15 m aperture diameter, 120° rim angle, and 12° tilt angle. During 2001–2002, the system was tested using oil (above 200 °C) as heat transfer fluid (HTF). Later it was switched to a 'water only' system more suitable for cooking and easy to use; it produced steam at 150 °C.

- Europe: The European Commission funded Phase 1 of a research project (1996–1999) that developed a prototype [21] producing hot air at 850 °C in Crete, Greece. The original concept aimed to produce 1 MW, but the prototype developed in Phase 1 was downsized. The prototype consisted of a reflective surface with a radius of curvature of 30 m (a 47° wide segment in the north-south direction and a 60° wide segment in the east-west direction) that supplied heat to the 35 kWe solar-gas turbine. The volumetric receiver with a secondary concentrator followed the concentrated solar flux at a distance of 14.7 m from the surface of the mirrors. As two-thirds of the mirror costs were the civil works associated with the construction of the bowl, a revised concept was conceived for Phase 2, which included a flat fixed Fresnel mirror. The most relevant prototypes to date are summarized in Table 1.

Table 1. The largest SRTA prototypes to date.

Site	Crosbyton, TX, USA	Marseilles, France Recife, Brazil	Haifa, Israel	Auroville, India	Crete, Greece
Photo		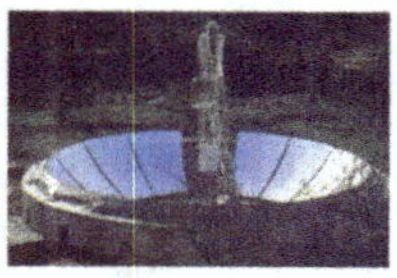			
Date Size HTF/ temp Status	1970s 19.7 m aperture diameter Steam at 538 °C Decommissioned	1970s 10 m aperture diameter Gilotherm TH oil at 330 °C Decommissioned	1970s 10 m aperture diameter Steam at 300 °C Decommissioned	1990s 15 m aperture diameter Low pressure Steam Still in operation	1990s 30 m curvature Air at 850 °C Decommissioned

As shown in the Crete project, large bowls installed at relatively high latitudes require huge civil works, while Fresnel configurations have the potential to be cheaper and easier to build. Figure 3 shows how a series of concentric spheres of different radius continue to concentrate the solar flux into a single focal line that passes through the center common to all the spheres and points to the sun. This is the basis for the design of Fresnel-type SRTAs. Larbi [22] and Sánchez [23] studied the optical design characteristics of different Fresnel-type configurations trying to reduce the investment in reflector infrastructure with minimum reduction in energy delivery and concentration ratio. However, so far, no relevant prototype has been built, but only small models (see Figure 3).

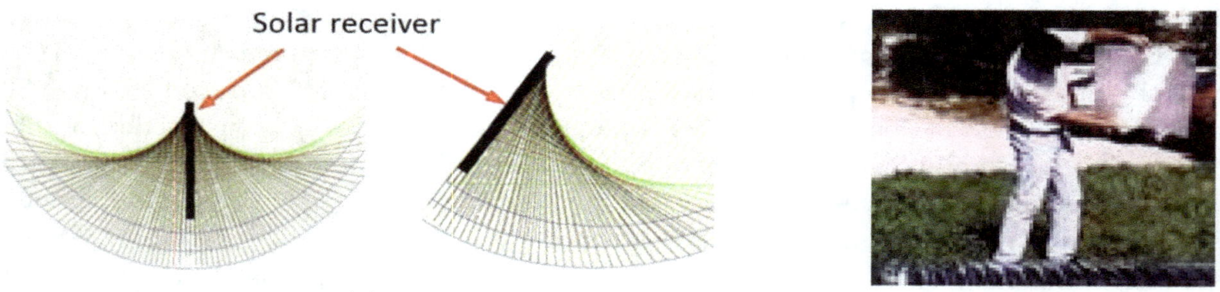

(a) (b)

Figure 3. The Fresnel approach for spherical concentrators [23]. (**a**) Raytracing for concentric spheres; (**b**) Model built by Sánchez in the early 2000s to prove the Fresnel concept.

SRTAs have not only been developed for solar electricity production, but different developments [8,12,15] have focused on solar heat production for a variety of uses. The MOSAIC configuration could also be applied for these purposes, especially to supply solar heat to industrial installations.

3. Materials and Methods

The MOSAIC project investigates a new concept of SRTA. The static reflector is not a continuous bowl, but a semi-Fresnel design is proposed. The proposed cable-based tracking system is also new. These innovations are intended to reduce costs. Furthermore, it is expected that in the future, its deployment in the field will be easy and fast thanks to the modularity of the concept. In summary, the system pursues the following objectives:

- Low costs because of the fixed solar field. Eliminating the drives of thousands of heliostats eliminates the most expensive elements of the solar field and those that require the greatest resources for their maintenance. In addition, the proposed Fresnel configuration limits the main drawbacks of previous SRTA configurations (huge civil works costs, high wind loads, etc.).
- High concentration ratios and high thermal efficiency of the system. The system allows concentration ratios much higher than those of the parabolic trough collectors and maximum concentrations in the upper part of the receiver close to those of the solar towers.
- High operating temperatures allow high efficiency of the thermodynamic cycle that is fed by the system. Additionally, higher operating temperatures reduce the size of storage tanks and improve the cost-effectiveness of the thermal storage system, contributing to lower electricity costs.
- Scalable plants to any power output as a result of the modularity of the concept. The maximum size is not limited, and it is possible to think of plants with large power blocks that are more efficient and large centralized thermal storage systems that are more cost-effective. If needed, smaller plants can also be designed to produce electricity or to supply heat to industrial processes.

In turn, to achieve all these potential advantages (low costs, high efficiencies, and cheap energy storage), a system that moves the receiver must be developed, which introduces important challenges to be solved. The next section also describes other possible drawbacks of SRTAs and the strategy followed to minimize them. A 3D view of a MOSAIC module is shown in Figure 4.

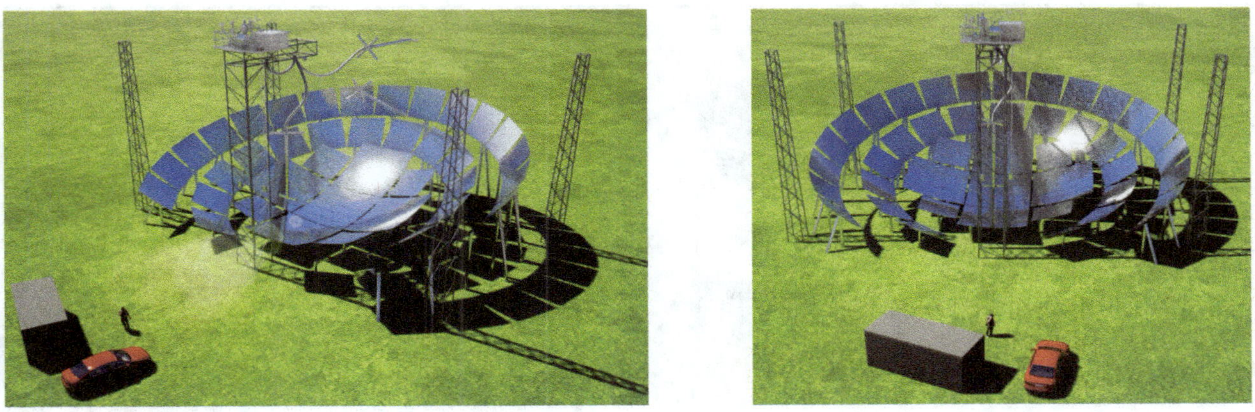

Figure 4. 3D view of a MOSAIC module.

4. Results and Discussion

Given the novelty of the MOSAIC approach, there are specific features and challenges that need to be further studied, tested, and validated. In order to gain the required experience, a complete prototype has been designed and is being built in Spain. The following describes the configuration of the system and of the various components, their main characteristics, and the reasons that led to the choice of that particular design

4.1. Solar Field.

As in other SRTAs, the solar field is fixed. This opens up great savings potential, as no drives, wiring, or trenching are required for the solar field. However, previous projects [21] have shown that these savings can be compromised when trying to build systems of relevant size for large-scale power plants.

A key parameter to define in order to optimize an SRTA is its tilt angle, which depends mainly on the latitude where it will be installed. As presented in a previous study [7], most interesting latitudes are above 30°, which can lead to rather high tilt angles in other to minimize cosine effect.

High tilt angles are not desirable for large SRTAs, as they result in higher and steeper solar field structures that are more exposed to strong winds (concentrators are fixed and cannot be moved to the stow position). This requires high, heavy-duty structures or huge excavations. Elevated parts can also also be a problem regarding accessibility for maintenance and repair.

The MOSAIC project addresses this problem through a Fresnel configuration. To ensure that most of the solar field is close to the ground, even with minimal or no excavation, the concentrator uses a set of concentric spheres. In addition, the discontinuities between spheres allow access corridors to the entire solar field. The long continuous ring-shaped surfaces also open up the possibility of developing automatic cleaning devices that slide over the mirrors, thus simplifying cleaning, which is the most resource-consuming maintenance operation in a CSP plant.

The optimal configuration must balance the costs of the system and its optical efficiency. As explained by Sánchez [23], this economic optimum is not achieved with a flat Fresnel configuration but with a "semi-Fresnel" configuration (see Table 2).

Table 2. Analysis of possible SRTA configurations and their effect on the annual contribution of each area [23]. The annual contribution graphs represent the kWh/m^2 provided by each mirror seen from the center of the spheres (aperture plane).

	'Actual Sphere'	**'Flat Fresnel'**	**'Semi-Fresnel'**
Configuration	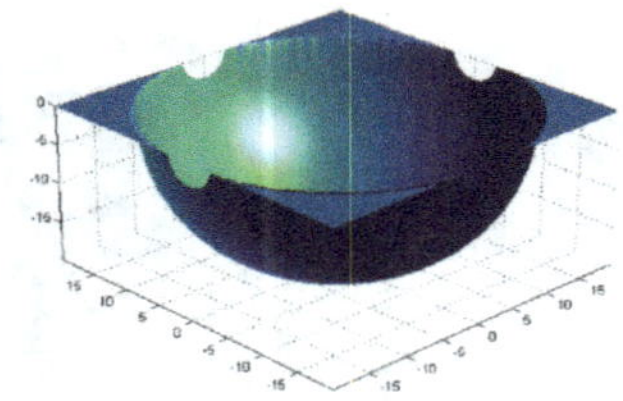	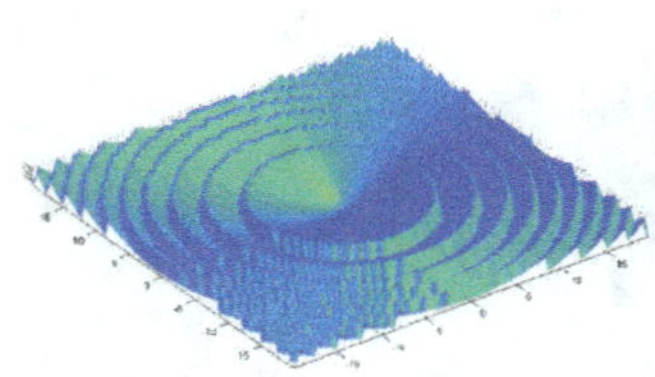	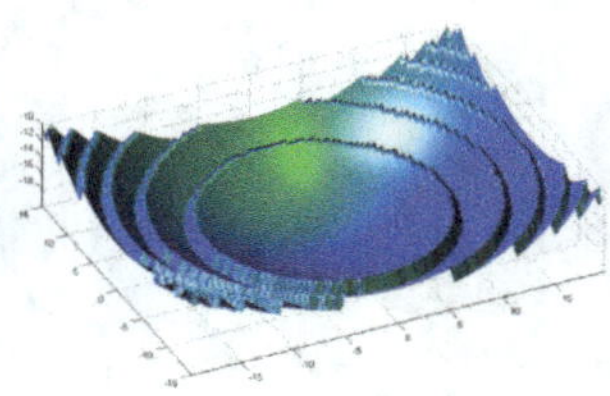
Annual contribution	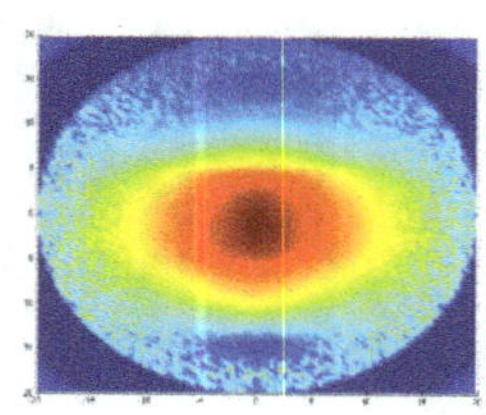	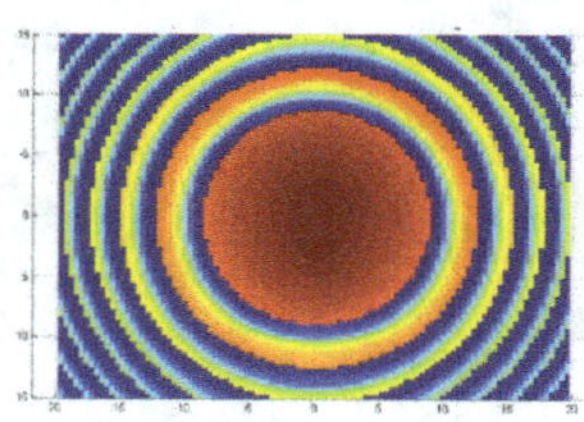	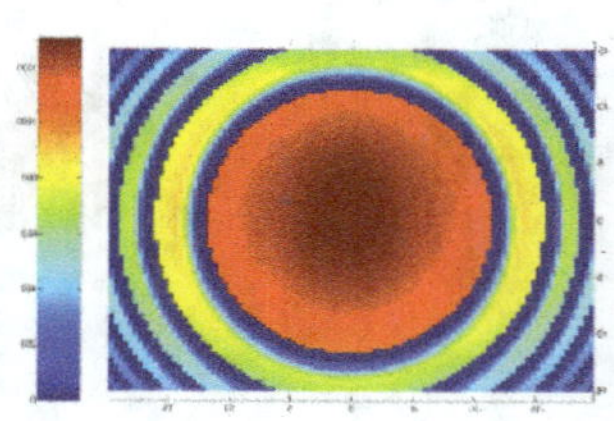

Depending on the latitude, terrain characteristics, expected wind speeds, etc., a different optimal module configuration can be defined. A MOSAIC module will always comprise a central bowl and additional rings corresponding to increasingly large concentric spheres separated by corridors (see Figure 5). Note that the descriptions will consider plants located in the northern hemisphere.

Table 2 shows that each mirror contributes unevenly. The mirrors in the center provide more energy. The mirrors on the top (north side of the module) will require higher supporting structures but will contribute more in winter while the lower mirrors (south side) will contribute more in summer. All this must be considered when deciding which mirrors will be implemented to obtain a balanced production throughout the year at minimum cost. Finally, practical considerations such as the width of the passageways have been considered to define the optimal configuration.

(a) (b)

Figure 5. Definition of the solar field for a MOSAIC module: (**a**) Schematic cross-section showing a central bowl and three rings beside a cylindrical receiver; (**b**) Annual contribution for a given configuration and tilt angle. The southern mirrors contributing less have been removed.

The design process defines the optimal size, the tilt angle, the number of curvatures (spheres), the width of the rings, the part of the collector to be excavated, and the part to be installed above ground level, the mirror areas to be implemented, etc. Given the cross-influences of the different variables and the practical constraints of design, the optimization process is not linear but iterative.

The first analysis led to modules with an aperture radius of 20 m, providing a maximum thermal power of around 500 kWt. Larger modules were also considered; these designs included a central bowl and three additional rings (see Figure 5b). The tilt angle would be 7.1° at latitudes of 30.5°.

The mirrors themselves also have particular requirements and impose additional restrictions. SRTAs require spherical mirrors, i.e., mirrors that are curved in two axes. Even in the case of relatively large spherical modules, their curvatures will be relatively strong compared to the facets of heliostats. This implies a greater complexity in their design and manufacture, which, in turn, limits the maximum size of the spherical mirror to be manufactured.

Auroville's solar bowl [15] addressed this problem by discretizing the reflective surface into a large number of small flat mirrors (see Figure 6). This slightly limits the concentration ratio, but above all, greatly increases the installation and canting time required. The Crete project [21] fixed thin (flexible) mirrors on a spherical concrete surface, where they were shaped, but ended up corroding in a short time. In addition, the mirrors, once attached, cannot be readjusted if required during their lifetime.

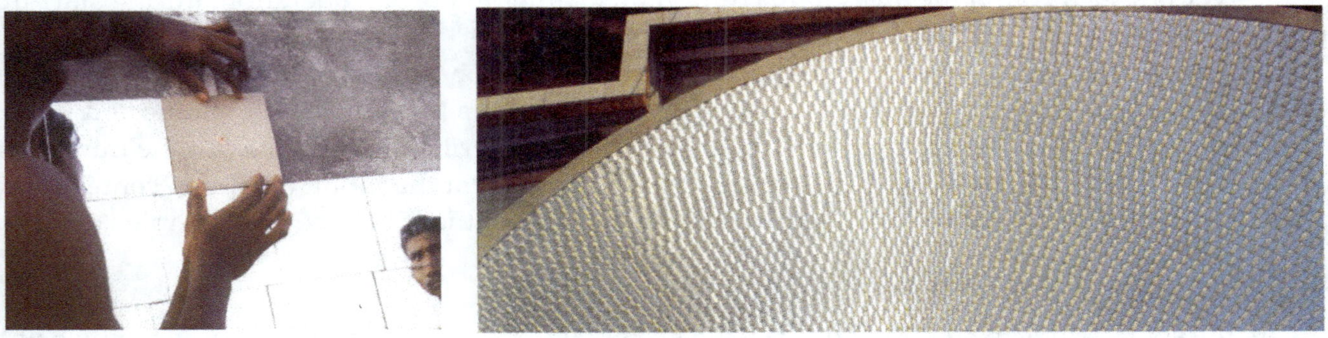

(a) (b)

Figure 6. Small facets glued to the spherical concrete bowl in Auroville, India [15]: (**a**) Gluing and canting process; (**b**) Mirrors seen from the center of the sphere.

In addition, hot spots (points of high concentration) will appear on the mirrors due to secondary reflections. Taking all this into account, Rioglass has developed tailored mirrors for the MOSAIC concept that have passed the laboratory tests. The mirrors to be manufactured have a surface of 1 x 1 m and a spherical geometry with the radius of the corresponding sphere. Accordingly, the width of the rings can only have discrete values (e.g. 1, 2, or 3 m).

With all this in mind, a prototype of appropriate size was defined to validate the concept and is under construction in Sangüesa, Navarra, Spain (latitude 42.59°). It includes a central bowl, and two incomplete outer rings tilted 15°. In total, there will be 600 mirrors of 1 m^2 (mirrors from the south that contribute less will be eliminated). The radii of curvature of the corresponding spheres are 15, 16.1, and 17.9 m, and together, they provide a peak thermal power close to 300 kWt. The aperture diameter of the system will be 30 m.

In order to optimize manufacturing costs, we have also aimed to use identical structures for different areas of the solar field. To this end, all mirrors will be installed in 3 x 3 or 5 x 5 mirror modules, which will then be lifted into position (see Figure 7). Depending on wind conditions and soil characteristics, future plants could implement a partially buried solar field. For the prototype, all mirrors will be installed in structures above ground level, which will facilitate any re-shaping or maintenance work, and access to the back of the mirrors.

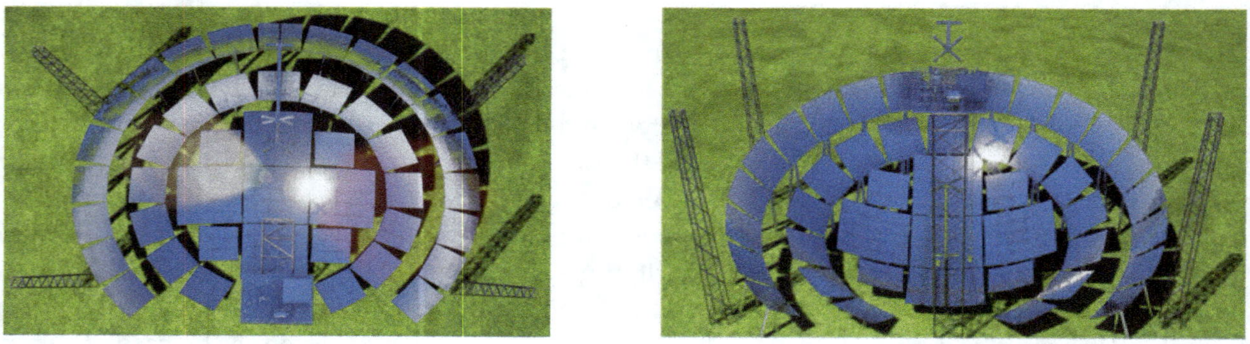

Figure 7. Solar field configuration to be implemented in the prototype. The outer rings are composed of 9-mirror modules (3 × 3), while the central bowl includes 25-mirror modules (5 × 5) and 9-mirror modules (3 × 3): (**a**) Top view; (**b**) Front view.

Another remarkable effect of SRTAs is that they do not have the ability to defocus the solar field. That is, a concentrated solar flux will always exist unless mirrors are covered. In contrast, the distribution of concentrated solar flux is known and is fixed for each day and time of the year. Therefore, it has to be guaranteed that no one can access this elevated area and that no element except the receiver passes through the areas of maximum flux, either in normal operation, during start-up, and shutdown, in emergencies or during maintenance operations.

One last key advantage of the proposed configuration should be highlighted. All other CSP plants require complex assembly systems and optical devices to regularly verify that the reflected flux is correct. In contrast, the optical quality of the SRTAs is easily verifiable on-site, as all the mirrors in the field have a common focal point. Therefore, a person placed at that point (common center of the spheres) can validate the entire solar field in a single step and without the need for complex devices.

4.2. Tracking System.

As the solar field is fixed, only one tracking system will be needed for each module, which will move the receiver to track the sun. This can lead to a reduction in investment costs, as well as fewer failures and maintenance operations.

In the past, SRTA tracking systems have relied on heavy and rigid structures to ensure a fixed point in the center of the sphere, where a beam is supporting the receiver pivots during its movement. They used a tripod-type configuration, such as those described in Figure 8. This tripod must be rigid to keep the center of rotation of the receiver in place. The arm holding the receiver and the receiver itself must also be rigid enough to ensure precise positioning, even for the highest elevations of the receiver (sunrise or sunset). The weight and cost of these stiff structures increase quadratically to the size of the system. In fact, projects implementing large solar bowls such as those in Crosbyton or Crete showed the high cost of the structures required to support the receiver.

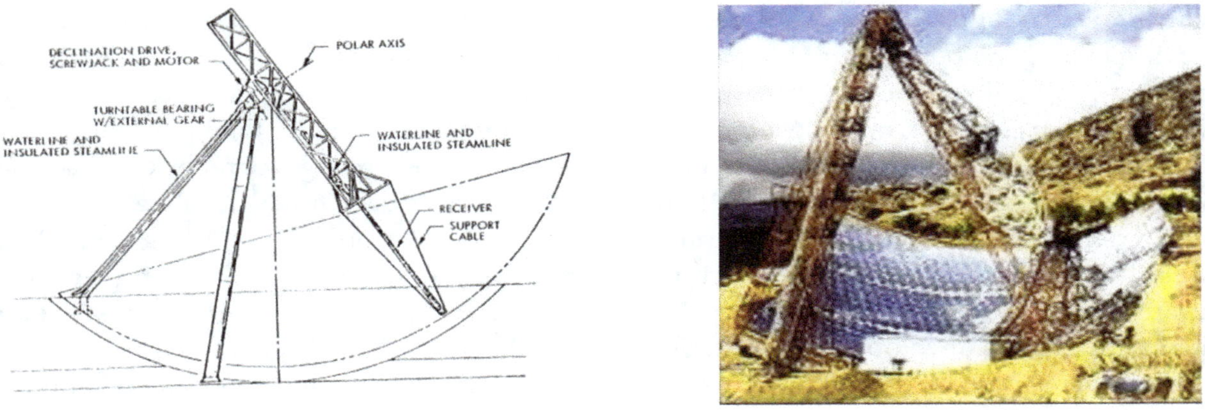

(a) (b)

Figure 8. Tripod-type structures supporting the tracking system, (**a**) the design for the system in Crosbyton as described in [24]; (**b**) Installation in Crete that included a tripod and an additional structure rolling over the spherical surface.

In the MOSAIC project, a new approach is adopted. The receiver will be suspended by several metallic cables, which will define its position in the air above the fixed mirror. These cables will be pulled from four light towers to correctly position and orient the receiver during the day.

The use of cables to operate the receiver had already been proposed [12,17], but those designs included a tripod to hold the receiver (see Figure 9). That is, the cables did not hold the receiver but simply pulled it to make it pivot around the center of the sphere. Therefore, they did not use the full potential of the cable drives.

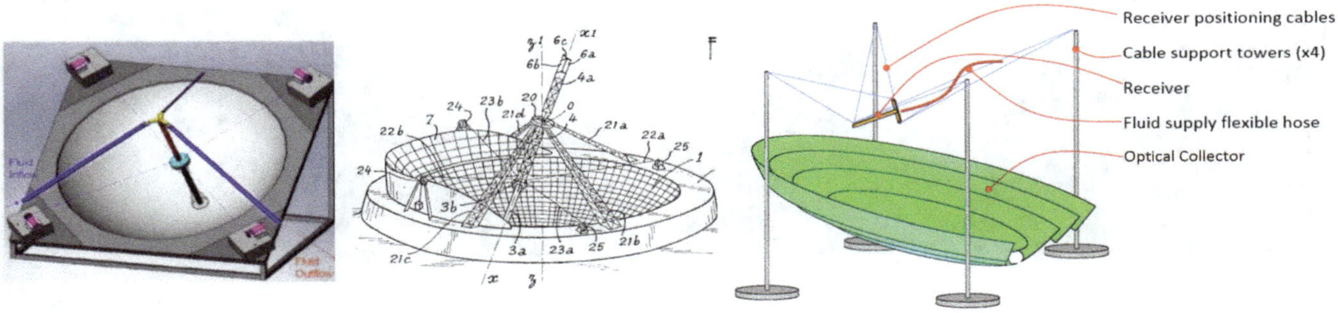

(a) (b) (c)

Figure 9. Proposed solution and previous approaches combining tripod-type structures and cable traction: (**a**) solar heat production system proposed by Cohen [12]; (**b**) system patented by Authier [17]; (**c**) sketch of the proposed solution.

The use of cables to drive and hold the receiver opens up new opportunities for cost reduction. Cable-driven handling solutions have inherent advantages, such as the ability to store cables on reels, provide large workspaces, relatively low moving masses, or low manufacturing costs. However, the accurate positioning of the element to be moved presents several challenges due to the compliance of the cables and supports, which operate under considerably different tensions depending on the position of the moving part, as well as due to the uncertainties of the nominal geometry of the plant. Therefore, a closed-loop controlled system is required to position the receiver in the desired poses accurately.

Unlike solar tower systems, the entire solar field of a module behaves like a single concentrator (it produces a solar flux that is defined by the position of the reflectors, the location of the site, and the solar time) and a single receiver must be positioned. As a result, closed-loop controlled systems can be implemented, as long as the actual position of the receiver can be determined. The proposed solution includes a position closed-loop control system based on artificial vision. Kortaberria et al. anticipated a possible implementation in a previous paper [25]; this allows for additional savings to be made since the tracking errors can be measured and corrected. That is, it is possible to relax the requirements related to tracking units, structure rigidity, and foundations and to use cheaper systems. This control software, although complex, once developed, will not add much cost to the total plant budget.

Regarding the system kinematics, defining the settings for positioning a parallel kinematic system, like the current one, is not a straightforward problem, but it involves complex non-linear mathematics. In order to define the optimal configuration (the height and position of the pulling points), and to ensure that the system is capable of reaching all the required workspace, avoiding singularities, and minimizing the required pulling forces, Matlab®models have been developed and subsequently validated with models developed in Adams®.

The receiver should cover a wide range of positions, but not all positions provide the same energy or are equally accessible. The receiver must be aligned with the sun and the center of the sphere during tracking. Early and late in the day, this implies higher positions and nearly horizontal orientations. This places higher requirements on the tracking system, which makes the system more expensive. On the other hand, in the morning or afternoon, the system provides less energy [7]. Therefore, the workspace has been optimized for latitudes between 30° and 40°discarding non-economic positions.

Figure 10 shows the final design, which includes 4 towers and 8 actuated cables. A 'parking' position has also been added. The required cameras for the closed-loop control system will be installed in the pulling towers. Active targets will also be integrated into the receiver's support arms.

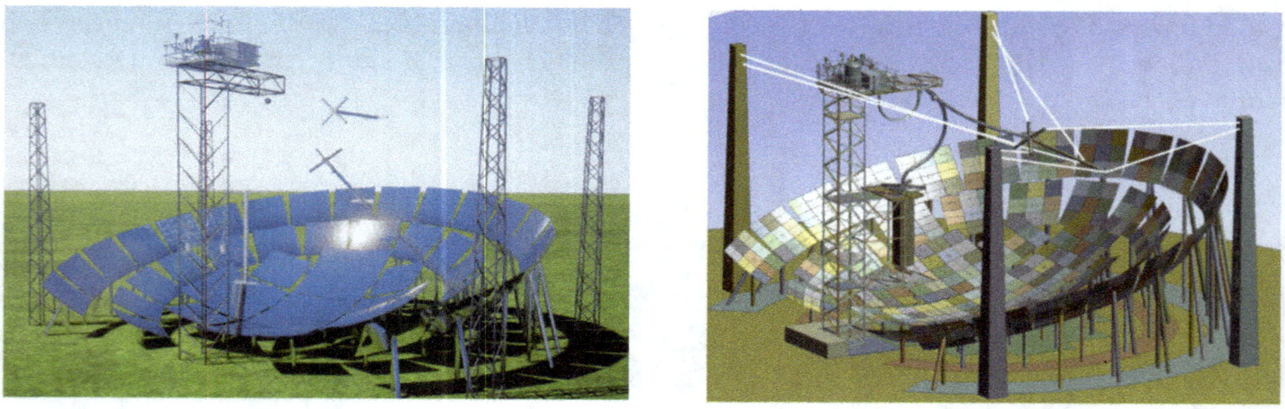

(a) (b)

Figure 10. Proposed tracking configuration: (**a**) Receiver at noon positions for two different seasons; (**b**) schematic representation of the cables that hold and move the receiver.

Another feature of this approach is that the cables allow the receiver to be easily brought to the floor for any maintenance task, thus reducing costs and increasing system availability.

Another challenge of this solution is how to get the HTF to/from a moving receiver in all the required positions. Senior Flexonics is developing a customized, flexible hose for the MOSAIC concept (see Figure 10). Preliminary tests on a full-scale hose prototype showed that the hose could reach all required positions.

4.3. Receiver

Previously suggested receivers [15–17,26] included tubes wound around a cylinder (materialized or not) or a bundle of tubes placed according to the generating lines of that cylinder, [17] or added conical shapes at the top end (see Figure 11), or even used volumetric receivers [21] placed in the zone of maximum flow. Transparent covers (see Figure 11c) were also proposed to minimize convection losses. However, the reflected rays impinge very parallel to the receiver in the areas of maximum flux, and therefore, such a cover may be counterproductive.

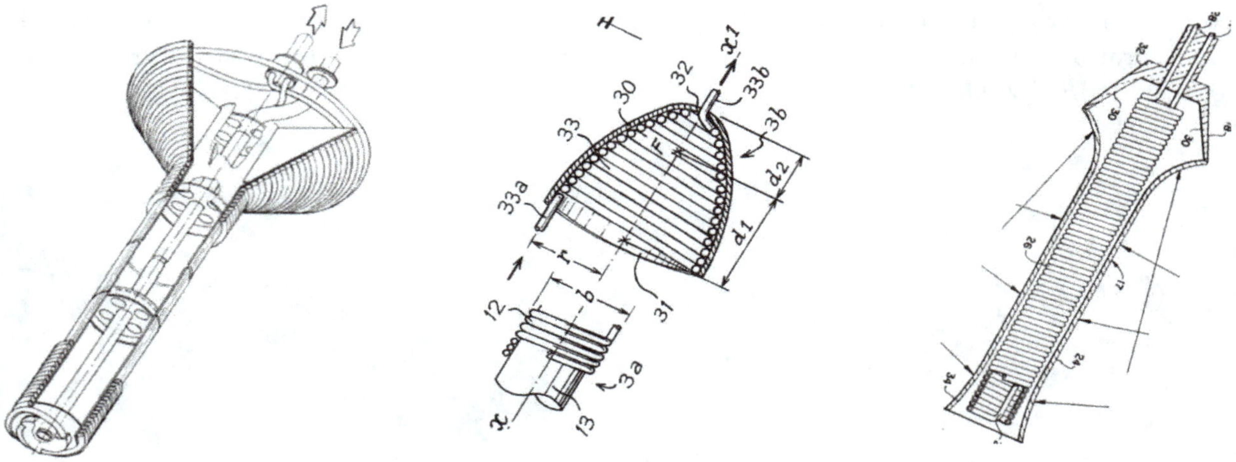

(a) (b) (c)

Figure 11. Receivers made of spiral tubes: (**a**) Design implemented in Auroville, India [15]; (**b**) design proposed by Authier [17] including a pre-heater (3a) and a high concentration heater on top (3b); (**c**) receiver patented by Steward [26], including a transparent envelope (24).

To facilitate thermal storage, the MOSAIC concept will use liquids as HTF, preferably molten salts. Optical and thermal-fluid-dynamic analyses of different receiver configurations (see Figure 12) were carried out. A ray-tracing simulation model developed in Tonatiuh has allowed the determination in detail of the incident flux map on the receiver, for each time of the year. This flux information has been used for the design of the receiver, which has been carried out in Modelica®. For each configuration, the influence of different design parameters on thermal efficiency, pumping losses or thermal stress has been analyzed. In addition, other practical considerations such as the manufacturability and drainage of the HTF were also considered. Details of the design process and the models developed will be included in [27].

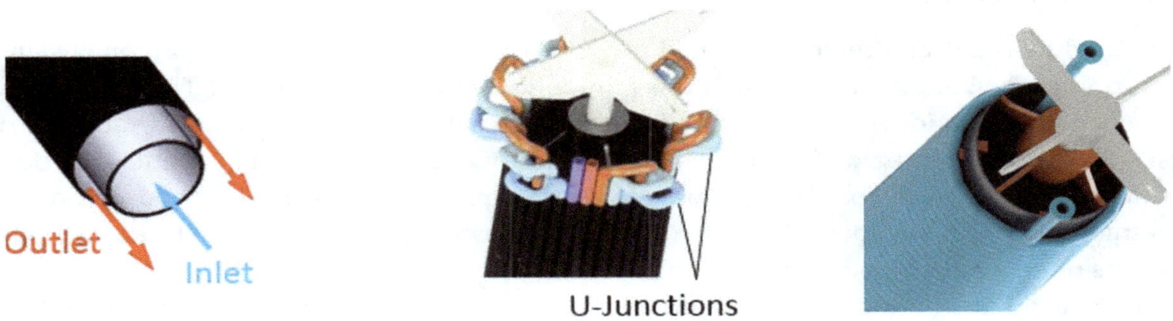

(a) (b) (c)

Figure 12. Different configurations analyzed: (**a**) Annular configuration; (**b**) tubes placed according to the generating lines of a cylinder; (**c**) coil type configuration.

As shown in [7], the solar flux incident on the surface of the receiver is very uneven and varies throughout the day and from one day to another. However, unlike tower power plants, where the flux distribution on the receiver has high uncertainty, the SRTA solar field is fixed, and the theoretical flux distribution is known in advance. This is a guarantee for the safety of the receiver, as this calculated value can never be exceeded.

As a result of this study, a coil type configuration was selected (see Figure 13). It includes a bundle of three parallel helical pipes made of Inconel Alloy 625. The 1-inch tubes are wound on a cylindrical surface at the bottom and a 40° conical surface at the top. Taking into account manufacturing constraints and the spillage due to mirror inaccuracies, the outer diameter of the cylindrical part of the receiver has been fixed to 0.3 m. In order to maximize efficiency, HTF enters the receiver on the lowest end located closest to the mirror surface and leaves the receiver at the top as it is the zone of maximum flux concentration. To increase solar absorptance, the receiver will be painted with Pyromark®High Temperature Paint 2500 Flat Black.

(a) (b)

Figure 13. Receiver design: (**a**) Main dimensions; (**b**) 3D view.

Previous implementations used 'complete' receivers (bottom of the receiver is close to the mirror). However, the radiation received at the receiver increases from the mirror to the top of the receiver. A shorter receiver will be cheaper, lighter, and easier to move. What is more, it will have lower thermal and pumping losses. After an analysis of the costs, thermal losses, and intercepted radiation [27], 40% of the receiver closest to the mirror will be removed.

4.4. Modular Plant Configuration

Compared to a central receiver CSP plant, a MOSAIC plant will have more components and more kilometers of piping, which is more similar to a parabolic through the STE plant. This means that more attention needs to be paid to the analysis of losses (thermal and pumping), and that there may be more risks of leakage or freezing if molten salts are used. On the contrary, all modules are built from similar components, so it is possible to replicate the different modules from standardized elements. The tracking towers can also be industrialized and pre-assembled, allowing fast deployment in the field. All this offers enormous cost reduction potential.

In addition, adjacent modules can share the tracking towers. This way, each tower will support the cables of the four adjacent modules, resulting in additional savings.

5. Conclusions

The MOSAIC concept proposes a new SRTA approach for a high-concentration modular plant to produce large scale solar electricity or heat. Compared to other CSP approaches, MOSAIC presents potential advantages (e.g., lower solar field costs, easier maintenance, high efficiencies, cheap energy storage, tailored power capacity) that could lead to more competitive CSP plants.

SRTA systems have demonstrated their applicability for solar heat collection at intermediate and high temperatures, but their large-scale implementation has been poorly studied. Compared to previous SRTA configurations, MOSAIC also addresses the main drawbacks of previous fixed mirror systems. Nevertheless, this new configuration presents uncertainties and challenges that need to be overcome. This article presents the advantages and challenges of this new concept and the measures proposed to overcome them. Some of the most relevant new features of the system are:

- The semi-Fresnel design of the solar field reduces civil works costs and wind loads, thus addressing the main drawbacks of previous SRTA configurations.
- The cable-based tracking system that replaces the numerous heliostat drives offers cost reduction potential (CAPEX and OPEX).
- A modular configuration that allows the development of large, cost-efficient plants, as well as smaller plants for electricity or heat production.

In the coming months, the construction of a large prototype (>300 kWt) will be completed at the CENER facilities (Spain). The results measured and the lessons learned during the implementation and operation of the prototype will make it possible to validate the concept, weigh up the hypotheses and propose improvements for future versions. Consequently, the proposed design for future commercial plants may be redefined.

Author Contributions: Conceptualization, C.V. and M.S.; Methodology, C.V. and M.S.; Validation, S.H. and A.B.; Formal Analysis, D.O.; I.P.; A.P. and S.H.; Investigation, C.V.; S.H.; D.O.; I.P.; A.P.; M.S. and A.B.; Writing-Original Draft Preparation, C.V.; Writing-Review & Editing, S.H. and M.S.; Project Administration, C.V.; Funding Acquisition, C.V. and M.S. All authors have read and agreed to the published version of the manuscript.

Acknowledgments: We would like to express our sincere gratitude to our colleagues John Harper and Gilles Guigan for their warm welcome at Auroville and their valuable advice.

References

1. McPherson, M.; Mehos, M.; Denholm, P. Leveraging concentrating solar power plant dispatchability: A review of the impacts of global market structures and policy. *Energy Policy* **2020**, *139*, 111335. [CrossRef]
2. Crespo, D. STE Can Replace Coal, Nuclear and Early Gas as demonstrated in an Hourly Simulation over 4 Years in the Spanish Electricity Mix. *AIP Conf. Proc.* **2019**, *2126*, 130003. [CrossRef]
3. Rashid, K.; Safdarnejad, S.M.; Ellingwood, K.; Powell, M. Techno-economic evaluation of different hybridization schemes for a solar thermal/gas power plant. *Energy* **2019**, *181*, 91–106. [CrossRef]
4. Rashid, K.; Mohammadi, K.; Powell, K. Dynamic simulation and techno-economic analysis of a concentrated solar power (CSP) plant hybridized with both thermal energy storage and natural gas. *J. Clean. Prod.* **2020**, *248*, 119193. [CrossRef]
5. Rovense, F.; Reyes-Belmonte, M.A.; Gonzalez-Aguilar, J.; Amelio, M.; Bova, S.; Romero, M. Application of un-fired closed Brayton cycle with mass flow regulation and particles based thermal energy storage systems for CSP. *AIP Conf. Proc.* **2019**, *2126*, 030047. [CrossRef]
6. Bhargava, K.R.; Grossb, F.; Schramek, P. Life Cycle cost optimized heliostat size for power towers. *Energy Proc.* **2014**, *49*, 40–49. [CrossRef]
7. Villasante, C.; Pagola, I.; Pena, A.; Sanchez, M.; Olarra, A.; Gomez-Acedo, E.; Herrero, S. MOSAIC, a new CSP plant concept for the highest concentration ratios at the lowest cost. *AIP Conf. Proc.* **2019**, *2126*, 060008. [CrossRef]
8. Steward, W.G.; Kreith, F. Stationary concentrating reflector/tracking absorber solar energy collector: Optical design characteristics. *Appl. Opt.* **1975**, *14*, 1509–1512. [CrossRef] [PubMed]
9. Dirks, J.A.; Williams, T.A.; Brown, D.R. Performance and Cost Implications of the Fixed Mirror, Distributed Focus (FMDF) Collector. *J. Sol. Energy Eng.* **1992**, *114*, 254–259. [CrossRef]
10. Goodman, J.H. *Solar Concentrating Architectonics; Solar Bowl Architectonics and Interior Heliostats Architectonics*; c/o HANDance Designs: Walton: Spring Green, WI, USA, 1993.
11. Adams, W. *Solar Heat-A Substitute for Fuel in Tropical Countries for Heating Steam Boilers and Other Purposes*; NYC Public Library; N.U.C. #NA 0065154; Education Society's Press: Byculla, Bombay, India, 1878; p. 112.
12. Cohen, S.; Grossman, G. Development of a solar collector with a stationary spherical reflector/tracking absorber for industrial process heat. *Sol. Energy* **2016**, *128*, 31–40. [CrossRef]
13. Berland, J.J.G. Machine Solaire à Air Atmosphérique. French Patent No. FR635283 (A), 3 Decembar 1928.
14. Kreider, J.F. Thermal performance analysis of the stationary reflector/tracking absorber (SRTA) solar concentrator. *ASME J. Heat Transf.* **1975**, *97*, 451–456. [CrossRef]

15. Guigan, G.; Harper, J. Auroville Solar Bowl Concentrator for Community Scale Steam Cooking. Report of the Project Funded by the Government of India, May 2008. Available online: https://research.auroville.org/system/papers/attachments/000/000/623/original/Auroville_Solar_Bowl_Concentrator_for_the_Communit..Cooking_Auroshilpam_2008.pdf (accessed on 30 January 2020).
16. O'Hair, E.A.; Simpson, T.L.; Green, B. Results from Operation of the Crosbyton Solar Bowl. In Proceedings of the 8th ASME Solar Energy Division Conference, Anaheim, CA, USA; Ferber, R.R., Ed.; American Society of Mechanical Engineers: New York, NY, USA, 1986; pp. 205–209.
17. Authier, B. Capteur D'energie Solaire a Miroir Spherique Fixe. French Patent No. FR2365085 (A1), 14 April 1978.
18. Grossman, G.; Fruchter, E.; Kreith, F. An experimental investigation of a stationary reflector/tracking absorber solar collector at intermediate temperatures. *J. Sol. Energy Eng.* **1982**, *104*, 340–344.
19. Experimental Set-up for a SRTA at Technion-Israel Institute of Technology. Available online: https://www.youtube.com/watch?v=WBb7hDGUSng&t=182s (accessed on 1 April 2020).
20. Harper, J. *Construction of a Stationary Spherical Reflector for Rural Use*; Final report of the project sponsored by Tata Energy Research Institute; Tata Energy Research Institute: Auroville, Tamil Nadu, India, 1982.
21. Kenna, J. Construction and Testing of a Pilot Solar thermal power station using an innovative mirror concept. Report for JOR3CT960046 project (1996–1999) funded under FP4-NNE-JOULE C Program. Available online: http://cordis.europa.eu/publication/rcn/1452_en.html (accessed on 30 January 2020).
22. Larbi, A.B. A new design of a (3D) Fresnel collector with fixed mirrors and tracking absorber. *J. Sol. Energy Eng.* **2000**, *122*, 63–68. [CrossRef]
23. Sanchez, M. Potential of Optimized Non-Tracking Mirror Concentrators for Distributed Solar Applications. In Proceedings of the SolarPACES Conferences, Oaxaca, Mexico, October 2004.
24. Hariharan, S. Simulation of the receiver in a fixed-mirror distributed focus solar power system. Ph.D. Thesis, Texas Tech University, Lubbock, TX, USA, August 1981.
25. Kortaberria, G.; Gomez-Acedo, E.; Molina, J.; Tellaeche, A.; Minguez, R. Theoretical accuracy assessment of model-based photogrammetric approach for pose estimation of cylindrical elements. *Meas. Sci. Technol.* **2019**, *30*, 1–17. [CrossRef]
26. Steward, W.G. Receiver for solar energy. US Patent No. U.S.4173968 (A), 13 November 1979.
27. Zanino, R.; Cagnoli, M.; Falsig, J.J.; Pagola, I.; Peña, A.; Savoldi, L.; Villasante, C. Design and Analysis of the Helical Receiver for the MOSAIC Prototype Solar Bowl System. *J. Sol. Energy*. under review.

4

Novel Design of Primary Optical Elements based on a Linear Fresnel Lens for Concentrator Photovoltaic Technology

Thanh Tuan Pham [1,2], Ngoc Hai Vu [3,*] and Seoyong Shin [1,*]

1 Department of Information and Communication Engineering, Myongji University, Gyeonggi-do 17058, Korea; pttuan1412@gmail.com
2 Institute of Applied Material Science, Vietnam Academy of Science and Technology, Ho Chi Minh City 70072, Vietnam
3 Electrical and Electronics Engineering, Phenikaa University, Hanoi 1000, Vietnam
* Correspondence: hai.vungoc@phenikaa-uni.edu.vn (N.H.V.); sshin@mju.ac.kr (S.S.);

Abstract: In this paper, we present a design and optical simulation of a novel linear Fresnel lens. The lens can be applied to a concentrator photovoltaic (CPV) system as a primary optical element (POE) to increase the concentration ratio and improve the uniformity of irradiance distribution over the receiver. In addition, the CPV system can use the proposed lens as a concentrator without involving a secondary optical element (SOE). The designed lens, which is a combination of two linear Fresnel lenses placed perpendicular to each other, can collect and distribute the direct sunlight on two dimensions. The lens is first designed in the MATLAB program, based on the edge ray theorem, Snell's law, and the conservation of the optical path length, and then drawn in three dimensions (3D) by using LightTools™. Furthermore, in order to optimize the structure and investigate the performance of the lens, the ray tracing and the simulation are also performed in LightTools™. The results show that the newly designed lens can achieve a high concentration ratio of 576 times, a high optical efficiency of 82.4%, an acceptable tolerance of 0.84°, and high uniform irradiance of around 77% for both horizontal and vertical investigation lines over the receiver.

Keywords: linear Fresnel lens; concentrator photovoltaic (CPV) system; high concentrator ratio; uniform irradiance distribution

---

1. Introduction

In recent years, the development of photovoltaic technology has brought about significant improvement in the efficiency of solar cells. The latest record of direct conversion of sunlight into electricity of multi-junction solar cells reaches 46%, as published by the Fraunhofer Institute, Germany [1]. Therefore, solar energy [2–4] becomes a promising resource to completely replace fossil energy in the future. However, the high cost of multi-junction solar cells [5–7] leads to the price of the photovoltaic system being high. An effective way to reduce the cost per watt value is to cut down the size of the required cell area by using a cheaper optical system in concentrator photovoltaic (CPV) technology.

The CPV [8–11] is a potential technology to promote the popularity of solar energy by reducing the cost of the photovoltaic system [12]. Generally, the CPV consists of an optical part and multi-junction solar cells with high efficiency. The optical part operates as a concentrator to converge sunlight in a large area to a small receiver (multi-junction solar cells). The concentrator used in CPV is usually a mirror or a lens [13–15]. The most popular lens used in CPV system is the Fresnel lens, due to its compactness, low cost with high transparency, and availability for a high concentration ratio.

The Fresnel lens [16–18] used in CPV systems can be a linear Fresnel lens or a regular Fresnel lens (just called Fresnel lens) with circular grooves. In the framework of CPV technology, the Fresnel lens is applicable for a high concentration ratio (> 100 times) while the linear Fresnel lens is suitable to aim at a regular or low concentration ratio [19]. The Fresnel lens can focus the sunlight on two dimensions so that it easily reaches a high concentration ratio. Therefore, it is usually applied to CPV systems using high-efficiency multi-junction solar cells. On the other hand, the linear Fresnel lens is enabled only to focus the sunlight on one dimension, so it is normally used for solar thermal technology [20–22].

The Fresnel lens has been widely used as a primary optical component (POE) in CPV systems. However, a natural property of the Fresnel lens is that it always creates a hot spot in the center of the solar cell [23–25]. Thus, it distributes the sunlight over the receiver in a non-uniform manner, which reduce the efficiency of the solar cell. In addition, solar cells exposed to non-uniform irradiance distribution with a hot spot in the center can be degraded quickly by a high thermal effect if the cooling system is not good enough.

An effective way to improve the uniformity of irradiance over the solar cells is to use a combination consisting of a primary optical element (POE) and a secondary optical element (SOE) component [26–28]. Generally, the SOE has two functions in the optical part of the CPV system. Firstly, it helps to redistribute the sunlight over the receiver to improve the uniformity of irradiance distribution. Secondly, it helps to increase the sunlight tracking tolerance in the CPV system. Therefore, the combination of POE and SOE is the normal choice for the CPV system. The SOE usually has to operate under harsh conditions with high irradiance and high temperatures, demanding that its design and properties are of high quality to ensure its lifetime is equal to that of the CPV system [29,30].

In CPV technology, the optical system, consisting of the POE and the SOE, accounts for from 12% to 18% of the total system cost, in which the SOE cost is about 25–45% of the optical part depending on the concentration level (number of SOEs/m^2), the material of the SOE, and the technology [31–35]. Assume that a module carries 100 SOEs per square meter (in a geometry concentration ratio around 1000 times and solar cells size 1 cm^2), about 370,000 SOEs per MWp are needed under 30% module efficiency and 900 W/m^2 irradiances for concentration standard operating conditions [36], which is a huge number of SOEs. Therefore, it is undoubtedly economically efficient to have a CPV system that can work without employing SOEs. Several efforts have been made to design an optical CPV system without an SOE. Zhuang et al. [37] designed a hybrid Fresnel lens for generating uniform irradiance. In the research, the Fresnel lens with high concentration consists of two parts. The center part of the lens is designed based on refraction phenomena, while the outer area is designed based on internal reflection phenomena. Languy et al. [24] introduced a method to deal with chromatic aberration, in which the Fresnel lens grooves are made by polymethyl methacrylate and polycarbonate to enhance the concentration and achieve the spectrally uniform flux on the receiver. Leutz et al. [38] proposed a non-imaging Fresnel lens that had an improvement in the acceptance angle by using the edge ray principle method. Gonzalez et al. [39] used the Simultaneous Multiple Surface (SMS) method to increase the acceptance angle of the linear Fresnel lens. In the study, the lens has a small concentration ratio (less than 100 times), so it is suitable for low CPV systems (LCPV). In addition, there are many more research studies to design optical systems for CPV systems without SOEs, such as Noburo et al. [40], Akisawa et al. [41], Ryu et al. [42], etc. Although there are many studies on designing the concentrator of the CPV system by modifying the Fresnel lens [43–45], several disadvantages still need to be resolved.

In this paper, we propose a novel design of the linear Fresnel lens, which is able to collect and distribute the sunlight on two dimensions over the receiver. The lens structure consists of two faces: the upper groove surface and the lower groove surface, which are linear Fresnel lenses arranged perpendicular to each other. Each linear lens focuses sunlight on one dimension so that the whole novel lens, consisting of two linear Fresnel lenses, will focus sunlight on two dimensions. The linear Fresnel lenses used as upper and lower surfaces are designed by using the conservation of the optical path length and the edge ray theorem. The shape of the newly designed lens is constructed in the

MATLAB program. The data from MATLAB are then applied to LightTools™ software to draw the lens in three dimensions (3D). The ray tracing is used to optimize the lens structure. The simulation is performed to estimate the efficiency of the novel lens, which works as a concentrator in the CPV system.

The remainder of the paper is organized as follows: in Section 2, the basic idea of the design of the newly linear Fresnel lens is indicated. Section 3 describes in detail the process of building the grooves of the upper surface, while Section 4 demonstrates the construction procedure for the grooves of the lower surface. Section 5 presents the simulation results and discussion. Finally, the brief concluding remarks and future work are included in Section 6.

2. The Basic Idea

The conventional structure of the linear Fresnel lens consists of a series of linear grooves that act as individual refracting surfaces bending parallel light rays into a common focal line. In the structure, the lens collects and distributes sunlight on one dimension, so the lens usually has a small or regular geometry concentration ratio (smaller than 100 times). To increase the concentration level, the series of the grooves is revolved around an axis that is orthogonal to the lens at the center instead of extruding along an axis orthogonal to a cross-section of the grooves in the linear Fresnel lens. Figure 1 shows the way to build two kinds of Fresnel lenses, distributing sunlight on one or two dimensions. By revolving the grooves around an axis, the Fresnel lens can easily achieve a high concentration ratio. However, this structure type always distributes sunlight non-uniformly over the receiver leading to the appearance of a hot spot that negatively affects the efficiency and the lifetime of the multi-junction solar cells. A new idea is proposed to overcome these issues, in which a combination of two linear Fresnel lenses placed perpendicular to each other is presented. Figure 2 shows the basic structure of the novel linear Fresnel lens.

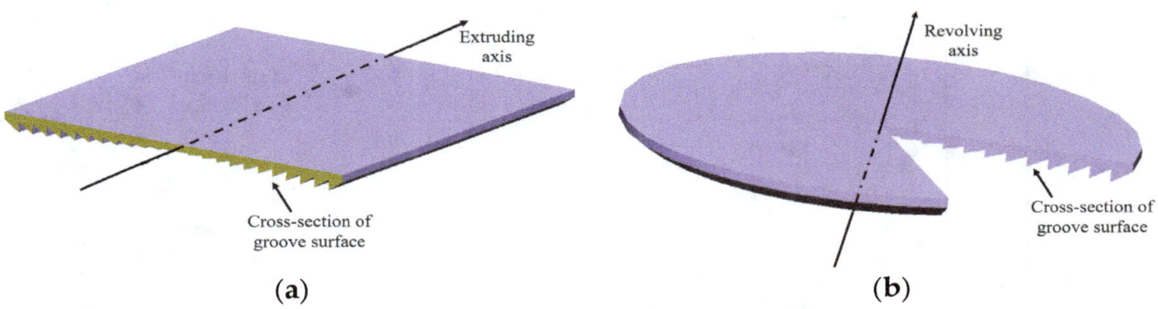

Figure 1. Two types of Fresnel lenses constructed from a cross-section of the grooves with (**a**) extruding axis and (**b**) revolving axis.

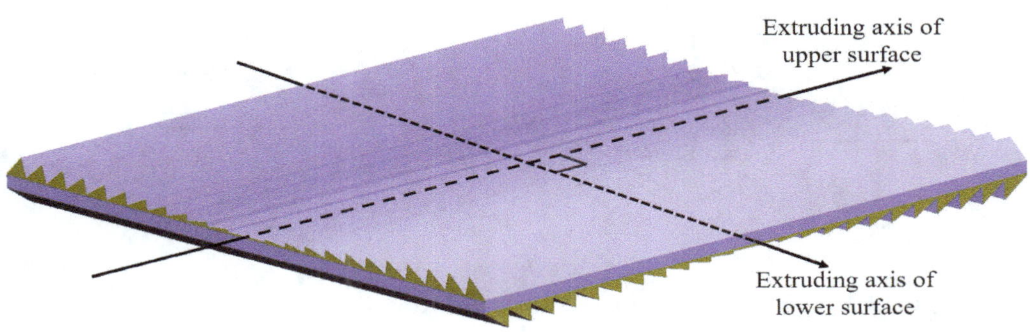

Figure 2. The structure of the novel Fresnel lens consisting of two surfaces placed perpendicular to each other.

The lens shape is constructed in the MATLAB program with the flowchart of design process shown in Figure 3. In the process, initial parameters must be chosen to build the lens. The initial parameters can be receiver size, distance between lens and receiver, size of grooves, geometry concentration ratio target, etc. Based on these parameters, the lens is constructed step by step from the center to the edge of the lens or vice versa, and from the upper surface to the lower surface. In addition, the structure of the lens has to be optimized to get a uniform irradiance distribution over the receiver. Such a challenge can be overcome by using a design with multi-focal points instead of a design with only one focal point, as in the case in the conventional design. In the novel design, each groove has its own focal point to distribute the sunlight uniformly over the receiver, leading to the whole lens being able to distribute the sunlight uniformly over the target area. Therefore, the estimation of focal point positions is important to obtain uniform irradiance distribution. The technique to estimate the positions of focal points is changed slightly in the design of the upper and the lower surfaces that will be described in detail in following sections.

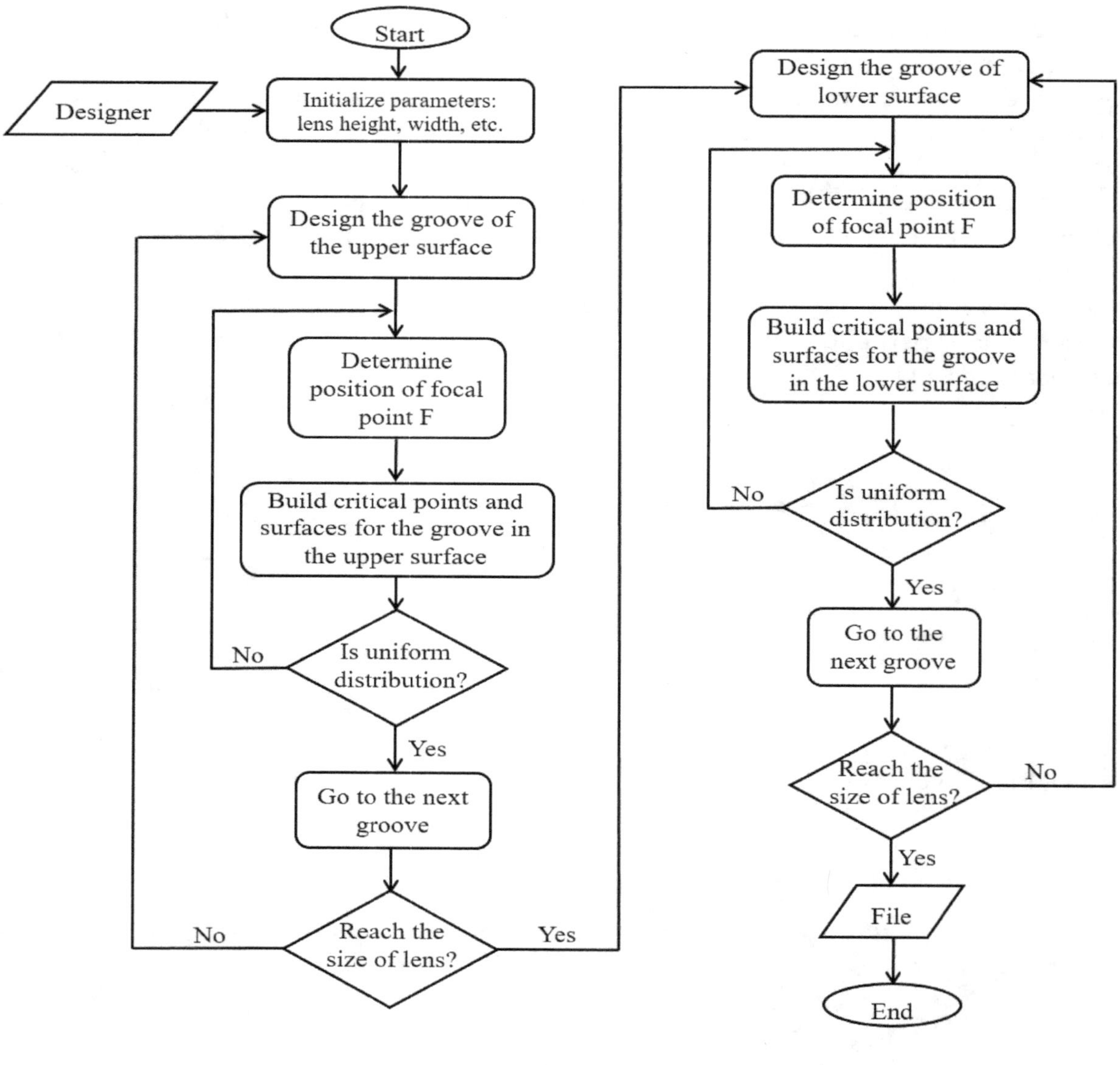

Figure 3. The flowchart of the lens designed in the MATLAB program.

Generally, direct sunlight is not perfect parallel so that the direct sunlight comes to the Earth surface with an angle about 0.265°, namely angular aperture of sunlight or solar angle [43]. As a result, the bundle of direct sunlight comes to every point of Fresnel lens under angle 0.265° with cone shape as example in Figure 4. However, it is much simpler to design the lens if we just use normal incident. Therefore, in this paper, the novel lens is designed by using the normal incidents.

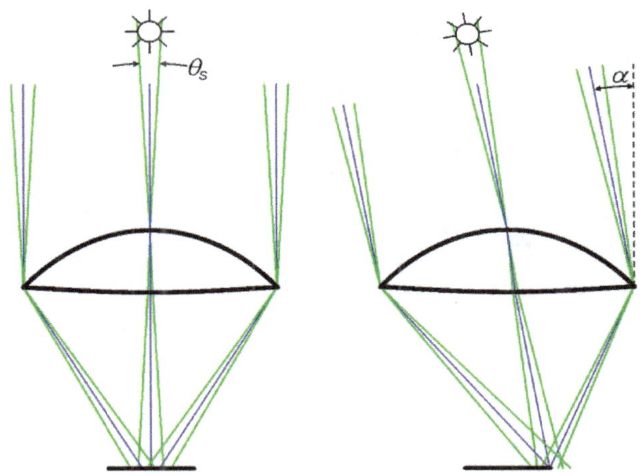

Figure 4. Angular aperture of sunlight θ_S and acceptance angle α.

In this study, the novel linear Fresnel lens is designed by using a combination of Snell's law, conservation of pathlength, and the edge ray principle. The edge ray principle is defined as a design method, in which the light rays coming from the edges of the source are redirected towards the edges of the receiver, this will ensure that all light rays coming from the inner points in the source will end up on the receiver. In this method, the image does not need to form, the only goal is to transfer the light from the source to the target. Edge ray principle is illustrated in Figure 5.

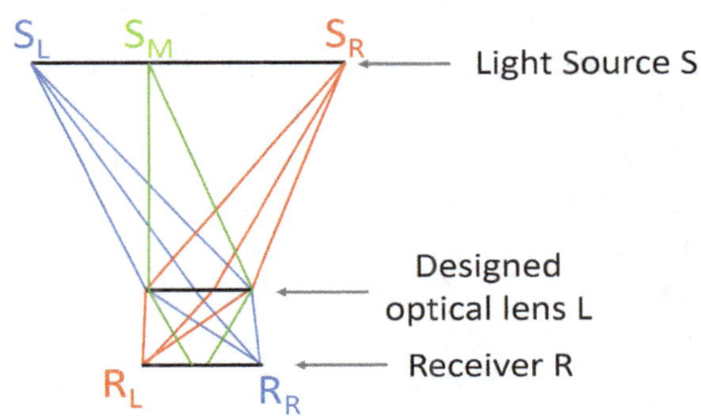

Figure 5. The basic of the edge ray principle, in which the illumination energy from light source S is transferred to receiver R without needed image formation.

Based on the edge ray principle, the ray coming to the edge of the designed Fresnel lens should be redirected to reach the extreme point of the receiver [46]. However, in our design, the novel linear Fresnel lens comprises two grove surfaces: the upper surface and lower surface, in which they are placed perpendicular to each other. Each groove surface consists of some grooves, which play main role to guide the direct sunlight to the receiver. To build the novel lens, we tried to design groove by groove. Each groove is designed by using Snell's law, the conservation of pathlength, and edge ray principle. Therefore, the bundle of rays coming to each groove consists of left edge ray and right edge ray that come to the groove at left edge point and right edge point of groove, respectively. Additionally, every ray between the left edge ray and right edge ray should be come to somewhere on the receiver. In another word, each groove plays role as a mini lens so that we can use the edge ray principle to design every groove. When the design process for one groove has been completed, the same procedure is repeated to build the next groove. The edge ray principle applied to design one groove is illustrated in Figure 6, in which the left edge ray, the right edge ray, and an example of ray between left and right edge rays are shown.

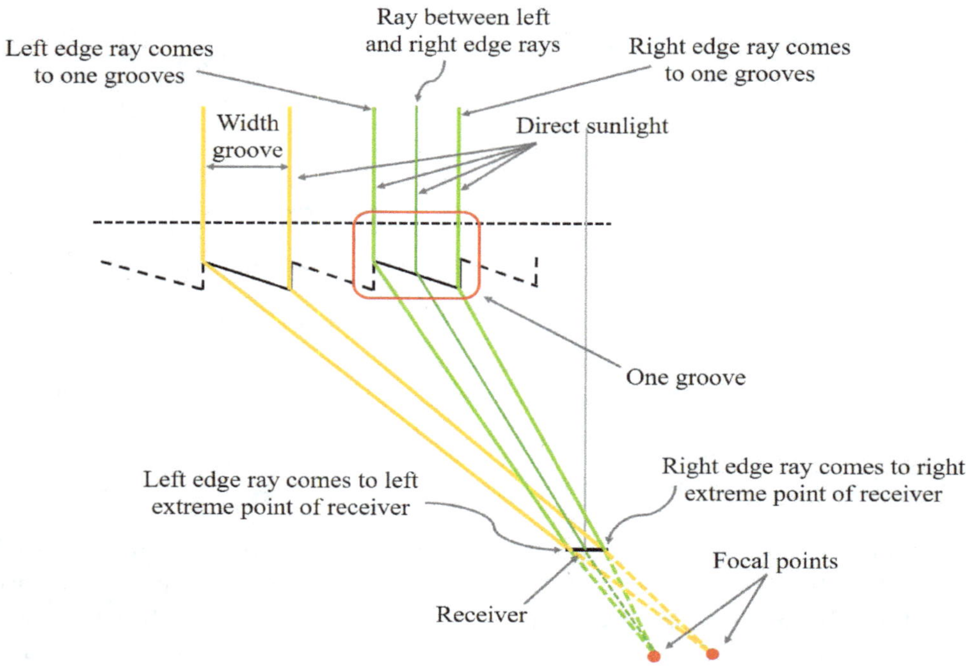

Figure 6. The edge ray principle is applied to design one groove, in which the left edge ray, the right edge ray, and an example of ray between left and right edge rays coming to one groove are shown.

Figures 7 and 8 show the difference between the conventional distribution and the new technique for sunlight distribution. In the new technique, the bundle of sunlight coming over one groove has the left and the right edge rays that go to the left and the right extreme sides of the receiver, respectively. The left and the right edge rays intersect with each other at an unreal point *F* (Figure 8), which can be either beneath the receiver or above the groove, depending on the size of the receiver and the groove. If the groove size is smaller than that of the receiver, the intersection point *F* will be above the groove. Otherwise, the intersection point *F* will be under the receiver side. Furthermore, the estimation of the focal point of each groove is important. A good focal point can help the bundle of sunlight to be distributed uniformly over the receiver. The bundle of rays coming to one groove is converged at the focal point *F* so that all the rays can reach the receiver. As a result, the direct sunlight arriving at one groove is distributed uniformly from the left to the right extreme sides of the receiver.

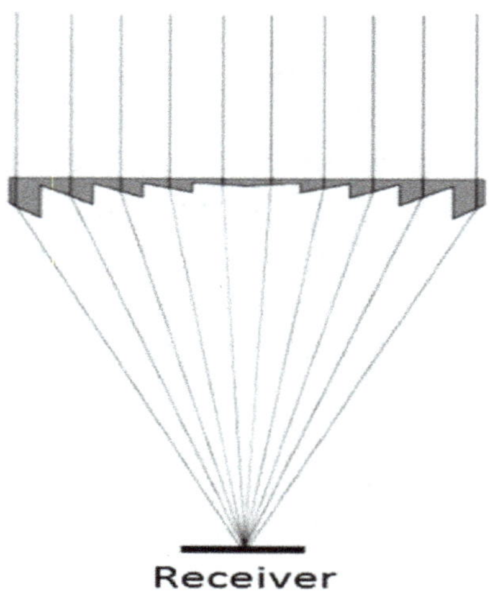

Figure 7. The irradiance distribution in the conventional design of the Fresnel lens, in which there is only one focal point at the center of the receiver.

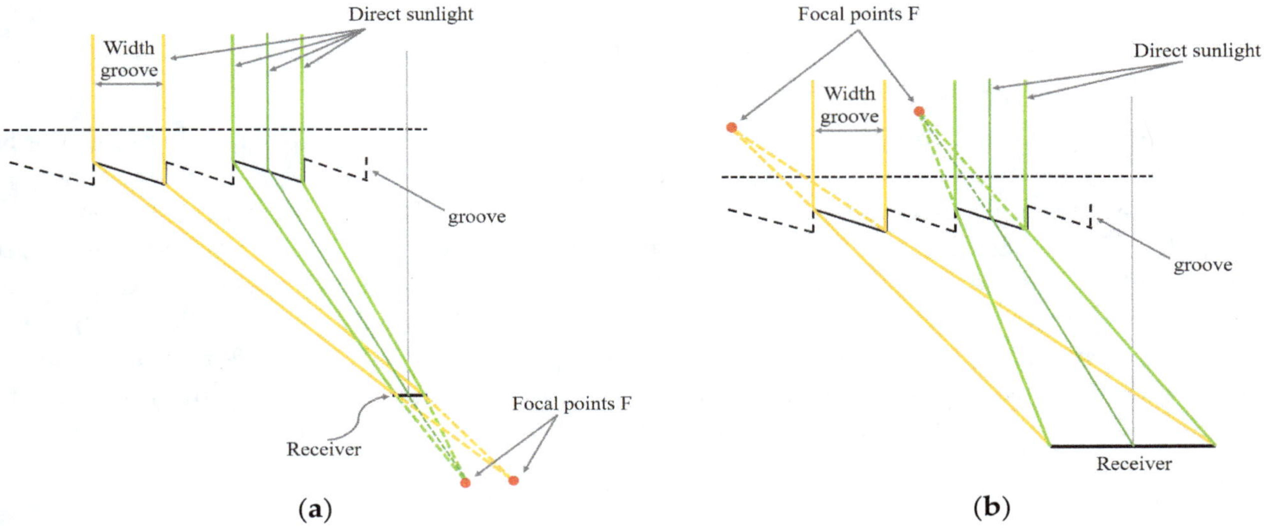

Figure 8. The new technique to distribute the sunlight uniformly over the receiver using multi-focal points that can be either (**a**) under the receiver or (**b**) above the lens depending on the size of the receiver and the grooves. Note: The figures are not drawn to scale.

3. Design of the Linear Fresnel Lens as an Upper Groove Surface

The novel linear Fresnel lens is built by a combination of a linear Fresnel lens with an upper groove surface, namely upper surface, and a linear Fresnel lens with a lower groove surface, namely lower surface. The task of the two surfaces is to distribute the direct sunlight. Each surface focuses the sunlight on one dimension (on a focal line). However, when two surfaces are placed perpendicular to each other to become a new structure, the lens will focus the sunlight on two dimensions. As a result, although the linear Fresnel lenses are used to collect the sunlight, a high concentration ratio can be achieved by the proposed structure. In this section, the design of the upper surface is described in detail.

In the proposed design, each groove is designed to distribute the sunlight uniformly over the half or the whole of the receiver (Figure 9) because of the symmetry of the structures of the lens and the receiver. In addition, the size of the grooves is chosen to be bigger than that of the receiver, thus the focal point of each groove lies beneath the receiver. In Figure 9, each groove is constructed by the Cartesian surface S_1 as input and the flat surface S_2 as output. The bundle of rays coming to one groove is refracted at the input surface and then exits the lens at the output surface. Finally, the bundle of rays converges at the focal point of the groove, in which the left edge ray reaches the left extreme point X_1 of the receiver while the right edge ray passes through the receiver center I_0. After that, two edge rays intersect with each other at the focal point F of the groove. All rays that lie between the left and the right edge rays also converge at the focal point F due to the conservation of optical path length. As a result, all rays of the bundle coming to one groove reach the left- or the right-half of the receiver, depending on the groove position. The design procedure of one groove is described with reference to Figure 9 as follows:

Step 1: Initialize parameters such as the width of the output beam w, the thickness of the lens d, the position of points P_0 and P_1, the height of the lens, etc. The width of the output beam w is defined as the distance between the left and the right edge rays that exit from the output surface S_2.

Step 2: The left edge ray coming to the point P_1 is first refracted by angle β_1 and then approach surface S_2 at P_2. After that, the ray is refracted again by angle β_2 and finally reaches the extreme point X_1 of the receiver. Because the position of P_1 is known in advance, the position of P_2 can be calculated as follows:

$$\tan \beta_2 = \frac{X_1 - xP_2}{yP_2}, \tag{1}$$

$$n \sin \beta_1 = \sin \beta_2, \tag{2}$$

$$\tan \beta_1 = \frac{xP_2 - xP_0}{yP_1 - yP_0}, \tag{3}$$

where X_1 is the position of the left extreme point of the receiver, n is the refractive index of the lens, xP_0, yP_0, yP_1, xP_2 are the positions of P_0, P_1, P_2 on the x-axis and the y-axis, respectively, β_1 and β_2 are the refractive angles of the left edge ray at the input surface S_1 and the output surface S_2, respectively.

Step 3: In this design, P_3 is the limit point of the surface S_1, which is a Cartesian oval surface, so it is difficult to find out the position of the point P_3. In contrast, the point P_5 is on the exit surface S_2, which is a flat surface, so it is easy to estimate the position of P_5 by using the position of P_2 and the width of output beam w as Equation (4). Therefore, the position of P_5 is calculated before estimating the position of P_3. In addition, the position of P_5 on the y-axis is equal to the height of the lens, thus the position of P_5 can be estimated totally as follows:

$$xP_5 = xP_2 + w. \tag{4}$$

Step 4: Depending on the size of the receiver and the grooves, the left and the right edge rays can intersect with each other at the position beneath the receiver or above the groove. In this study, we assume that the size of the groove is bigger than that of the receiver, thus the intersection position is beneath the receiver. The position of the focal point F, which is also the intersection point of the left and the right edge rays, can be calculated by using straight lines P_5I_0 and P_2X_1.

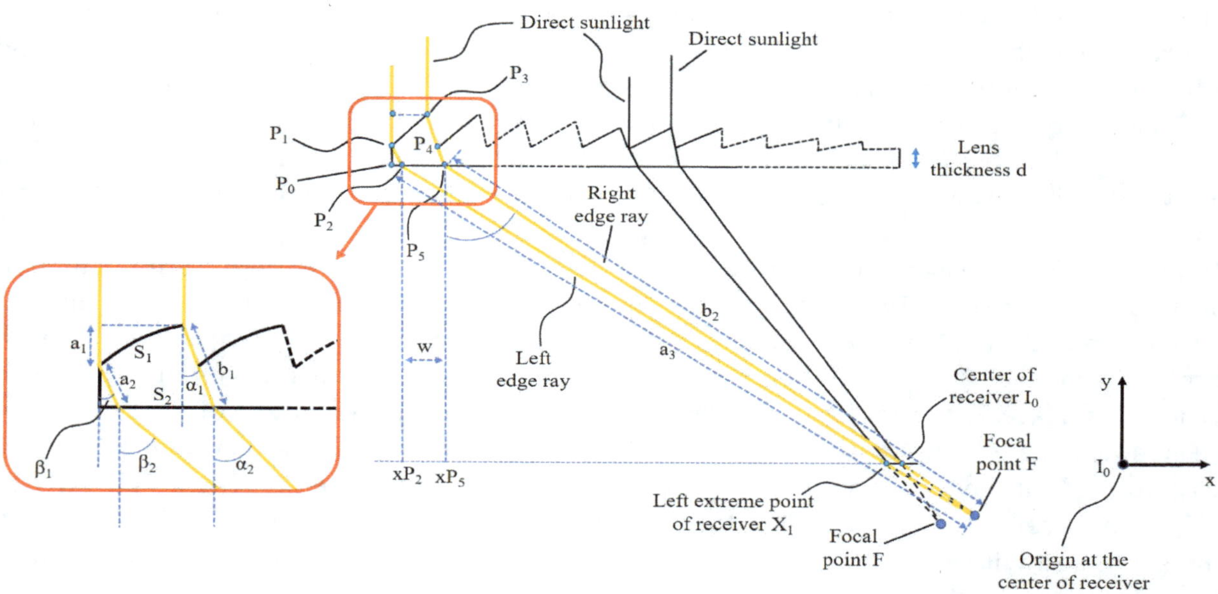

Figure 9. The way to distribute the direct sunlight in each groove of the linear Fresnel lens as an upper groove surface.

Step 5: On the right edge ray of the groove, the position of P_3 can be calculated by using the conservation of the optical path length and Snell's law. The right edge ray coming to the groove at the surface S_1 is refracted by angle α_1 and then refracted again by angle α_2 at the surface S_2 to approach the receiver center. Therefore, the position of P_3 can be calculated by using the equations as follows:

$$\tan \alpha_1 = \frac{xP_5 - xP_3}{yP_3 - yP_5}, \tag{5}$$

$$n \sin \alpha_1 = \sin \alpha_2, \tag{6}$$

$$a_1 + na_2 + a_3 = nb_1 + b_2, \tag{7}$$

where xP_3, yP_3, xP_5, and yP_5 are the positions of P_3 and P_5 on the x-axis and the y-axis, a_1, a_2, and a_3 are optical path lengths of the left edge ray, b_1 and b_2 are the optical path lengths of the right edge ray.

Step 6: All rays arriving at the groove between P_1 and P_3 have to focus on the focal point F of the groove. Furthermore, they also have to satisfy the conditions in Step 5. Therefore, choosing one point on the surface S_2 helps to find out one corresponding point on the surface S_1 by using the conditions in Step 5. As a result, the input surface S_1 can be built completely by using a large number of rays between two edge rays in this step.

Step 7: In this design, P_4 can be calculated by using the lens thickness d and the condition of P_4 lying on the straight line P_3P_5. When the design process for one groove is finished, the same procedure is repeated to build the next groove and so on until the size of the lens is reached. In the new process, P_4 in the current groove will be P_1 in the next one.

In brief, this section has described the procedure to design the upper surface of the novel linear Fresnel lens in detail, in which each groove designed in seven steps is defined by critical points P_0, P_1, P_2, P_3, P_4, and P_5 and two surfaces S_1, and S_2.

4. Design of the Linear Fresnel Lens as a Lower Groove Surface

In a similar way to the design of the upper surface, each groove of the Fresnel lens as the lower groove surface is constructed based on the refraction phenomena, the edge ray theorem, and the conservation of the optical path length. However, the structure of the grooves in the lower surface is different from the structure of the grooves in the case of the upper surface. Each lower groove consists of an input surface, which is flat, and an output surface, which is a Cartesian oval surface. Therefore, the sunlight distribution is also slightly different. The bundle of rays coming to one groove is refracted only one time at the output surface to travel to the receiver. The left edge ray has to reach the left extreme point of the receiver while the right edge ray has to pass through the center of the receiver. After that, the left edge ray intersects with the right edge rays at the focal point of the groove, which is an unreal point beneath the receiver. According to the conservation of the optical path length, all rays between the two edge rays are refracted at the output surface to converge at the focal point of the groove so that all the rays reach the receiver. In this way, the sunlight is distributed uniformly over the receiver by one groove. All the grooves distribute sunlight uniformly so that the whole lens also distributes sunlight uniformly. In this design, the structure of the output surface is important to get uniform irradiance distribution over the receiver. The design process of one groove is described in detail with reference to Figure 10 as follows:

Step 1: Initialize parameters such as the width groove, the lens thickness, the size of the receiver, the height of the lens, etc. Furthermore, the positions of P_1, P_2, P_3, and P_4 are known in advance by using the value of the width groove w and the lens thickness d in Figure 10. Therefore, the work remaining is to find out the position of P_5 and then estimate the Cartesian oval surface S_2.

Step 2: The left edge ray of the ray bundle comes to the groove at P_1. Because the input surface S_1 of the groove is a flat surface, the left edge ray approaches P_2 without any refraction. At the P_2 position, the left edge ray is refracted to travel to the left extreme point X_1 of the receiver. By using this condition, the normal vector at P_2 is calculated using the equations as follows:

$$n \sin \beta_1 = \sin(\beta_2 + \beta_1), \tag{8}$$

$$\tan \beta_2 = \frac{X_1 - xP_2}{yP_2}, \tag{9}$$

where n is the refractive index, β_1 and β_2 are the refractive angles of the left edge ray at the exit surface S_2, xP_2 and yP_2 are the positions of P_2 on the x-axis and the y-axis, respectively.

Step 3: The position of P_5 is estimated by using two conditions. Firstly, the right edge ray, which exits the groove at P_5, has to approach the center I_0 of the receiver and then intersects with the left edge ray at the focal point F of the groove. Secondly, the optical path length of the left edge ray has to be equal to that of the right edge ray. Therefore, the position of P_5 is calculated by using the following equations:

$$\tan \alpha_1 = \frac{I_0 - xP_5}{yP_5}, \tag{10}$$

$$na_1 + a_2 = nb_1 + b_2, \tag{11}$$

where I_0 is the center of the receiver, α_1 is the refractive angle of the right edge ray at P_5, xP_5 and yP_5 are the positions of P_5 on the x-axis and the y-axis, a_1, a_2, b_1, and b_2 are the optical path lengths of the left and the right edge rays, respectively.

Step 4: All rays between the left and the right edge rays have to converge on the focal point F due to the conservation of the optical path length. In this way, the bundle of rays coming to the groove can be distributed uniformly over the half of the receiver from the center I_0 to the extreme point X_1. By using individual rays coming to the input surface S_1 between the left and the right edge rays, the corresponding exit points at the output surface S_2 can be calculated based on the condition of the optical path length. Therefore, the output surface S_2 can be built by calculating the output points of a large number of individual rays between the left and the right edge rays.

Step 5: When the design process for one groove has been completed, the same procedure is repeated to build the next groove. In the new process, P_3 and P_4 in the current groove will be P_1 and P_2 in the next one, respectively.

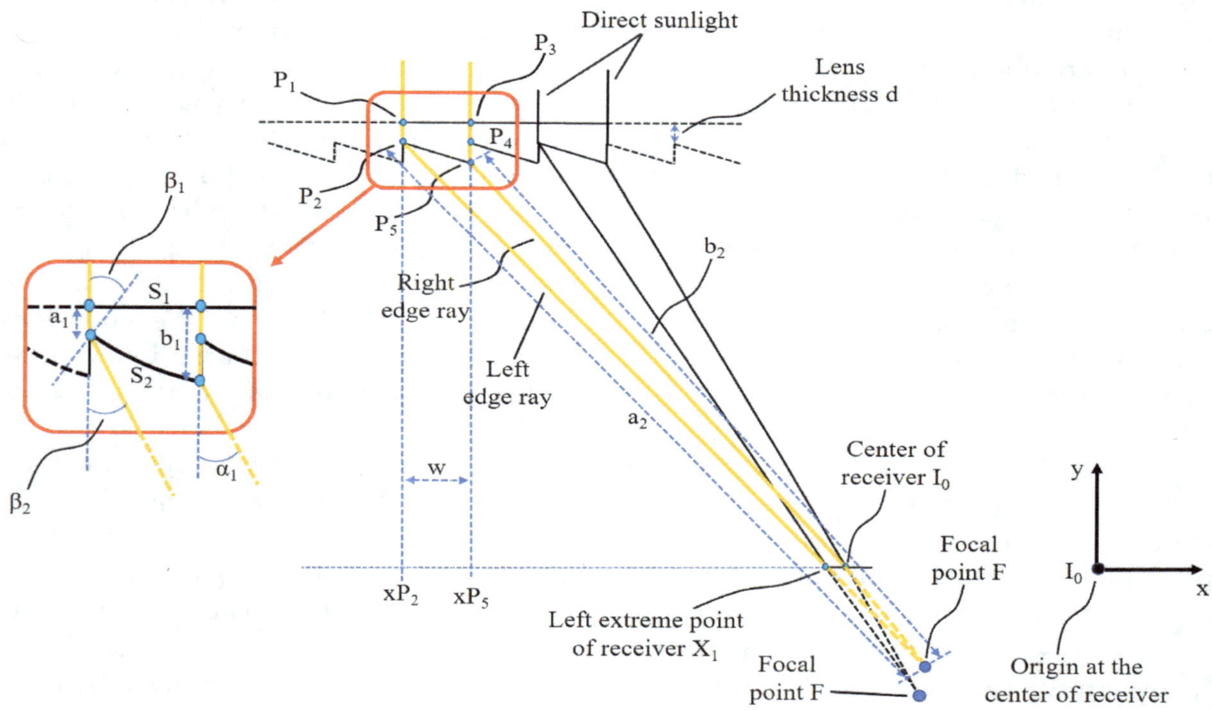

Figure 10. The distribution of direct sunlight in each groove of the linear Fresnel lens as a lower groove surface.

In conclusion, this section has described in detail the design process of the lower surface of the novel lens. The lower surface consists of a series of grooves that focus sunlight onto multi-focal points to improve the uniformity of irradiance distribution over the receiver. Each groove is constructed from a flat input surface and a Cartesian oval output surface.

5. Performance and Discussion

In this study, the novel designed lens is applied to CPV technology with the goal of increasing the performance and decreasing the cost of the photovoltaic system. In terms of performance, there are three aspects to improve the CPV system. Firstly, the optical efficiency can be increased thanks to the performance without a secondary optical element (SOE). Secondly, the new structure helps to distribute the sunlight uniformly over the receiver. Thirdly, the shape of irradiance distribution is a square area which brings about a better match with the shape of the solar cells. Furthermore, the cost can be reduced by the improvement of the performance and the simple process to produce the novel Fresnel lens.

The CPV system is built in MATLAB and then simulated in LightTools™ to estimate the performance of the novel linear Fresnel lens. Parameters are chosen to facilitate application in real conditions. Therefore, a receiver with the size of 10 × 10 mm is chosen to represent a multi-junction solar cell, which usually has a square shape with a width of between 5 mm and 10 mm in commercial products [47]. Moreover, the concentration ratio of CPV using highly efficient multi-junction solar cells has a range of from 300 to 1000 times, depending on the design of the CPV system [10]. Concentrated sunlight can help to increase the efficiency of the solar cells. However, the loss due to the electrode resistance and the series resistance of the solar cells is boosted by the squared function of the photocurrent. As a result, the efficiency of solar cells increases to the maximum value and then decreases regardless of the fact that concentration ratio is still increasing. In the current CPV technology, the optimum concentration ratio can be different and will usually be from 400 to 600 times depending on the quality of the solar cells and the photovoltaic system [48]. Therefore, a size of 240 × 240 mm is chosen for the simulation of the designed lens. With the chosen size, the concentration ratio calculated by the equation $C = \frac{D_{lens}^2}{d_{receiver}^2}$ is 576 times, where D_{lens} is the width of the designed lens, and $d_{receiver}$ is the width of the receiver. These parameters are chosen to be suitable for CPV in the current technology. However, they can be changed easily depending on the aim of the design.

The novel linear Fresnel lens is designed in the MATLAB program and then the data from MATLAB are applied to Lighttools™ to draw the lens in three dimensions (3D). The ray tracing technique is used to estimate the performance of the design process. Furthermore, the simulation process is also performed using Lighttools™ to find out the optimum structure and the optical properties of the lens. Table 1 shows the parameters for ray tracing and simulation of the lens in Lighttools™.

Table 1. The parameters of the designed linear Fresnel lens for ray tracing and simulation.

Design Parameters	Values
Dimension of lens	240 × 240 mm
Height of lens	240 mm
F-number	1
Dimension of receiver	10 × 10 mm
Number of grooves	24
Material	Polymethyl methacrylate (PMMA)
Thickness	5 mm
Concentration ratio	576 times
Wavelength for ray tracing	550 nm
Spectrum for light source	390–1000 nm
Power of light source	20 kW
Number of rays for ray tracing	2,000,000 rays
Refractive index	1.492 at 550 nm

Figure 11 shows the ray tracing of the linear Fresnel lens using only one groove surface. The sunlight is focused on one dimension as a focal line. However, a combination of these linear Fresnel lens structures can distribute sunlight on two dimensions as indicated in Figure 12.

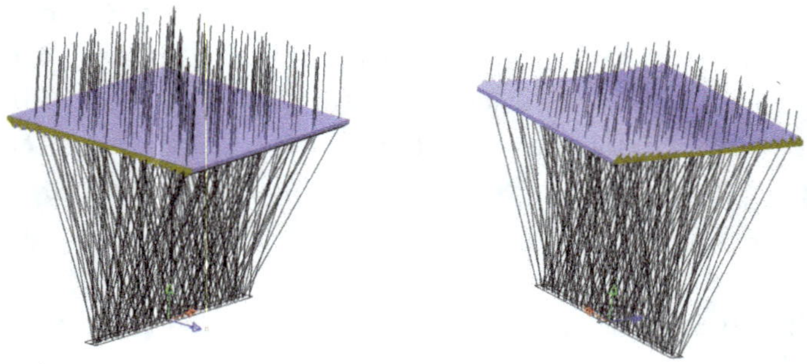

(a) (b)

Figure 11. The ray tracing of sunlight convergence on one dimension onto the focal line using the linear Fresnel lens with (**a**) the lower groove surface and (**b**) the upper groove surface.

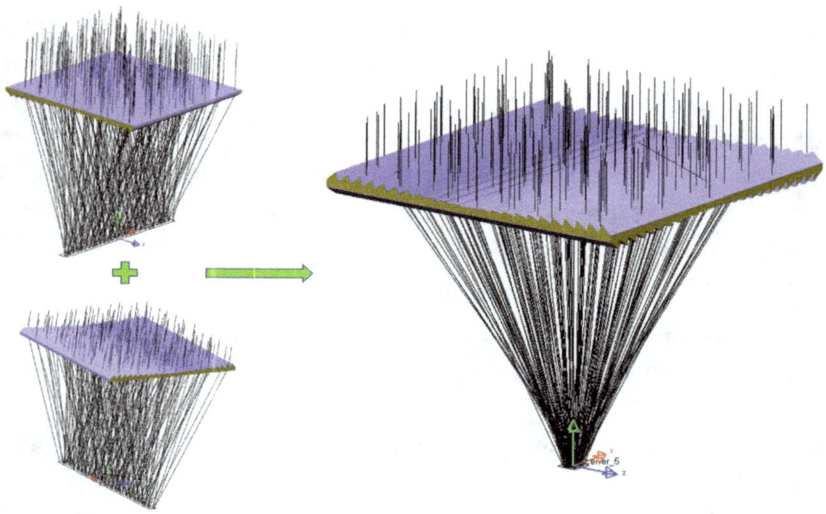

Figure 12. The ray tracing of sunlight convergence on two dimensions using the novel Fresnel lens with two groove surfaces (upper and lower surfaces) placed perpendicular to each other.

Besides the structure of the lens shown in Figure 12, Figure 13 shows other combinational structures of individual linear Fresnel lenses which are placed perpendicular to each other. The main advantage of these structures is simplicity; however, their optical loss is bigger than that of the structure shown in Figure 12 because of the sunlight passing through four interfaces in these structures.

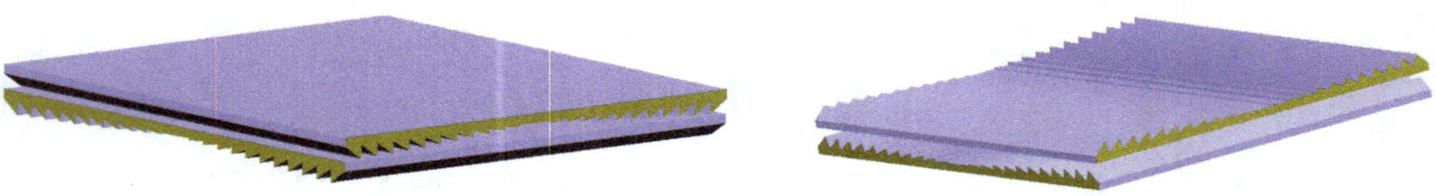

(a) (b)

(c) (d)

Figure 13. Four structures of the POE using a combination of two individual Fresnel lenses that are constructed from (**a**) two lower groove surfaces, (**b**) two upper groove surfaces, (**c**) one lower groove surface above one upper groove surface, and (**d**) one upper groove surface above one lower groove surface.

In term of simulation, a system is built in LightTools^{TM} to represent a CPV system, in which the novel designed Fresnel lens, which is square with size of 240 × 240 mm, collects and distributes direct sunlight over the square receiver with size of 10 × 10 mm. Figure 14 shows the irradiance distribution over the receiver of the system's simulation process. The result indicates that the distribution area is a square that helps to match better with the shape of solar cells. In addition, there is no hot spot at the center of the solar cell and the irradiance distribution is quite uniform in the area of the receiver. The uniformity can be calculated by using the Equation (12):

$$U = 100 - \frac{max - min}{average} \times 100\%, \tag{12}$$

where *max* and *min* are the maximum and the minimum value of the irradiance on the solar cell, respectively, *average* is the mean value of the irradiance on the whole solar cell. The ideal uniformity is 100%. The simulation results of the proposed design are quite good, with the horizontal uniformity reaching 78.6% while the vertical uniformity approaches 76.4%.

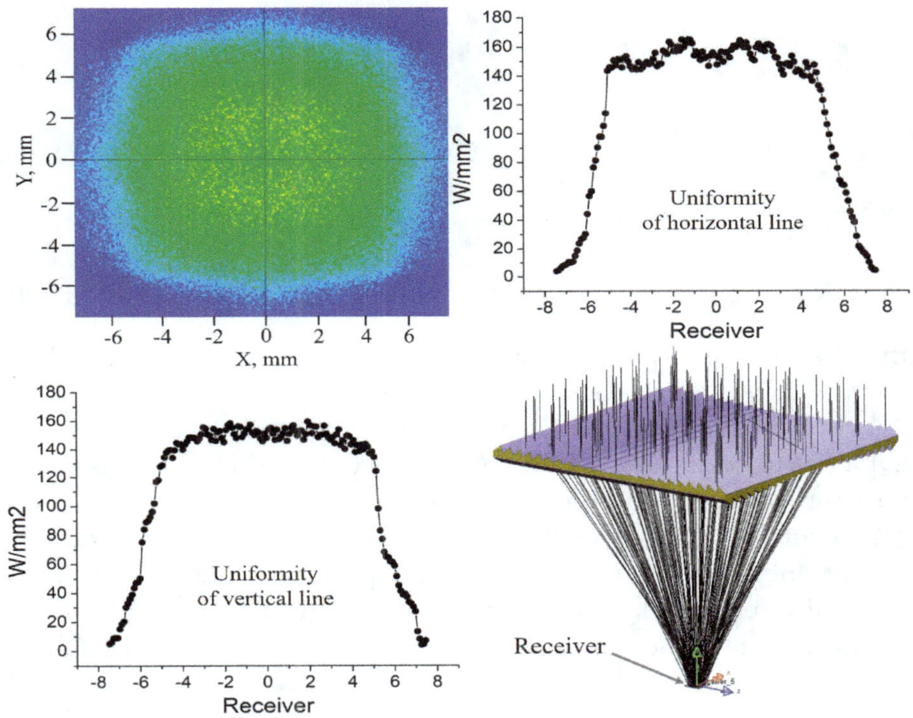

Figure 14. Ray tracing of the novel designed lens, irradiance distribution of the sunlight over the receiver, and the uniformity on the horizontal and vertical investigation lines.

The optical efficiency is an important index of the CPV system and it is defined by the ratio of the irradiance reaching the receiver over the irradiance coming to the designed lens. The efficiency of the conventional Fresnel lens working as a POE of the CPV system without a SOE is usually around 90% because of the optical loss on the interference surfaces. Although the efficiency is high, the uniformity of the conventional Fresnel lens is low, and the concentrated light is not match well to square multi-junction solar cell. Therefore, a combination of a POE and a SOE is necessary for the CPV system with a high concentration ratio. Nevertheless, the combination leads to the optical efficiency being reduced by around 10% more. In this design, the input power coming to the novel lens is 20 kW while the power reaching the receiver is 16.489 kW. Therefore, the optical efficiency is 82.445%. There are two reasons that affect the optical efficiency of the novel designed Fresnel lens. Firstly, each groove of the upper surface has a small loss area as shown in Figure 15. Secondly, the thickness of the lens is not constant because of the internal impact between the upper surface and the lower surface in the design process. Although the efficiency of the novel lens is lower than that of the conventional Fresnel lens, it is still considered a good result compared to the optical system of CPV using the combination of a POE and a SOE, which has optical efficiency around 80%.

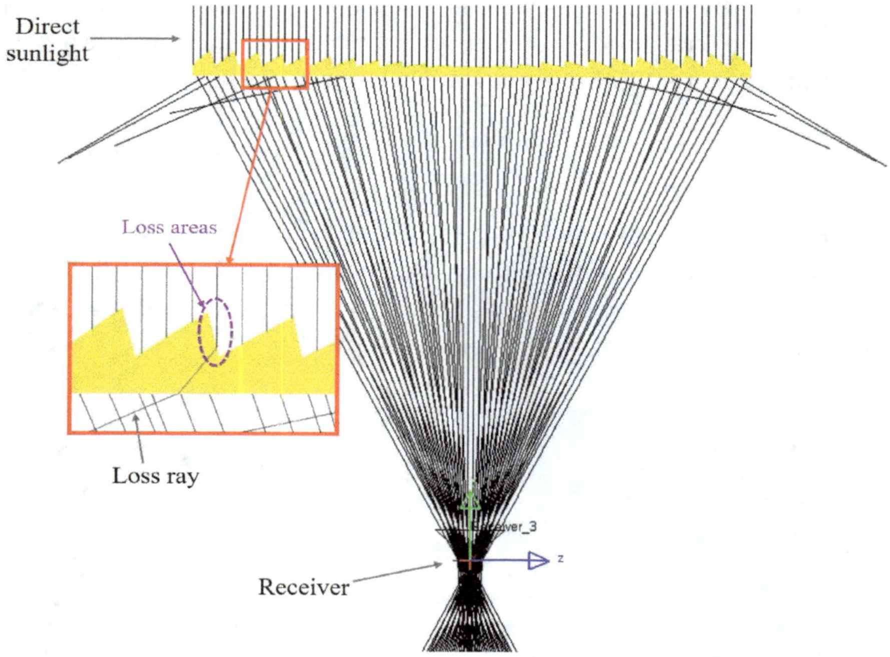

Figure 15. Ray tracing of the upper surface and the loss areas on the upper surface.

In addition, the tolerance (acceptance angle) is also an important index for the CPV system. The acceptance angle in the CPV system is defined as the incident angle at which the solar power over the receiver drops to 90% of its maximum [8,49]. The requirement for the acceptance angle to apply to the real condition is that it has to be larger than the solar angle of 0.265° [43]. In this design, the acceptance angle is investigated and the result in Figure 16 shows that it is about 0.84°, which is an acceptable value for a CPV system without a SOE, thanks to the development of sunlight tracking technology [50]. The concentration acceptance product (CAP) is also an importance parameter for the CPV system. The CAP parameter is meaningful if we know the acceptance angle at a given concentration, we can use CAP parameter to estimate the acceptance angle that we would obtain with different concentration.

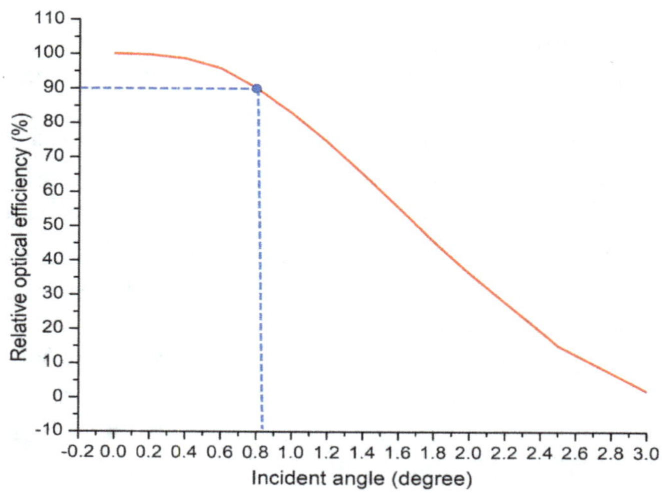

Figure 16. The relative optical efficiency depends on the incident angle of the direct sunlight.

In our design, the acceptance angle is $\alpha_a = 0.84°$ with the concentration ratio is $C_g = 576$ times so that the CAP = $C_g^{0.5}\sin(\alpha_a) = 0.352$. With the value of CAP, we can estimate that the concentration is about 406 times if the we increase the acceptance angle from 0.84° to 1°. Generally, the concentration ratio of CPV using highly efficient multi-junction solar cells has a range of from 300 to 1000 times, depending on the design of the CPV system. Concentrated sunlight can help to increase the efficiency of the solar cells. However, the loss due to the electrode resistance and the series resistance of the solar cells is boosted by the squared function of the photocurrent that was studied by Xu et al. [48].

As a result, the efficiency of solar cells increases to the maximum value and then decreases although concentration ratio is still increasing. In the current CPV technology, the optimum concentration ratio can be different and will usually be from 400 to 600 times depending on the quality of the solar cells and the photovoltaic system. Therefore, the results show the concentration ratio by 406 times with the acceptance angle by 1° that can be a good potential candidate to consider being an optical concentrator system of CPV system. To more clearly, the results of the designed linear Fresnel lens are used to compare to other study results from other research groups. In comparison, all concentrator systems have the same entry aperture area 625 cm^2 and the same acceptance angle ($\alpha_a = 1°$). The following table shows the comparison of concentration ratio C_g and f_{number} with the same acceptance angle while Figure 17 shows the size comparison of multi-junction solar cells and secondary optical element with the same acceptance angle.

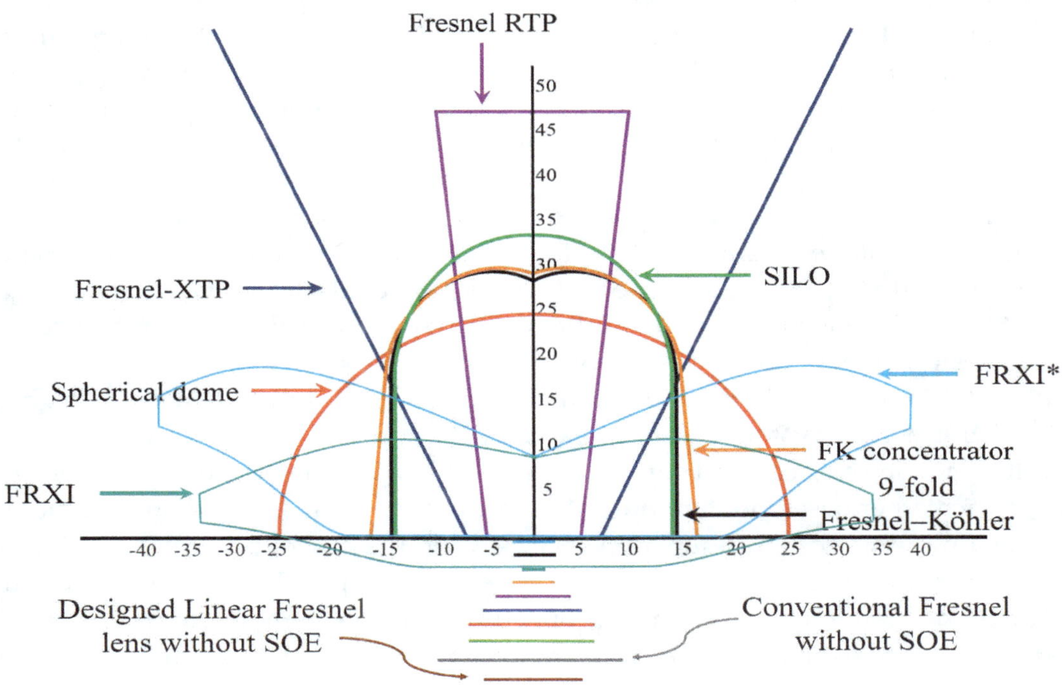

Figure 17. Cross section views of some types of SOEs and size comparison of these optical concentrator systems with the same acceptance angle. The structures and designs of these optical concentrator systems are described in detail in Miñano et al. [51] and Mendes-Lopes et al. [52].

Based on comparison in Table 2 and Figure 17, if we need a super high concentration for the CPV system, F-RXI type of optical concentrator system is the best choice because of highest concentration ratio. In contrast, if we take care about the compact property of CPV system, RTP type of optical concentrator system is the best choice because of smallest f_{number}. However, depending on the structure of CPV, designer can choose a suitable type of optical concentrator system. As above-mentioned, if designer choose a medium concentration ratio and f_{number} to avoid facing with serious issues of high temperature and the loss due to the electrode resistance and the series resistance, our design can be a good potential choice.

Table 2. Comparison of concentration ratio and f_{number} with the same acceptance angle between our proposed design and other designs that described in Minano et al. [51].

Concentrator Type	f_{number}	Concentration Ratio C_g	Acceptance Angle α_a
No SOE	1.5	104	1°
Spherical dome	1.5	257	1°
SILO	1.2	248	1°
Our proposed design	1.0	406	1°
XTP	1.3	425	1°
RTP	0.85	677	1°
FK	1.0	1057	1°
9-Fold	1.0	1062	1°
F-RXI	1.4	2300	1°
F-RXI*	1.4	1750	1°

There are some studies about designing CPV system without SOE that is published in literatures before. Gonzalez et al. [39] used the Simultaneous Multiple Surface (SMS) method to increase the acceptance angle a uniformity of irradiance distribution of the linear Fresnel lens without SOE. The results in this study show that the uniformity and acceptance are improved much, however, the structure of the designed Fresnel lens is just suitable for low concentrator photovoltaic (LCPV) because of low concentration ratio (less than 100 times). Another study comes from Ryu et al. [42]. In this design, the author tried to design an optical concentrator system with high uniformity irradiance distribution without SOE. The flat Fresnel lens was built by some parts; each of them collects and distributes the sunlight to different places on the receiver. This structure obtains high uniformity with acceptance angle greater 1° for all of concentration ratio from 9 to 121 times. Although the results show that the uniformity is good with mild acceptance angle, the concentration ratio is small so that this structure is not suitable for HCPV system. A prototype of flat Fresnel lens designed by Pan et al. [43] was introduced by using the idea to change all pitches of the Fresnel concentrator focusing on the different position of the receiver. The designed Fresnel lens obtained high concentration ratio around 1018 times, but the acceptance is small 0.305°. Therefore, it is difficult to apply the designed Fresnel lens to CPV system without in real conditions. Comparing to all these types of optical concentrator system without the SOE abovementioned, our novel designed linear Fresnel lens has good properties, which help our designed lens can become a potential choice for HCPV system.

Generally, a CPV system consist of three main components: optical concentrator, high effective solar cells, and sun tracking system. Assume that all CPV systems use the same the best commercial version of multi-junction solar cell and sun tracking system in the market so we can compare between CPV systems to each other based on their optical concentrator systems. Therefore, the module cost will differ depending on the size of multi-junction solar cell and the cost of production of Fresnel lens and SOE. However, the estimation of the cost of CPV system is complex because the cost of the working CPV system depends on the reliability and the effective of that system in real conditions. To simplify the comparison, a comparison among all types of optical concentrators in Table 2 is just focused on the size of multi-junction solar cell and the cost of production. In term of multi-junction solar cell, the concentration ratio is higher, the size of solar cell is smaller so that the module cost is cheaper. In this area, F-RXI is the best choice. However, a concentration ratio from 300–600 is a good choice in real conditions because of considering electrical loss by series loss and the temperature issue. Therefore, our designed linear Fresnel lens is a good candidate. In addition, the optical concentrator system based on our designed linear Fresnel lens does not need to use SOE so that the cost of module is decreased significantly comparing to other types of optical concentrators. In term of production cost of the lens, there are some technologies to manufacture Fresnel lens or Primary optical component (POE) for CPV system, such as the embossing method (Zhang et al. [53]), molding method (Huang et al. [54]), and extrusion method (Benz et al. [55]). The conventional linear Fresnel lens can be manufactured easily by these methods. However, our designed linear Fresnel lens has a special structure, in which two groove surfaces is placed perpendicular to each other so that the extrusion method cannot be used to manufacture. Fortunately, the embossing method is a mature method to manufacture conventional

Fresnel lens can be applied to make our designed linear Fresnel without any addition special skill so that the production cost of the novel linear Fresnel lens is similar that of conventional linear Fresnel lens. In conclusion, the cost of bigger multi-junction solar cells in our designed concentrator system comparing to FK, 9-Fold, F-RXI, F-RXI* types of optical concentrators can be compensated by removing SOE. Moreover, although our designed linear Fresnel lens has a special structure, which the extrusion method could not apply to manufacture, we can use the mature embossing method to manufacture easily so that the production cost of POE is similar that of other types of optical concentrators. To more clearly, a prototype of the designed linear Fresnel lens is needed to manufacture to apply to CPV system. After that, a comparison about performance and cost production can be implemented with more accuracy between our designed optical concentrator and other types of other optical concentrators.

6. Conclusions

In this study, the design of the novel linear Fresnel lens is presented in detail. The lens has two groove surfaces (upper and lower groove surfaces) placed perpendicular to each other. Each surface focuses direct sunlight on one dimension so that the whole lens can collect and distribute the sunlight on two dimensions, thereby, increasing the concentration ratio. Furthermore, a new technique using multi-focal points to distribute the sunlight uniformly over the receiver is proposed, in which each groove has its own focal point on which all rays arriving at the groove converge. The position of the focal point for each groove is determined by using the edge ray theorem, Snell's law, and the conservation of the optical path length so that the bundle of rays can be distributed uniformly from the center to the extreme point of the receiver. The simulation results show that the irradiance distribution is a square area that helps the concentrated sunlight to match better to the shape of the solar cells. In addition, the irradiance distribution is quite uniform with the uniformity reaches around 77% for both horizontal and vertical investigation lines. An acceptance angle with an acceptable value of about 0.84° is also investigated in this research to choose or design a suitable sun tracking system for the CPV. All these factors substantially help to enhance the performance and decrease the cost of the CPV system for massive installation in real conditions. Therefore, a concentrator based on the novel structure of the linear Fresnel lens can help to promote the development of high performance and effective cost of the CPV system generation.

In future research, the prototype of the designed lens will be produced and then experiments using that prototype as a concentrator in the CPV system can be carried out to check the performance of the designed lens. The results of the experiment will be compared to the simulation results to understand and optimize the novel lens and the CPV system.

Author Contributions: All authors contributed extensively to the work presented in this paper. T.T.P. proposed the research idea to design a concentrator for a CPV system. S.S. supervised the design and simulation process of the system. T.T.P., N.H.V., and S.S. together discussed setting up the conditions for the simulation in the LightTools software. T.T.P. carried out the design and the simulation, collected the dataset, and wrote the manuscript. S.S. and N.H.V. contributed to the manuscript revision.

References

1. Fraunhofer, I. New world record for solar cell efficiency at 46% French-German cooperation confirms competitive advantage of European photovoltaic industry. *Press Release* **2014**, *26*. Available online: http://www.ise.fraunhofer.de/en/press-and-media/press-releases/press-releases-2014/new-world-record-for-solar-cell-efficiency-at-46-percent (accessed on 26 March 2019).
2. Abbott, D. Keeping the energy debate clean: How do we supply the world's energy needs? *Proc. IEEE* **2010**, *98*, 42–66. [CrossRef]

3. Khamooshi, M.; Salati, H.; Egelioglu, F.; Hooshyar Faghiri, A.; Tarabishi, J.; Babadi, S. A review of solar photovoltaic concentrators. *Int. J. Photoenergy* **2014**, *2014*. [CrossRef]
4. Mendoza, B. Total solar irradiance and climate. *Adv. Space Res.* **2005**, *35*, 882–890. [CrossRef]
5. Chukwuka, C.; Folly, K.A. Overview of concentrated photovoltaic (CPV) cells. *J. Power Energy Eng.* **2014**, *2*, 1. [CrossRef]
6. Morales-Acevedo, A. *Solar Cells: Research and Application Perspectives*; InTech: Rijeka, Croatia, 2013. [CrossRef]
7. Tanabe, K. A review of ultrahigh efficiency III-V semiconductor compound solar cells: Multijunction tandem, lower dimensional, photonic up/down conversion and plasmonic nanometallic structures. *Energies* **2009**, *2*, 504–530. [CrossRef]
8. Buljan, M.; Mendes-Lopes, J.; Benítez, P.; Miñano, J.C. Recent trends in concentrated photovoltaics concentrators' architecture. *J. Photonics Energy* **2014**, *4*, 040995. [CrossRef]
9. Luque, A.; Hegedus, S. *Handbook of Photovoltaic Science and Engineering*; John Wiley & Sons: Hoboken, NJ, USA, 2011. [CrossRef]
10. Pérez-Higueras, P.; Fernández, E.F. *High Concentrator Photovoltaics: Fundamentals, Engineering and Power Plants*; Springer: Berlin, Germany, 2015. [CrossRef]
11. Shanks, K.; Senthilarasu, S.; Mallick, T.K. Optics for concentrating photovoltaics: Trends, limits and opportunities for materials and design. *Renew. Sustain. Energy Rev.* **2016**, *60*, 394–407. [CrossRef]
12. Parida, B.; Iniyan, S.; Goic, R. A review of solar photovoltaic technologies. *Renew. Sustain. Energy Rev.* **2011**, *15*, 1625–1636. [CrossRef]
13. Chong, K.-K.; Lau, S.-L.; Yew, T.-K.; Tan, P.C.-L. Design and development in optics of concentrator photovoltaic system. *Renew. Sustain. Energy Rev.* **2013**, *19*, 598–612. [CrossRef]
14. Dreger, M.; Wiesenfarth, M.; Kisser, A.; Schmid, T.; Bett, A.W. Development and investigation of a CPV module with Cassegrain mirror optics. *AIP Conf. Proc.* **2014**, *177*, 177–182.
15. Rumyantsev, V.D. Solar concentrator modules with silicone-on-glass Fresnel lens panels and multijunction cells. *Opt. Express* **2010**, *18*, A17–A24. [CrossRef] [PubMed]
16. Cucco, S.; Faranda, R.; Invernizzi, F.; Leva, S. Analysis of a Fresnel lenses concentrator. In Proceedings of the Power and Energy Society General Meeting, Manchester Grand Hyatt, San Diego, CA, USA, 22–26 July 2012; pp. 1–8.
17. Erismann, F. Design of a plastic aspheric Fresnel lens with a spherical shape. *Opt. Eng.* **1997**, *36*, 988–992. [CrossRef]
18. Tien, N.X.; Shin, S. A Novel Concentrator Photovoltaic (CPV) System with the Improvement of Irradiance Uniformity and the Capturing of Diffuse Solar Radiation. *Appl. Sci.* **2016**, *6*, 251. [CrossRef]
19. Xie, W.; Dai, Y.; Wang, R.; Sumathy, K. Concentrated solar energy applications using Fresnel lenses: A review. *Renew. Sustain. Energy Rev.* **2011**, *15*, 2588–2606. [CrossRef]
20. Bahaidarah, H.M.; Tanweer, B.; Gandhidasan, P.; Rehman, S. A combined optical, thermal and electrical performance study of a V-trough PV system—Experimental and analytical investigations. *Energies* **2015**, *8*, 2803–2827. [CrossRef]
21. Du, B.; Hu, E.; Kolhe, M. Performance analysis of water cooled concentrated photovoltaic (CPV) system. *Renew. Sustain. Energy Rev.* **2012**, *16*, 6732–6736. [CrossRef]
22. Rosell, J.; Vallverdu, X.; Lechon, M.; Ibanez, M. Design and simulation of a low concentrating photovoltaic/thermal system. *Energy Convers. Manag.* **2005**, *46*, 3034–3046. [CrossRef]
23. Espinet, P.; Algora, C.; Rey-Stolle, I.; Garcia, I.; Baudrit, M. Electroluminescence characterization of III–V multi-junction solar cells. In Proceedings of the Photovoltaic Specialists Conference, 2008. PVSC'08, San Diego, CA, USA, 11–16 May 2008; pp. 1–6.
24. Languy, F.; Fleury, K.; Lenaerts, C.; Loicq, J.; Regaert, D.; Thibert, T.; Habraken, S. Flat Fresnel doublets made of PMMA and PC: Combining low cost production and very high concentration ratio for CPV. *Opt. Express* **2011**, *19*, A280–A294. [CrossRef]
25. Leutz, R.; Suzuki, A. *Nonimaging Fresnel Lenses: Design and Performance of Solar Concentrators*; Springer: Berlin, Germany, 2012; Volume 83.
26. Jing, L.; Liu, H.; Zhao, H.; Lu, Z.; Wu, H.; Wang, H.; Xu, J. Design of novel compound Fresnel lens for high-performance photovoltaic concentrator. *Int. J. Photoenergy* **2012**, *2012*. [CrossRef]

27. Sharaf, O.Z.; Orhan, M.F. Concentrated photovoltaic thermal (CPVT) solar collector systems: Part I–Fundamentals, design considerations and current technologies. *Renew. Sustain. Energy Rev.* **2015**, *50*, 1500–1565. [CrossRef]
28. Ullah, I.; Shin, S.-Y. Development of optical fiber-based daylighting system with uniform illumination. *J. Opt. Soc. Korea* **2012**, *16*, 247–255. [CrossRef]
29. Chen, Y.-C.; Chiang, H.-W. Design of the secondary optical elements for concentrated photovoltaic units with Fresnel lenses. *Appl. Sci.* **2015**, *5*, 770–786. [CrossRef]
30. Victoria, M.; Domínguez, C.; Antón, I.; Sala, G. Comparative analysis of different secondary optical elements for aspheric primary lenses. *Optics Express* **2009**, *17*, 6487–6492. [CrossRef] [PubMed]
31. Horowitz, K.A.; Woodhouse, M.; Lee, H.; Smestad, G.P.A.; Woodhouse, M.; Lee, H.; Smestad, G.P. A bottom-up cost analysis of a high concentration PV module. In *AIP Conference Proceedings*; AIP Publishing: Melville, NY, USA, 2015; Volume 1679, p. 100001.
32. Luther, J.; Bett, A.; Burger, B.; Dimroth, F. High concentration photovoltaics based on III–V multi-junction solar cells. In Proceedings of the 21st European Photovoltaic Solar Energy Conference, Dresden, Germany, 4–8 September 2006; pp. 2054–2057.
33. Messmer, E.R. CPV market evolution and the potential in cost reduction of CPV modules. In Proceedings of the 9th Conference on Concentrator Photovoltaic Systems (CPV-9), Miyazaki, Japan, 15–17 April 2013.
34. Philipps, S.P.; Bett, A.W.; Horowitz, K.; Kurtz, S. *Current Status of Concentrator Photovoltaic (CPV) Technology*; NREL (National Renewable Energy Laboratory): Golden, CO, USA, 2015.
35. Tomosk, S.; Wright, D.; Hinzer, K.; Haysom, J.E. Analysis of present and future financial viability of high-concentrating photovoltaic projects. In *High Concentrator PhotoVoltaics*; Springer: Berlin, Germany, 2015; pp. 377–400. [CrossRef]
36. Algora, C.; Rey-Stolle, I. *Handbook on Concentrator Photovoltaic Technology*; John Wiley & Sons: Hoboken, NJ, USA, 2016. [CrossRef]
37. Zhuang, Z.; Yu, F. Optimization design of hybrid Fresnel-based concentrator for generating uniformity irradiance with the broad solar spectrum. *Opt. Laser Technol.* **2014**, *60*, 27–33. [CrossRef]
38. Leutz, R.; Suzuki, A.; Akisawa, A.; Kashiwagi, T. Design of a nonimaging fresnel lens for solar concentrators. *Solar Energy* **1999**, *65*, 379–387. [CrossRef]
39. González, J.C. Design and analysis of a curved cylindrical Fresnel lens that produces high irradiance uniformity on the solar cell. *Appl. Opt.* **2009**, *48*, 2127–2132. [CrossRef]
40. Yamada, N.; Okamoto, K. Experimental measurements of a prototype high concentration Fresnel lens CPV module for the harvesting of diffuse solar radiation. *Opt. Express* **2014**, *22*, A28–A34. [CrossRef]
41. Akisawa, A.; Hiramatsu, M.; Ozaki, K. Design of dome-shaped non-imaging Fresnel lenses taking chromatic aberration into account. *Solar Energy* **2012**, *86*, 877–885. [CrossRef]
42. Ryu, K.; Rhee, J.-G.; Park, K.-M.; Kim, J. Concept and design of modular Fresnel lenses for concentration solar PV system. *Solar Energy* **2006**, *80*, 1580–1587. [CrossRef]
43. Pan, J.-W.; Huang, J.-Y.; Wang, C.-M.; Hong, H.-F.; Liang, Y.-P. High concentration and homogenized Fresnel lens without secondary optics element. *Opt. Commun.* **2011**, *284*, 4283–4288. [CrossRef]
44. Pham, T.T.; Vu, N.H.; Shin, S. Daylighting System Based on Novel Design of Linear Fresnel lens. *Buildings* **2017**, *7*, 92. [CrossRef]
45. Vu, N.H.; Pham, T.T.; Shin, S. LED Uniform Illumination Using Double Linear Fresnel Lenses for Energy Saving. *Energies* **2017**, *10*, 2091. [CrossRef]
46. Chaves, J. *Introduction to Nonimaging Optics*; CRC Press: Boca Raton, FL, USA, 2008.
47. Messmer, E.R. Solar Cell Efficiency vs. Module Power Output: Simulation of a Solar Cell in a CPV Module. In *Solar Cells-Research and Application Perspectives*; InTech: Rijeka, Croatia, 2013. [CrossRef]
48. Xu, Q.; Ji, Y.; Riggs, B.; Ollanik, A.; Farrar-Foley, N.; Ermer, J.H.; Romanin, V.; Lynn, P.; Codd, D.; Escarra, M.D. A transmissive, spectrum-splitting concentrating photovoltaic module for hybrid photovoltaic-solar thermal energy conversion. *Solar Energy* **2016**, *137*, 585–593. [CrossRef]
49. Yavrian, A.; Tremblay, S.; Levesque, M.; Gilbert, R. How to increase the efficiency of a high concentrating PV (HCPV) by increasing the acceptance angle to $\pm 3.2^\circ$. In *AIP Conference Proceedings*; AIP: College Park, MD, USA, 2013; pp. 197–200.
50. Racharla, S.; Rajan, K. Solar tracking system—A review. *Int. J. Sustain. Eng.* **2017**, *10*, 72–81. [CrossRef]

51. Miñano, J.C.; Benítez, P.; Zamora, P.; Buljan, M.; Mohedano, R.; Santamaría, A.J.O.E. Free-form optics for Fresnel-lens-based photovoltaic concentrators. *Opt. Express* **2013**, *21*, A494–A502. [CrossRef]
52. Mendes-Lopes, J.; Benítez, P.; Zamora, P.; Miñano, J.C.J.O.E. 9-fold Fresnel–Köhler concentrator with Fresnel lens of variable focal point. *Opt. Express* **2014**, *22*, A1153–A1163. [CrossRef] [PubMed]
53. Zhang, X.; Liu, K.; Shan, X.; Liu, Y.J.O.E. Roll-to-roll embossing of optical linear Fresnel lens polymer film for solar concentration. *Opt. Express* **2014**, *22*, A1835–A1842. [CrossRef]
54. Huang, C.-Y.; Chen, C.-C.; Chou, H.-Y.; Chou, C.-P. Fabrication of fresnel lens by glass molding technique. *Opt. Rev.* **2013**, *20*, 202–204. [CrossRef]
55. Benz, V.; Berkenkopf, M.; Fischer, U.; Lorenz, H.; Meier-Kaiser, M. Process for Manufacture of Fresnel Lenses. U.S. Patent 5,870,233, 9 February 1999.

5

Heat-Transfer Characteristics of Liquid Sodium in a Solar Receiver Tube with a Non-Uniform Heat Flux

Jing Liu [1], Yongqing He [1,*] and Xianliang Lei [2]

[1] School of Chemical Engineering, Kunming University of Science and Technology, Kunming 650500, China; Jedxy1037@gmail.com

[2] State Key Laboratory of Multiphase Flow in Power Engineering, Xi'an Jiaotong University, Xi'an 710049, China; xianlianglei@xjtu.edu.cn

* Correspondence: yqhe@kmust.edu.cn

Abstract: This paper presents a numerical simulation on the heat transfer of liquid sodium in a solar receiver tube, as the liquid sodium is a promising heat-transfer candidate for the next generation solar-power-tower (SPT) system. A comparison between three mediums—solar salt, Hitec and liquid sodium—is presented under uniform and nonuniform heat-flux configurations. We studied the effects of mass flow rate (Q_m), inlet temperature (T_{in}), and maximum heat flux (q_{omax}), on the average heat-transfer coefficient (h) and the friction coefficient (f) of the three mediums. The results show that the h of liquid sodium is about 2.5 to 5 times than other two molten salts when T_{in} is varying from 550 to 800 K, Q_m is 1.0 kg/s, and q_{omax} is 0.1 MW/m^2. For maximum heat fluxes from 0.1 to 0.3 MW/m^2, the h of liquid sodium is always an order of magnitude larger than that of Hitec and Solar-Salt (S-S), while maintaining a small friction coefficient.

Keywords: solar-power tower; liquid sodium; solar salt; Hitec; heat flux

---

1. Introduction

Compared to the parabolic trough, Fresnel and dish collectors, the solar-power-tower (SPT) plant has the remarkable advantages, such as lower electricity cost, large-scale power generation and higher efficient thermodynamic cycles [1–3]. The SPT is equipped with a large number of heliostats on the ground, each with a tracking mechanism that accurately reflects the reflection of sunlight onto the receiver at the top of a tall tower. The concentrating magnification on the receiver can exceed 1000 times. One typical arrangement of the SPT receivers is the external tubular receiver designed for Solar Two project, in which only half of the surface of the tube is exposed to solar irradiation. This may bring about many problems, such as aggravating the plastic deformation of the receiver tube, facilitating degradation of the selective absorptive coating and decreasing the allowable solar heat flux [4,5]. Since the nonuniform solar heat flux tends to cause the temperature inhomogeneity of the heat-transfer fluid (HTF) and, further, the thermal stress on the heat-transfer tubes, a much broader range of operational temperatures is required. The liquid metal as the promising candidate for the exposed cylindrical heat absorber of SPT has been proposed [6].

Nitrate salts have been used as HTFs and thermal storage mediums for decades in the concentrating SPT industry. The most commonly used HTFs are solar salt (S-S, 60% $NaNO_3$ and 40% KNO_3) and Hitec (53% KNO_3 + 40% $NaNO_2$ + 7% $NaNO_3$). Both the nitrate salt mixtures will decompose above 873 °K, which has seriously limited the overall efficiency in the SPT system. Furthermore, recent research efforts have shown that the nitrate salts are more suitable for use in parabolic trough systems due to their low working temperature [7]. However, the next-generation SPT systems require a higher incident peak flux and operating temperature. If the liquid metal is the heat-transfer fluid, it can provide an incident peak flux above 0.6 MW/m^2. Liquid sodium (Na), characterized by chemical stability at temperatures up to near 1173 °K has a lower melting point to 371 °K [8], as well as superior thermal conductivity and low Prandtl number. The material properties mentioned above can largely improve heat transfer when compared to conventional fluids such as oil or salt mixtures [9]. In fact, during the early years of the development of central receiver systems (CRSs), liquid sodium was one of the prominent HTFs under investigation. Indeed, several test projects have been developed, and the efficiency was obtained at 88%–96% in the 1980s [10]. However, the disadvantage of liquid sodium is its high combustibility when in contact with water even if no air. Fortunately, special and protective measures have also been developed from previous experiences [11]. Also, this technology has continued to be investigated by several institutions around the world [12–14]. Besides, further work has reported that liquid metals have attractive properties for CSP applications [15,16]. Recently, Amy et al. [17] demonstrated how a ceramic, mechanical pump that can be used to continuously circulate liquid metal at temperatures of around 1473 °K–1746 °K. This study solves the problem that collecting, transporting, storing liquid metal above 1300 °K brings.

The prediction of heat transfer of liquid metal has been the subject of many investigations. DeAngelis et al. [18] have examined using a liquid metal as heat-transfer fluid in conjunction with a receiver. It is feasible to reach temperatures of 1623 °K at greater than 90% efficiency. Boerema et al. [15] compared liquid sodium and Hitec, and the use of liquid sodium can achieve 57% absorber area reduction and 1.1% efficiency improvement. Even if liquid sodium is an excellent heat-transfer medium, the specific application background was not mentioned in the paper. Pacio et al. [19,20] summarized the current state-of-the-art of liquid metals (LMs) as HTFs in solar power plants.

Additionally, the liquid sodium (Na) was proposed as an efficient HTF to allow extending the design ranges, and able to contribute to the development of next-generation SPT. Matsubara et al. [21] studied the spanwise heat transport in turbulent channel flow with Prandtl numbers ranging from 0.025 to 5.0. It is regrettable that they do not compare the heat-transfer property with traditional heat-transfer fluid (e.g., solar salts). Rodríguez-Sanchez et al. [22,23] focused on the thermal, mechanical and hydrodynamic analysis associated to the nonuniformity of the heat flux, and they also considered the thermal field and thermal stresses along the solid wall [24].

A higher cost, more complex system and security issues are factors that have to be considered for the experimental research on liquid metal.

Therefore, numerical simulation is an effective research method [25,26]. In this work, we present a numerical simulation on the heat transfer of liquid sodium under nonuniform heat flux. A physical model of a single tube is established to investigate the heat-transfer performance of the receiver tube, and two other commonly used heat-transfer media, Hitec and solar salt, are also to be considered. First, we built solar heat-flux distributions on the whole absorber outer wall and circumferential variation of heat flux on the inner wall of the tube. Then, the heterogeneity of the temperature on the circumferential tangent plane and the solid wall is presented. Cloud images show the wall temperature distribution of the three HTFs. The temperature difference (Θ-Θ_{ref}) of the three HTFs along the circumferential angle is compared when q_{omax} is 0.1 MW/m^2, Re is ranging from 10,000 to 30,000, and inlet temperature is 550 °K. Last, the influence of three parameters, Q_m (1.0 to 3.0 kg/s), q_{omax} (0.1 to 0.3 MW/m^2), T_{in} (550 °K to 800 °K) on the heat-transfer characteristics of sodium, S-S and Hitec are discussed. This research can offer technical references for the design and construction of experimental facilities.

2. Mathematical model

2.1. Physical Model

The research background of this study is the heat receiver of the solar-power-tower system (Figure 1a). Since the collector is cylindrical and the fluid flows serpentinely in the collector tube, we investigate only the single collector tube (Figure 1b).

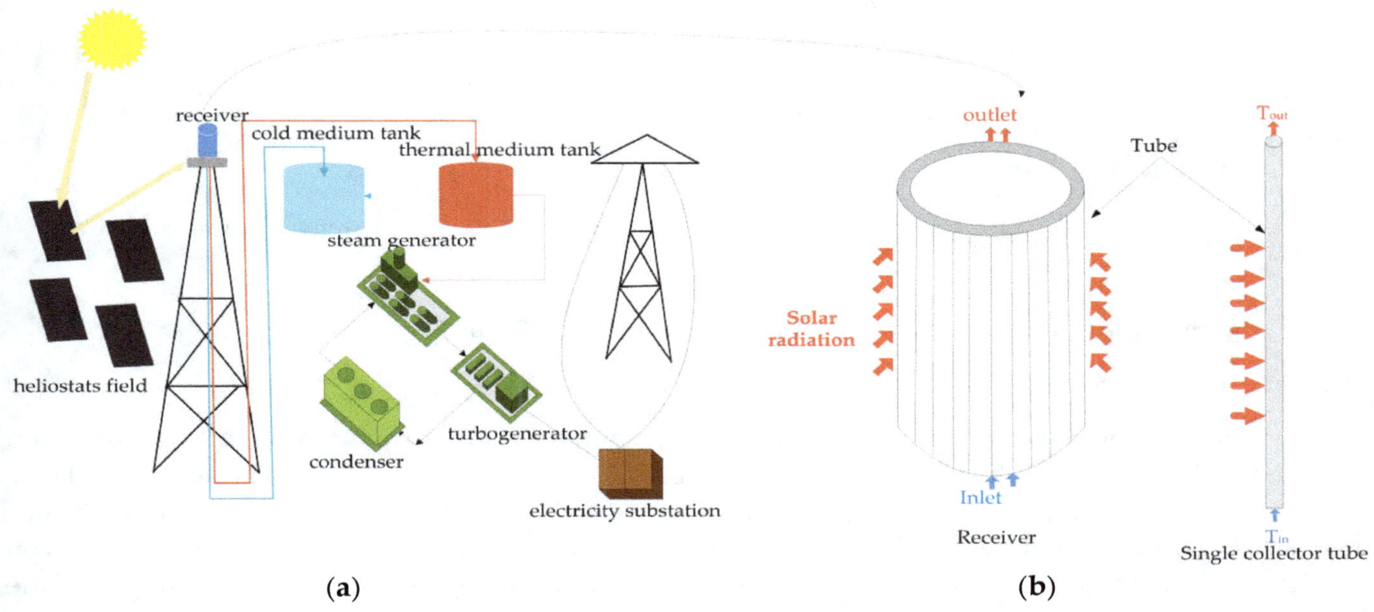

Figure 1. (**a**) Solar power tower system; (**b**) Exposed cylindrical heat receiver and a single receiver tube

We consider the conjugate heat-transfer problem of a receiver tube subject to inhomogeneous heat flux along the axial direction (z), circumferential direction (θ), and radial direction (r). The heat transport from the regional source (see Figure 2) contains three orthogonal components. The outer diameter (R) of the geometric model is 20 mm, and the tube length is 100R. The effect of different thermal boundary conditions on the heat-transfer performance of the collector tube is discussed.

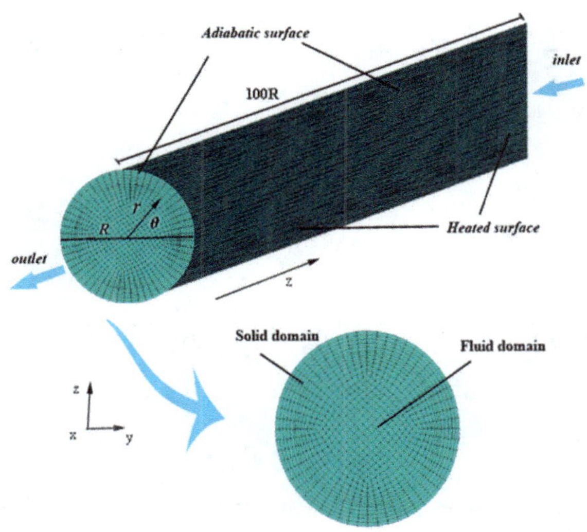

Figure 2. Schematic diagram of tower solar collector.

As shown in Figure 3, this paper deals with two heat-flux conditions of the turbulent-flow model:

(a) A cosine heat flux (see Equation (1)) [27] is imposed on one half of the wall of the tube, while the other half is considered adiabatic.

$$q = \begin{cases} q_{\text{omax}} \cdot e^{-\frac{9}{2}\cdot(z-1)^2} \cdot \cos\theta, & \cos\theta \geq 0 \\ 0, & \cos\theta < 0 \end{cases} \tag{1}$$

where the q_{omax} is the maximum heat flux on the wall of the collector tube, θ is the circumferential angle, and z is the length along the axial direction of the receiver tube.

(b) The heat flux to the bright side is constant, and the backlight side is an adiabatic.

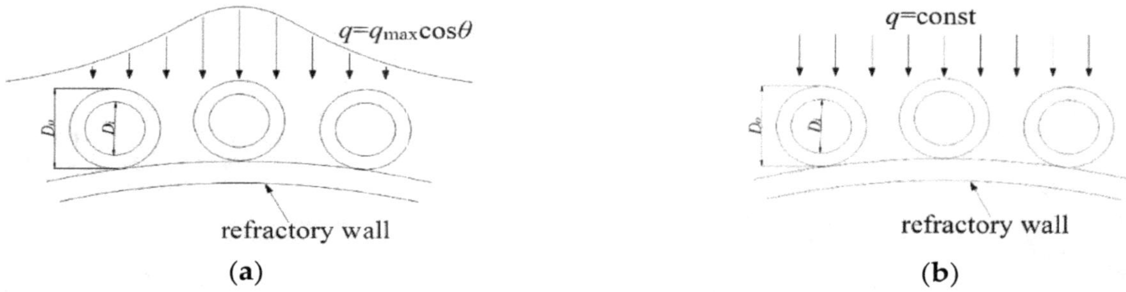

Figure 3. The heat flux conditions: (**a**) nonuniform heat flux (NHF) (**b**) uniform heat flux (UHF).

2.2. Governing Equations

A conjugate heat-transfer model is adopted to obtain the heat flux and temperature distribution in both the fluid and the tube wall. To simplify the model, the following assumptions are made: (1) the mediums are homogeneous and isotropic; (2) the liquid phase is in a state of local thermal equilibrium; (3) the heat loss at the outer wall of the receiver tube is not considered since this paper only focuses on the heat-transfer characteristic of the HTFs in the tube. For the heat loss on the outer surface of the tube, one can refer to the references [23,28].

The governing equations for continuity, momentum, energy and standard k–ε two-equation turbulence model for the incompressible Newtonian fluid can be expressed as follows:

Continuity equation:

$$\frac{\partial}{\partial \mathrm{x}}(\rho u_i) = 0, \tag{2}$$

Momentum equation:

$$\frac{\partial}{\partial x_i}(\rho u_i u_j) = -\frac{\partial p}{\partial x_i} + \frac{\partial}{\partial x_j}[(\mu_t + \mu) + (\frac{\partial u_i}{\partial x_j} + \frac{\partial u_j}{\partial x_i}) - \frac{2}{3}(\mu_t + \mu)\frac{\partial u_l}{\partial x_l}\delta_{\mathrm{ij}}] + \rho g_i, \tag{3}$$

Energy equation:

$$\frac{\partial}{\partial x_i}(\rho u_i T) = \frac{\partial}{\partial x_i}[(\frac{\mu}{\mathrm{Pr}} + \frac{\mu_t}{\sigma_t})\frac{\partial T}{\partial x_i}] + S_R, \tag{4}$$

k equation:

$$\frac{\partial}{\partial x_i}(\rho u_i k) = \frac{\partial}{\partial x_i}[(\mu + \frac{\mu_t}{\sigma_k})\frac{\partial k}{\partial x_i}] + G_k - \rho\varepsilon, \tag{5}$$

ε equation:

$$\frac{\partial}{\partial x_i}(\rho u_i \varepsilon) = \frac{\partial}{\partial x_i}[(\mu + \frac{\mu_t}{\sigma_\varepsilon})\frac{\partial \varepsilon}{\partial x_i}] + \frac{\varepsilon}{k}(c_1 G_k - c_2\rho\varepsilon), \tag{6}$$

where the turbulent viscosity μ_t and the production rate of k G_k are given by

$$\mu_t = c_\mu \rho \frac{k^2}{\varepsilon},\ G_k = \mu_t(\frac{\partial u_i}{\partial x_j} + \frac{\partial u_j}{\partial x_i}), \tag{7}$$

where the standard constants are employed, $c_\mu = 0.09$, $c_1 = 1.44$, $c_2 = 1.92$, $\sigma_k = 1.0$, $\sigma_\varepsilon = 1.3$, and $\sigma_T = 0.85$ [29].

2.3. Boundary Conditions

As a plurality of solar mirrors superposes the heat flux to the receiver tube wall, the radiation heat flux is highly nonuniform. The heat-flux expression is approximating the normal distribution function along the axial direction of the receiver tube and the cosine function distribution along the circumferential direction, as described in the Formula (1).

It is also convenient to define the total heat flux applied to the outer surface, and note that the energy conservation in the solid implies that:

$$Q = \int_0^{\pi} q_0(\theta) R_0 d\theta = 2 q_{omax} R_0, \quad (8)$$

where the total heat flux (Q) is kept constant for all the cases presented here, and R_0 = 20 mm.

Considering the wall temperature of the tube varies with the time and heat flux, to improve the accuracy and practicability of the numerical calculation, we adopted the formulas shown in Table 1 to evaluate the thermophysical properties of HTFs. The collector tube material is 316 L stainless steel, and its thermal conductivity (k_s) is 18.4 W/(m·K). Based on the assumptions, the boundary conditions are expressed as follows:

Table 1. Thermophysical properties of sodium [30], Solar salt and Hitec [7], where T is the fluid bulk temperature in Kelvin.

HTFs	Thermophysical Properties	Property Equation
Liquid sodium 371 °K–1255 °K	ρ/kg·m^{-3}	$219 + 275.32(1 - T/2503.7) + 511.58(1 - T/2503.7)^{0.5}$
	λ/W·m^{-1}·K^{-1}	$124.67 - 0.11381 \cdot T + 5.5226 \times 10^{-5} \cdot T^2 - 1.1842 \times 10^{-8} \cdot T^3$
	c_p/J·kg^{-1}·K^{-1}	$1658.2 - 0.84790 \cdot T + 4.4541 \times 10^{-4} \cdot T^2 - 2.9926 \times 10^6 \cdot T^{-2}$
	μ/P$_a$·s	$\ln(T) = -6.4406 - 0.3958\ln(T) + 556.835/T$
Solar salt 533 °K–873 °K	ρ/kg·m^{-3}	$2090 - 0.636(T - 273.15)$
	λ/W·m^{-1}·K^{-1}	$0.443 + 1.9 \times 10^{-4}(T - 273.15)$
	c_p/J·kg^{-1}·K^{-1}	$1443 + 0.172(T - 273.15)$
	μ/P$_a$·s	$2.2714 \times 10^{-2} - 1.2 \times 10^{-4}(T - 273.15) + 2.281 \times 10^{-7}(T - 273.15)^2 - 1.474 \times 10^{-10}(T - 273.15)^3$
Hitec 415 °K–808 °K	ρ/kg·m^{-3}	$-0.74(T - 273.15) + 2084$
	λ/W·m^{-1}·K^{-1}	$0.411 + 4.36 \times 10^{-4}(T - 273.15) + 1.54 \times 10^{-6}(T - 273.15)^2$
	c_p/J·kg^{-1}·K^{-1}	$1560 - (T - 273.15)$
	μ/P$_a$·s	$10^{2.7374}(T - 273.15)^{-2.104}$

(1) Fluid and the solid wall region

When the HTF flows around a stationary solid wall in a collector tube, where the solid wall is impermeable, the normal velocity should be satisfied $v_n = 0$. At the same time, the no-slip condition must be satisfied, and the tangential velocity $v_\tau = 0$. The heat-flux condition of the tube wall is:

$$q_w = -(\lambda \frac{\partial T}{\partial n}), \quad (9)$$

(2) The inlet and outlet temperature of the tube

The inlet velocity, pressure, and temperature of the tube line are formulated as follows, respectively.

$$T_{x=0} = T_0, \ u_{x=0} = u_0, \ p_{\text{inlet}} = p_0. \quad (10)$$

2.4. Numerical Methods

The governing Equations (2)–(7) are discretized by the finite volume method by using O-mesh and wall-dense nonuniform mesh. Moreover, the convective terms in momentum and energy equations are discretized with the second upwind scheme. The SIMPLE algorithm is used to ensure the coupling between velocity and pressure. The discretization of momentum, turbulent kinetic energy, energy, and dissipation rate are all second upwind schemes. The turbulence model is κ-ε model. The near-wall surface flow is solved by the standard wall function method, and all the non-dimensional number of near-wall y^+ is controlled by 30~60. The convergence criterion for the velocities and energy is that the maximum mass residual of the cells divided by the maximum residual is less than 10^{-5} and 10^{-7} for the continuity, momentum, and energy equation. Based on these methods, the performance of the receiver tube for the two models with different HTFs can be rapidly predicted [31].

2.5. Parameter Definitions

To predict the thermal and hydraulic characteristics, we define the time-averaged temperature in the cross-plane among the fluid as Θ, and the average temperature of the inlet and outlet of a receiver tube is named Θ_{ref}, also known as the qualitative temperature.

The Reynolds number and average Nusselt number in the receiver tube for the medium are given as [32].

$$Re = \frac{uD_i\rho}{\mu}, Nu = \frac{hR_i}{\lambda}, h = \frac{q}{\Delta t}, \tag{11}$$

where h is the average heat-transfer coefficient, q is the average heat flux in the tube, Δt the difference between the average temperature of the inner wall of the tube and the qualitative temperature.

The friction coefficient is defined as [33]:

$$f = \frac{\Delta p}{L}\frac{R_i}{(1/2)\rho u^2}, \tag{12}$$

where Δp is the pressure difference between the inlet and outlet of the receiver tube.

3. Model Verification and Cases Studied

3.1. Model Verification

We verified the calculation procedure by comparing them with empirical formulas and existing experimental data. Figure 4a exhibits the trend of Nu for S-S (Pr = 13) and Hitec (Pr = 50) as the Re ranging from 1×10^4 to 3×10^4 under the nonuniform heat flux on a receiver tube. The inlet temperature is set as the melting point for the three HTF, respectively. The result of the calculation is in good agreement with Dittus–Boetter correlation and Gnielinski correlation [7]. Figure 4b shows the variation tendency of Nu when the Pe is ranging from 65 to 203 for sodium (Pr = 0.01). The calculation results have a 2~6% difference with the Lyon-Martinelli equation correlation [34] and are consistent with the experimental data [30]. These results prove that the model and its calculation procedure is suitable and reasonable.

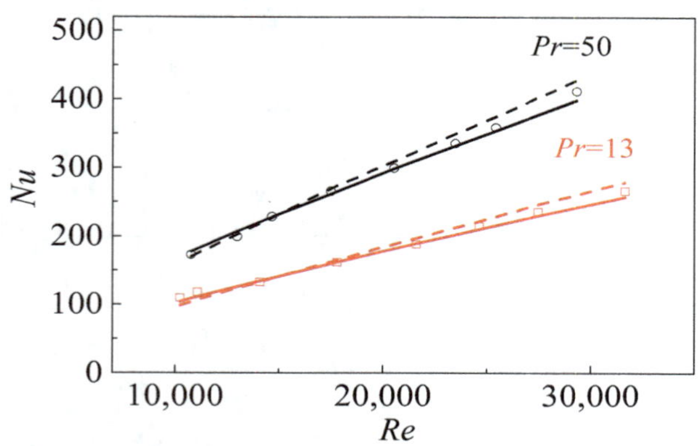

(a)

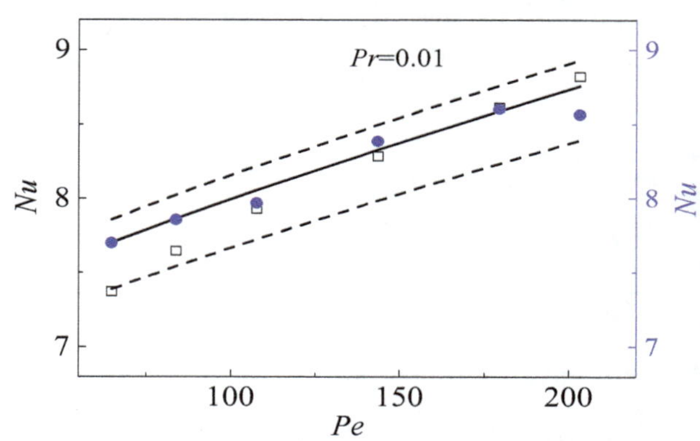

(b)

Figure 4. (**a**) Nu, as a function of Re, computed with inhomogeneous heating with WTR = 0.125. Red squares: Pr = 13. Black circles: Pr = 50. Solid lines: Dittus–Boetter correlation. Dashed lines: Gnielinski correlation. (**b**) Nu, as a function of the Péclet number, Pe. Solid lines: Lyon-Martinelli correlation (Nu = 7 + 0.028·$Pe^{0.8}$). Dashed lines refer to 2~6% from the correlation. Blue points: experimental data [34]. Black squares: numerical results.

3.2. Cases Studied

The main parameters used in the calculation cases are listed in Table 2. Among them, the mass flow rate varies from 1.0 to 3.0 kg/s, q_{omax} is from 0.1 to 0.3 MW/m^2, and the inlet temperature is from 550 °K to 800 °K. The temperature range is chosen between the melting point and the boiling point of the three mediums.

Table 2. The calculation parameters.

Test Condition	HTFs	Q_m (kg/s)	T_{in} (°K)	q_{omax} (MW/m^2)	h (W/m^2 K)
Case1	Sodium S-S Hitec	1.0~3.0	550	0.1	2989~14,089 2105~7318 2526~7947
Case2	Sodium S-S Hitec	1.0	550~800	0.1	23,822~21,520 4859~10,107 6047~11,664
Case3	Sodium S-S Hitec	1.0	550	0.1~0.3	75,960~75,753 5502~5581 6047~6167

4. Results and Discussion

4.1. Heat Flux

The heat-flux distributions on the whole absorber outer wall (case 1) are consistent with the result of Equation (1) shown in Figure 5.

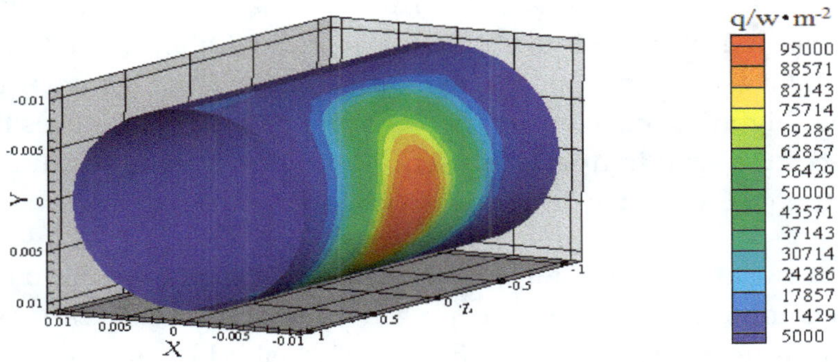

Figure 5. 3D heat-flux distribution on the outer tube's wall.

Figure 6 shows that the solar energy flux distribution, of different mediums on a cross section when Q_m is set as 1 kg/s, the inlet temperature is 550 °K, and q_{omax} is 0.1 MW/m^2. The distribution is unsymmetrical in the circumferential direction and is independent of the HTF in the receiver tube. It is noteworthy that the maximum heat flux through the inner wall is at the location 90°.

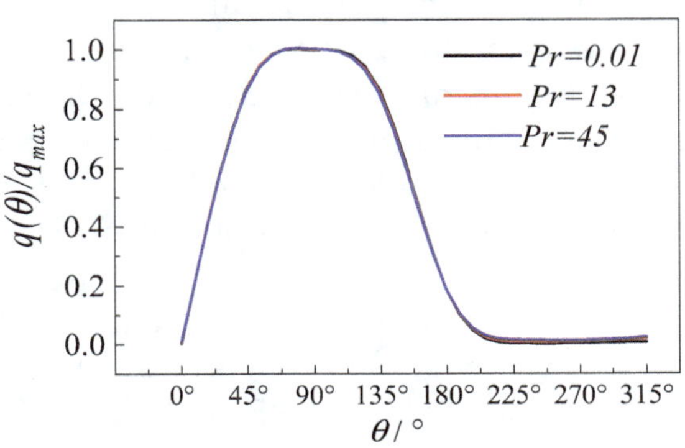

Figure 6. Circumferential variation of heat flux on the inner wall of the tube. The heat flux is non-dimensional with the maximum heat flux q_{omax} at the same WTR and Re.

4.2. Temperature Profile

The heat-flux distribution caused by inhomogeneous radiation inevitably influences the temperature distribution on the collector tube. The heterogeneity of the temperature on the circumferential tangent plane and the solid wall is shown in Figures 7 and 8. The coupled heat transfer between solid wall and fluid has been omitted in this section.

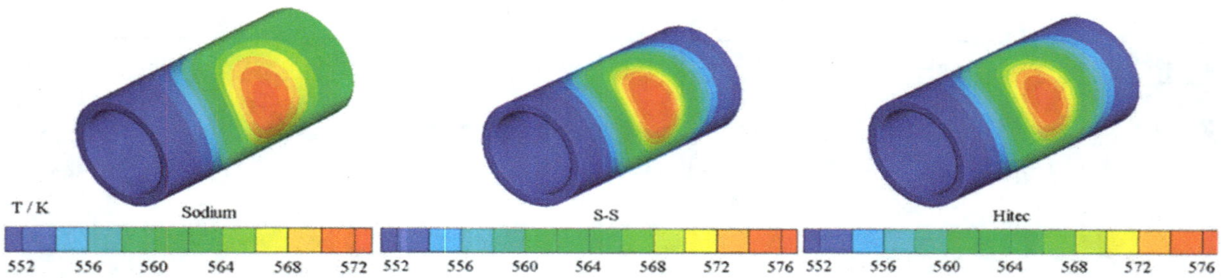

Figure 7. Temperature distribution on tube wall of sodium, S-S, and Hitec.

When the *Re* is 25,000, and the inlet temperature is 550 °K, the wall temperature distributions of the three HTFs are shown in Figure 7. It is found that the temperature distribution on the outer wall is extremely nonuniform around the tube. Compared with the other two HTFs, the temperature distribution on the solid wall of sodium is more uniform at the exit section. The outlet temperature is 561.08 °K, 550.93 °K, 551.06 °K on the area-weight-average from left to right of Figure 7. The tube wall temperature difference of sodium changes to 20 °K, solar salt changes to 24 °K and the Hitec changes to 24 °K after passing through two meters of the tube. Overall, for sodium, the temperature difference of the whole tube is the minimal, and the outlet temperature is higher than S-S or Hitec. The reasons for those are that high thermal conductivity of sodium, which enhances the turbulent flow, the heat transfer and reduces the wall temperature. Thus, in SPT projects, sodium is more suitable as a heat-transfer medium than S-S or Hitec, in terms of flow dynamics and heat transfer.

Figure 8 shows the variation of the temperature difference as a function of the circumferential coordinate. We made a comparison between the present NHF model calculations (solid-lines) and the results predicted by the uniform heat flux (UHF) model (dashed lines). The temperature difference (Θ-Θ_{ref}) means the difference in the temperature on the circumferential section and the qualitative temperature. The temperature difference (Θ-Θ_{ref}) of the three HTFs along the circumferential angle is compared when q_{omax} is 0.1 MW/m^2, *Re* is ranging from 10,000 to 30,000, and inlet temperature is set to 550 °K.

Note that the UHF model underestimates the inner temperature of the tube on the bright side while overestimates the backlight since the UHF model imposed the same heat flux on the bright side. Energy accumulation on both sides of the tube leads to two highest temperature points shown in Figure 8b,c whereas sodium has no obvious highest point. In the UHF model, the Θ-Θ_{ref} of sodium is lower than that in the NHF model, and the temperature difference is considerable as *Re* increasing. The results indicate that thermal conductivity still dominates even if the turbulence is significant when sodium is used as HTF. For the higher Prandtl number (S-S or Hitec), the average temperature in cross-plane of the adiabatic side must be equal to the bulk temperature. The reason is that the momentum transfer rate is higher than the heat-transfer rate, so the velocity surface is first established, and the heat-transfer delay. The changing range of Θ-Θ_{ref} is larger than sodium as *Re* increasing for S-S or Hitec. That shows that the flow and heat transfer of S-S or Hitec are greatly influenced by turbulence. As expected, the maximum temperature difference reached at the location of 90°, which agrees with the cosine effect expressed by the Equation (1).

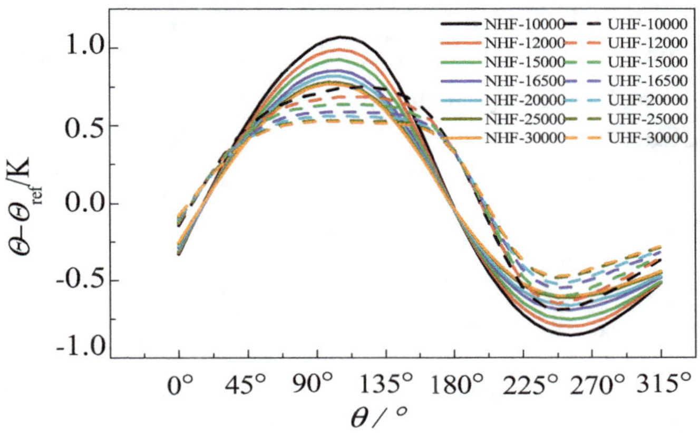

(**a**) Liquid sodium

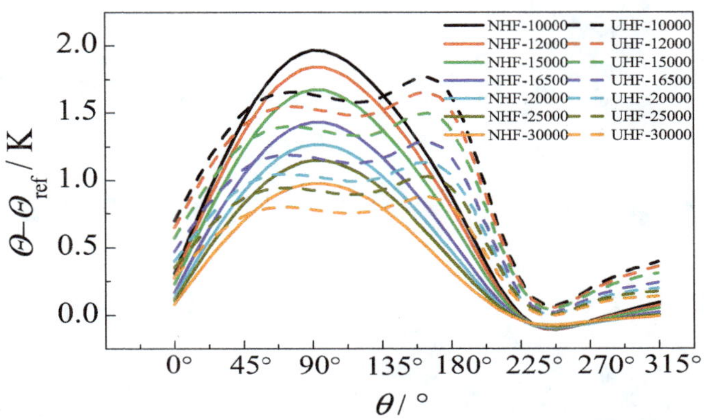

(**b**) Solar salt

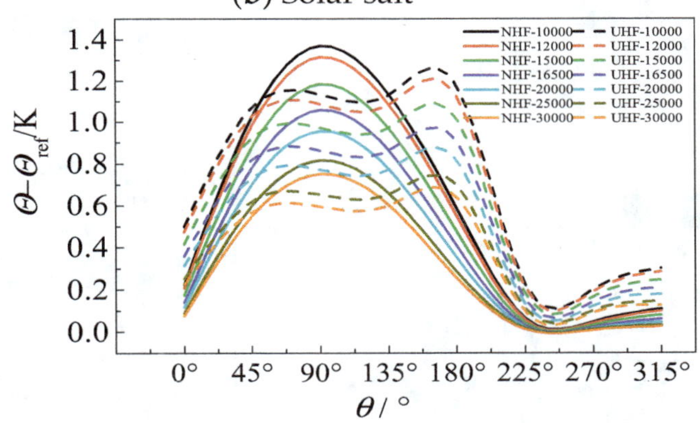

(**c**) Hitec

Figure 8. (**a**–**c**) Circumferential variation of the inner wall fluid temperature with the qualitative temperature for different Reynolds number at a fixed-inlet temperature (T_{in} = 550 °K). (**a**) Pr = 0.001 (**b**) Pr = 13 (**c**) Pr = 50. Solid lines: the present calculations; Dashed lines: homogeneous heat-flux model (HHFM).

The value of Θ-Θ_{ref} is slightly dependent on the Re but strongly affected by the Pr. In general, the Θ-Θ_{ref} is increasing as the Pr increases, decreasing as the Re increases. Comparing Figure 8a–c, the temperature difference of S-S at the location $\theta = 90°$ is 1.93 times higher than sodium, and Hitec is 1.35 times higher that of sodium when Re is 10,000. The temperature difference of S-S is 1.29 times higher than sodium, and Hitec is 2.0 times sodium when Re is 30,000. The results reflect that sodium has a good effect on solving the uneven temperature caused by nonuniform heat flux, which may reduce the risk of hot spots and, thus, reduce pipe stresses.

Another concern is the inner wall temperature, also called film temperature, which could lead to the degradation of the HTFs. The inner wall temperature distribution in NHF and UHF models are shown in Figure 9a–c when *Re* is 25,000, and the inlet temperature is 550 K. From Figure 9a–c, the average temperature of the fluid in the tube is: 562.8 °K, 554.9 °K, 554.7 °K. Figure 9a–c show that one of the most significant advantages of sodium over S-S and Hitec is that the temperature distribution is much more uniform on both the inner and outer surface of the tube.

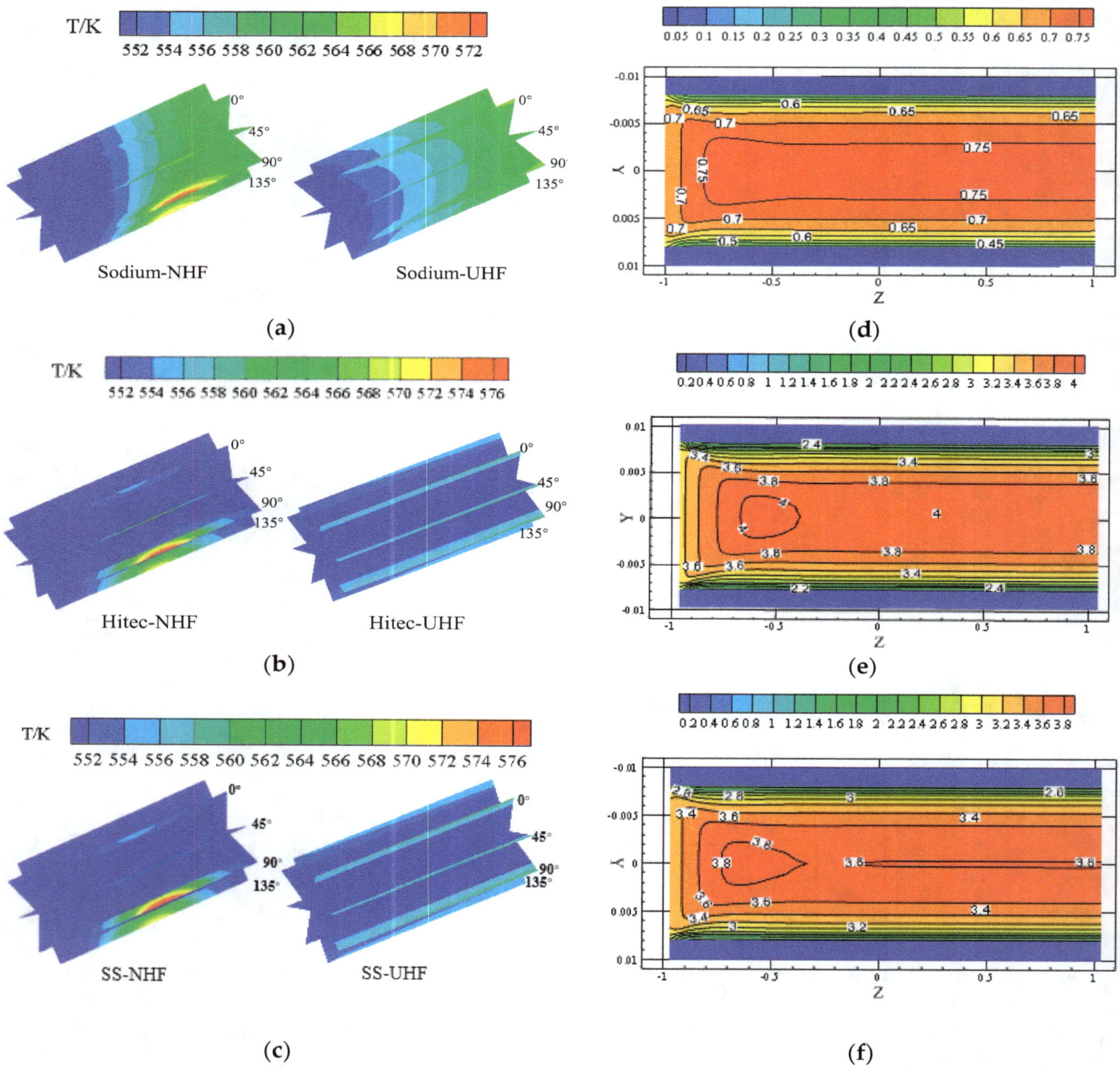

Figure 9. (**a–c**) Temperature distribution of three kinds of media under two models and (**d,e**) velocity distribution under NHF model.

Obviously, for the NHF model, there is a heat spot on the outer wall from 0° to 180°, especially at 90°. At the same time, the UHF model is idealistic. But, the highest heat spot temperature of sodium is the lowest, while S-S and Hitec are same. There is a considerable temperature gradient when the sodium flows in the receiver tube. Besides, not far from the entrance, the inner fluid temperature is higher and more uniform than S-S or Hitec flow in the tube.

Corresponding velocity distribution is shown in Figure 9e,f. The velocity of sodium is the lowest, only 0.19 times the other two mediums. At a distance of 0.2 m from the inlet, the sodium flow reached fully developed, and the boundary layer is thinner, while the other boundary conditions remained unchanged. It can be explained that the molecule heat conduction of sodium dominates in fully developed turbulent heat transfer. The heat diffusion of the medium with lower Prandtl number (sodium) is much higher than the momentum transfer diffusivity.

4.3. Effects of Mass Flow

Figure 10 shows the dependence of h, f of the three HTFs on Q_m, which varies from 0.2 to 1.4 kg/s. The values of maximum heat flux and inlet temperature are set to 0.1 MW/m^2 and 550 °K, respectively.

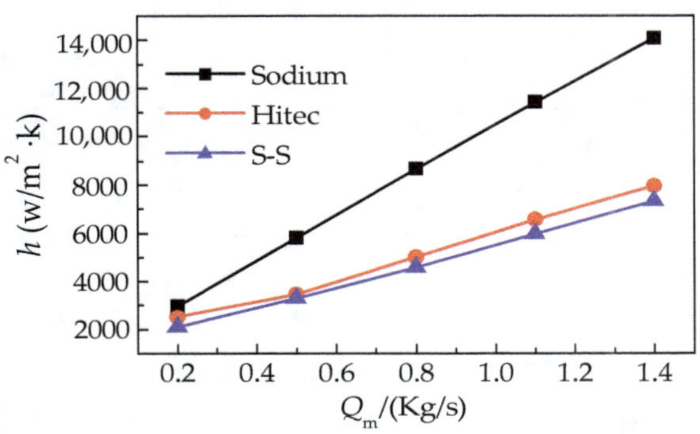

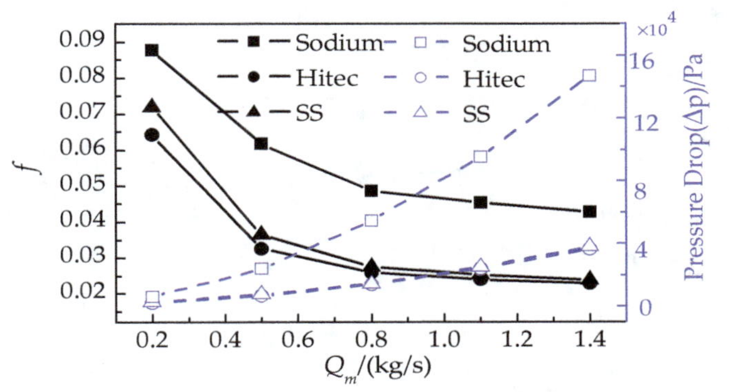

Figure 10. Effects of Q_m on the heat-transfer performance of the HTFs in the tube: (**a**) Q_m-h, (**b**) Q_m-f-Δp.

The average heat-transfer coefficient of all working fluids increases with the mass flow. When the Q_m increases by 0.3 kg/s, the h of S-S, Hitec and sodium increases by an average of 1303, 1355, 2775 W/m^2·°K, respectively. The variation of Q_m has a significant influence on the heat-transfer characteristic of the HTFs, especially on sodium. For sodium, the higher the Q_m, the better the heat-transfer coefficient. The reason is that the average temperature difference is getting smaller increasingly with the rise of Q_m when the maximum heat flux is 0.1 MW/m^2. However, the law of change in the average friction coefficient is reversed, all decreasing as the flow rate increases. When the Q_m is higher than 0.8 kg/s, the friction coefficient decreases slowly. S-S and Hitec, with high heat-transfer rate and low-pressure drop, are exceptionally beneficial for heat transfer. The results suggest that the Q_m should be higher than 0.8 kg/s to obtain greater heat-transfer performance. However, a larger mass-flow rate represents more considerable pump pressure and more substantial heat absorption capability. Besides, without increasing the solar thermal input, it is possible to cause the reduction of fluid temperature and increase the corresponding heat loss.

4.4. Effects of Inlet Temperature

The physical properties of the HTF are primarily affected by the inlet temperature (T_{in}). The typical operating temperatures for the three HTFs are shown in Table 3. The available temperature range is between the melting point and the boiling point. The trends of h, f for the three HTFs have been shown in Figure 11. T_{in} is varying from 550 °K to 800 °K, Q_m is 1.0 kg/s, and q_{omax} is 0.1 MW/m^2.

From Figure 11a, the heat-transfer coefficient of sodium is larger by about 2.5 to 5 times than the other two kinds of molten salt and decreases slowly with the temperature rising. The reason for the above unique heat-transfer feature of sodium is its high conductivity, and it decreases with increasing temperature even in turbulent flow. As shown in Figure 11c, the conductivity of sodium is 118, 154 times Hitec and S-S at 550 °K, respectively. Furthermore, the h of Hitec or S-S increases linearly with the increase of temperature whereas sodium is the opposite. The h of Hitec has not changed significantly between 750 °K and 800 °K because its boiling point is 808 °K. The greater the h, the smaller the f when the mass-flow rate is constant for the three HTFs. Besides, from Table 3, the energy-storage capacity (ρC_p) of Hitec or S-S is two times more than the sodium.

Table 3. Comparison of the physical properties for different HTFs proposed for CRS applications. The physical properties are evaluated at 550 °K, 1 bar. melting point = m.p, normal boiling point (n.b.p). Sources: [35–37].

Candidate HTF	m.p °K	n.b.p °K	ρC_p kJ/m^3·°K
Sodium	371	1255	1166
Solar salt	533	873	2498
Hitec	415	808	2411

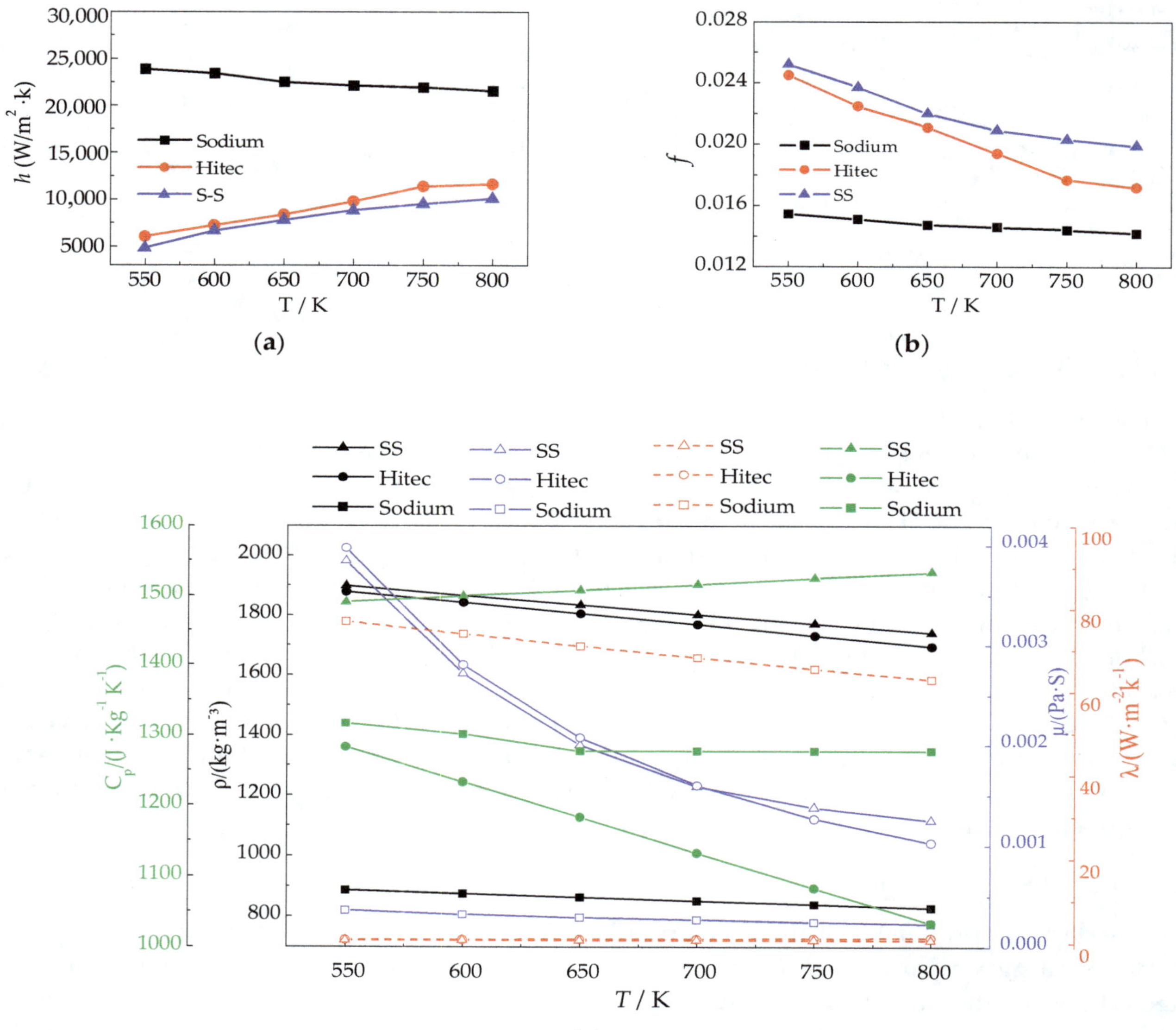

Figure 11. Effects of T_{in} on the performance of tube: (**a**) T-h; (**b**) T-f; (**c**) T-ρ-μ-C_p-λ.

Based on the above conclusions, we can select sodium as the heat-transfer medium in the collector tube on the top of the SPT, while the Hitec or S-S as the circulating medium or storage medium, respectively.

4.5. Effects of Heat Flux on the Outside Surface

Generally, the heat flux on the outside surfaces of the receiver tubes in SPT plants is quite limited to give a reasonable lifetime. The maximum heat flux allowed (allowable flux density), q_{omax}, has progressively grown for many years, such as 0.35 MW/m^2 in the Solar One plant and 0.8 MW/m^2 in the Solar Two plant. Besides, the q_{omax} is a crucial parameter for the receiver design, since it is directly related to the heliostat field cost, which involves the number of heliostats and the arrangement strategy [38,39].

Figure 12a,b show the variations of h, f of the three HTFs when q_{omax} varies from 0.1 to 0.3 MW/m^2. The values of mass flow rate and inlet temperature are set to 1.0 kg/s and 550 °K, respectively. The increased maximum heat flux does not affect the average heat-transfer coefficient of HTFs in the tube, and the h of sodium is invariably 12.56 times that of Hitec and 13.8 times that of S-S. At the same time, the flow of liquid sodium in the tube maintains a small friction coefficient compared to the other two working fluids (Figure 12b). Therefore, liquid sodium ensures economics while maintaining efficient system operation.

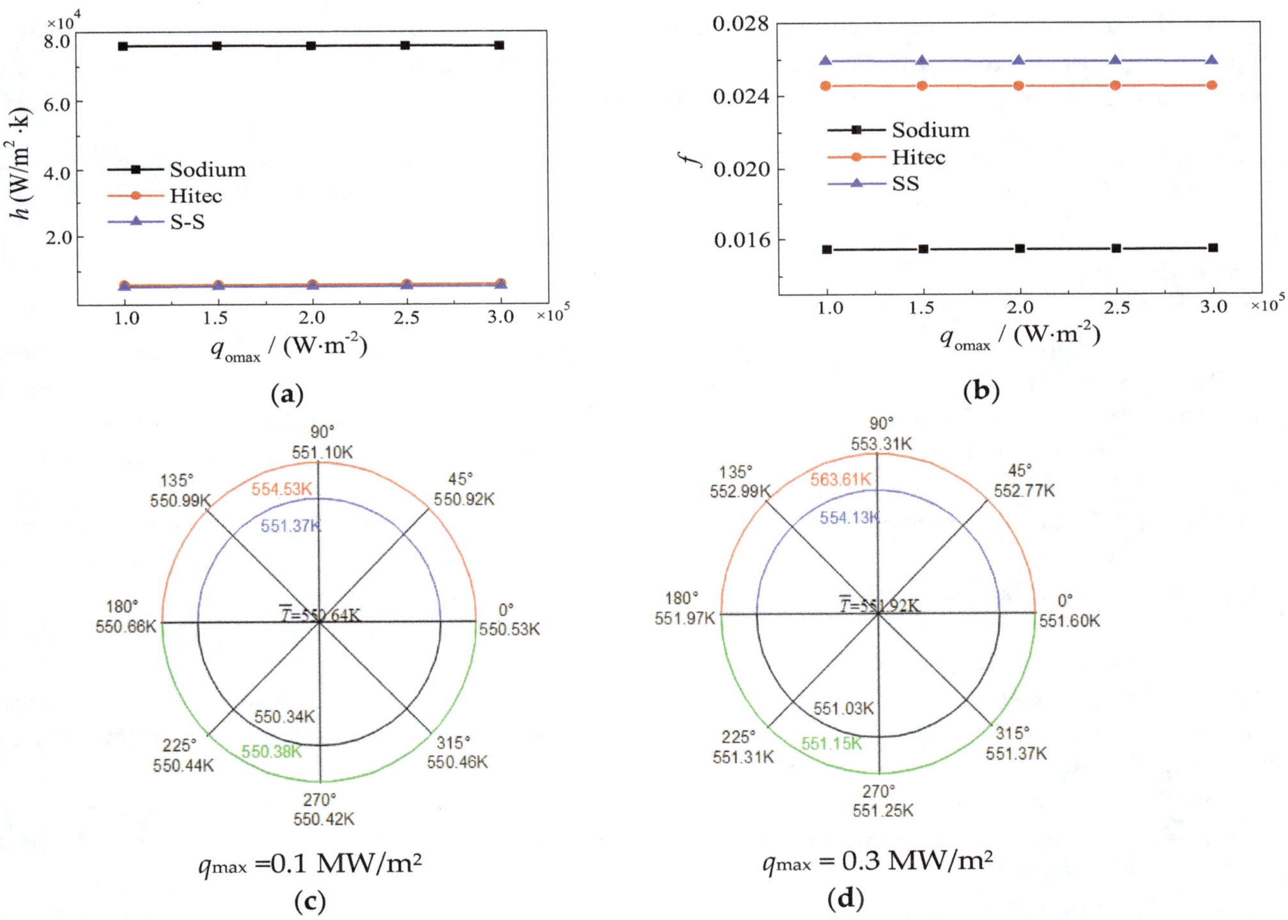

Figure 12. Effects of q_{omax}: (**a**) q_{omax}-h, (**b**) q_{omax}-f, (**c**,**d**) q_{omax}-T.

As shown in Figure 12c,d, when the maximum heat flux changed from 0.1 to 0.3 MW/m^2, the temperature of the outer wall increased from 554.53 °K to 563.61 °K while the average temperature of the circumferential increased by about 2 °K. At the same time, the temperature difference between the inner and outer walls is smaller than 4 °K in both cases. As a result, the thermal stress acting on the heat-transfer tubes is greatly reduced.

5. Conclusions

The heat-transfer characteristic of liquid sodium is compared with solar salt and Hitec in a heat-transfer tube for the SPT system. The calculations present the effects of operational parameters (Q_m, T_{in}, q_{omax}) on the thermal and thermo-hydraulic performance of three mediums under a nonuniform heat flux. The following conclusions are obtained:

(a) The tube wall temperature difference of sodium is 20 °K, which is 4 °K lower than other the two mediums when the *Re* is 25,000, and the inlet temperature is 550 °K. In either NHF model or UHF model, the temperature distribution of Sodium is more uniform and higher than S-S and Hitec, and the highest temperature on the hot spot is 4 °K lower than S-S or Hitec.

(b) Among these factors (Q_m, T_{in}, q_{omax}), the change of Q_m has an essential influence on the heat-transfer coefficient of HTFs. However, even in high turbulence, the thermal conductivity of Sodium still plays a crucial role. Meanwhile, the change of q_{omax} does not affect the average heat-transfer performance of the heat-transfer medium, but it has a significant influence on the temperature of the tube wall. Loading excessively high heat flux on the collector tube may cause many security issues in SPT system.

(c) Under the same boundary conditions, the heat-transfer performance of Sodium is one order of magnitude higher than that of S-S and Hitec at low temperature (550 °K~700 °K), and it is twice than the two kinds of molten salts at high temperature (700 °K +).

Author Contributions: Conceptualization, Y.H.; Data curation, J.L.; Formal analysis, J.L.; Funding acquisition, Y.H.; Investigation, J.L.; Methodology, Y.H. and X.L.; Validation, J.L.; Writing—original draft, J.L.; Writing—review & editing, J.L., Y.H. and X.L.

References

1. Sargent and Lundy LLC Consulting Group. *Assessment of Parabolic Trough and Power Tower Solar Technology Cost and Performance Forecasts*; Report No. NREL/SR-550-34440; NREL: Golden, CO, USA, 2003. [CrossRef]
2. García-Barberena, J.; Monreal, A.; Mutuberria, A.; Sánchez, M. Towards cost-competitive solar towers—Energy cost reductions based on decoupled solar combined cycles (DSCC). *Energy Procedia* **2014**, *49*, 1350–1360. [CrossRef]
3. Hinkley, J.; Curtin, B.; Hayward, J.; Wonhas, A.; Boyd, R.; Grima, C.; Tadros, A.; Hall, R.; Naicker, K.; Mikhail, A. *Concentrating Solar Power—Drivers and Opportunities for Cost-Competitive Electricity*; Report No. EP111647; CSIRO: Canberra, Australia, 2011. [CrossRef]
4. Roldán, M.I.; Monterreal, R. Heat flux and temperature prediction on a volumetric receiver installed in a solar furnace. *Appl. Energy* **2014**, *120*, 65–74. [CrossRef]
5. Chang, C.; Li, X.; Zhang, Q.Q. Experimental and Numerical Study of the Heat Transfer Characteristics in Solar Thermal Absorber Tubes with Circumferentially Non-uniform Heat Flux. *Energy Procedia* **2014**, *49*, 305–313. [CrossRef]
6. Romero, M.; González-Aguilar, J. Next generation of liquid metal and other high-performance receiver designs for concentrating solar thermal (CST) central tower systems. *Adv. Conc. Sol. Therm. Res. Technol.* **2017**, 129–154. [CrossRef]
7. Siegel, N.P.; Bradshaw, R.W.; Cordaro, J.B.; Kruizenga, A.M. Thermophysical property measurement of nitrate salt heat transfer fluids. In Proceedings of the ASME 2011 5th International Conference on Energy Sustainability, Washington, DC, USA, 7–10 August 2011; pp. 439–446. [CrossRef]
8. Pacio, J.; Singer, C.; Wetzel, T.; Uhlig, R. Thermodynamic evaluation of liquid metals as heat transfer fluids in concentrated solar power plants. *Appl. Therm. Eng.* **2013**, *60*, 295–302. [CrossRef]
9. Kearney, D.; Herrmann, U.; Nava, P.; Kelly, B.; Mahoney, R.; Pacheco, J.; Cable, R.; Potrovitza, N.; Blake, D.; Price, H. Assessment of a molten salt heat transfer fluid in a parabolic trough solar field. *J. Sol. Energy Eng.* **2003**, *125*, 170–176. [CrossRef]

10. Thomas, W.; Julio, P.; Luca, M.; Alfons, W.; Annette, H.; Wolfgang, H.; Carsten, S.; Georg, M.; Jürgen, K.; Robert, S.; et al. Liquid metal technology for concentrated solar power systems: Contributions by the German research program. *AIMS Energy* **2014**, *2*, 89–98. [CrossRef]
11. Guidez, J.; Martin, L.; Chetal, S.; Chellapandi, P.; Raj, B. Lessons learned from sodium-cooled fast reactor operation and their ramifications for future reactors with respect to enhanced safety and reliability. *Nucl. Technol.* **2008**, *164*, 207–220. [CrossRef]
12. Singer, C.; Buck, R.; Pitz-Paal, R.; Müller-Steinhagen, H. Assessment of solar power tower driven ultrasupercritical steam cycles applying tubular central receivers with varied heat transfer media. *J. Sol. Energy Eng.* **2010**, *132*, 41010. [CrossRef]
13. Hering, W.; Stieglitz, R.; Wetzel, T. Application of liquid metals for solar energy systems. EPJ Web of Conferences. *EDP Sci.* **2012**, *33*, 03003. [CrossRef]
14. Bienert, W.B. The heat pipe and its application to solar receivers. *Electr. Power Syst. Res.* **1980**, *3*, 111–123. [CrossRef]
15. Boerema, N.; Morrison, G.; Taylor, R.; Rosengarten, G. Liquid sodium versus Hitec as a heat transfer fluid in solar thermal central receiver systems. *Sol. Energy* **2012**, *86*, 2293–2305. [CrossRef]
16. Kotzé, J.P.; von Backström, T.W.; Erens, P.J. NaK as a primary heat transfer fluid in thermal solar power installations. In Proceedings of the SolarPACES 2012 International Conference, Durban, South Africa, 27–29 March 2014.
17. Amy, C.; Budenstein, D.; Bagepalli, M.; England, D.; Deangelis, F.; Wilk, G.; Jarrett, C.; Kelsall, C.; Hirschey, J.; Wen, H.; et al. Pumping liquid metal at high temperatures up to 1,673 kelvin. *Nature* **2017**, *550*, 199–203. [CrossRef] [PubMed]
18. DeAngelis, F.; Seyf, H.R.; Berman, R.; Schmidt, G.; Moore, D.; Henry, A. Design of a high temperature (1350 °C) solar receiver based on a liquid metal heat transfer fluid: Sensitivity analysis. *Sol. Energy* **2018**, *164*, 200–209. [CrossRef]
19. Pacio, J.; Wetzel, T. Assessment of liquid metal technology status and research paths for their use as efficient heat transfer fluids in solar central receiver systems. *Sol. Energy* **2013**, *93*, 11–22. [CrossRef]
20. Pacio, J.; Fritsch, A.; Singer, C.; Uhlig, R. Liquid metals as efficient coolants for high-intensity point-focus receivers: Implications to the design and performance of next-generation CSP systems. *Energy Procedia* **2014**, *49*, 647–655. [CrossRef]
21. Matsubara, K.; Sakurai, A.; Miura, T.; Kawabata, T. Spanwise heat transport in turbulent channel flow with Prandtl numbers ranging from 0.025 to 5.0. *J. Heat Transf.* **2012**, *134*, 041701. [CrossRef]
22. Flores, O.; Marugán-Cruz, C.; Santana, D.; Garcia-Villalba, M. Thermal stresses analysis of a circular tube in a central receiver. *Int. Conf. Sol. PACES* **2014**, *49*, 354–362. [CrossRef]
23. Rodríguez-Sánchez, M.; Venegas-Bernal, M.; Marugán-Cruz, C.; Santana, D. Thermal, mechanical and hydrodynamic analysis to optimize the design of molten salt central receivers of solar tower power plants. In Proceedings of the International Conference on Renewable Energies and Power Quality (ICREPQ'13), Bilbao, Spain, 20–22 March 2013; pp. 128–133. [CrossRef]
24. Irfan, M.A.; Chapman, W. Thermal stresses in radiant tubes due to axial, circumferential and radial temperature distributions. *Appl. Therm. Eng.* **2009**, *29*, 1913–1920. [CrossRef]
25. Fritsch, A.; Uhlig, R.; Marocco, L.; Frantz, C.; Flesch, R.; Hoffschmidt, B. A comparison between transient CFD and FEM simulations of solar central receiver tubes using molten salt and liquid metals. *Sol. Energy* **2017**, *155*, 259–266. [CrossRef]
26. Flesch, J.; Niedermeier, K.; Fritsch, A.; Musaeva, D.; Marocco, L.; Uhlig, R.; Baake, E.; Buck, R.; Wetzel, T. Liquid metals for solar power systems. *IOP Conf. Ser. Mater. Sci. Eng.* **2017**, *228*, 012012. [CrossRef]
27. Gartshore, I.; Salcudean, M.; Hassan, I. Film cooling injection hole geometry: Hole shape comparison for compound cooling orientation. *AIAA J.* **2015**, *39*, 1493–1499. [CrossRef]
28. Rodríguez-Sánchez, M.R.; Soria-Verdugo, A.; Almendros-Ibáñez, J.A.; Acosta-Ibáñez, A.; Santana, D. Thermal design guidelines of solar power towers. *Appl. Therm. Eng.* **2014**, *63*, 428–438. [CrossRef]
29. Patankar, S. *Numerical Heat Transfer and Fluid Flow*; Hemisphere Publishing: Washington, DC, USA, 1980.
30. Mikityuk, K. Heat transfer to liquid metal: Review of data and correlations for tube bundles. *Nucl. Eng. Des.* **2009**, *239*, 680–687. [CrossRef]
31. Minkowycz, W.J.; Sparrow, E.M.; Murthy, J.Y. *Handbook of Numerical Heat Transfer*, 2nd ed.; John Wiley & Sons: New York, NY, USA, 2000.

32. Wu, M.; Li, M.; Xu, C.; He, Y.L.; Tao, W.Q. The impact of concrete structure on the thermal performance of the dual-media thermocline thermal storage tank using concrete as the solid medium. *Appl. Energy* **2014**, *113*, 1363–1371. [CrossRef]
33. He, Y.L.; Tao, W.Q. Convective heat transfer enhancement: Mechanisms, techniques, and performance evaluation. *Adv. Heat Transf.* **2014**, *46*, 87–186. [CrossRef]
34. Gnielinski, V. New equations for heat and mass transfer in turbulent pipe and channel flow. *Int. Chem. Eng.* **1976**, *16*, 8–16.
35. Zavoico, A.B. *Solar Power Tower Design Basis Document, Revision 0*; Report of Sandia National Laboratories: San Francisco, CA, USA, 2001. [CrossRef]
36. Foust, O.J. *Sodium-NaK Engineering Handbook*; Gordon and Breach Science Publishers: New York, NY, USA, 1972; pp. 32–38.
37. VDI-Gesellschaft Verfahrenstechnik und Chemieingenieurwesen (GVC). *VDI Heat Atlas*, 2nd. ed.; Springer: Berlin, Germany, 2010.
38. Vant-Hull, L.L. The role of "allowable flux density" in the design and operation of molten-salt solar central receivers. *J. Sol. Energy Eng.* **2002**, *124*, 165–169. [CrossRef]
39. Liao, Z.; Li, X.; Xu, C.; Chang, C.; Wang, Z. Allowable flux density on a solar central receiver. *Renew. Energ.* **2014**, *62*, 747–753. [CrossRef]

6

Design, Manufacturing and Characterization of Linear Fresnel Reflector's Facets

Diego Pulido-Iparraguirre [1,2,3], Loreto Valenzuela [4,*], Jesús Fernández-Reche [4], José Galindo [1] and José Rodríguez [4]

1 Plataforma Solar de Almería, Crta. Senes, km. 4.5, E04200 Tabernas, Spain
2 Universidad de Antofagasta, Avenida Angamos, Antofagasta 601, Chile
3 Universidad de Almería, Crta. De Sacramento s/n, E04120, La Cañada de San Urbano, E04120 Almería, Spain
4 CIEMAT-Plataforma Solar de Almería, Crta. Senes, km. 4.5, E04200 Tabernas, Spain
* Correspondence: loreto.valenzuela@psa.es.

Abstract: This paper presents a procedure for making facetted mirrors to use in linear Fresnel reflectors, considering the design of the transversal geometry, materials, and structure configuration. Four different assemblies of the structure that supports and shapes the mirror are documented and evaluated. An assembly that implies a curved, pleated aluminum rectangular plate with a thin silvered-glass mirror vacuum glued to the plate is defined as the optimal. The geometrical quality of the chosen mirror facet's configuration is accomplished by photogrammetry.

Keywords: solar energy; linear Fresnel collector; mirror facet; manufacturing process; geometrical quality

1. Introduction

Concentrated solar thermal technologies are interesting renewable energy systems to be incorporated in industrial heat processes [1]. There are two main technologies that fit in the range of temperature demanded by several industrial processes, i.e., parabolic-through collector (PTC) and linear Fresnel reflector (LFR) [2].

In the last four decades, in terms of concentrated solar power plants, PTC has been leading the energy production and the number of facilities installed over LFR [3]. From an optical point of view, a PTC has a higher optical performance compared to a LFR, explained by the transversal component of the beam solar radiation impinging on the LFRs reflectors aperture, which decreases the optical efficiency of the collector. However, LFR has a simpler concept related to the absence of metal-glass welds in the receiver tubes, commonly present in the receivers for PTCs, and they do not require flexible hoses or ball joints between adjacent collectors. Also, the use of land is more efficient with LFRs because shadows between neighboring collectors are reduced. Finally, the optical components of LFR are cheaper due to the nearly flat shape of the mirrors. All these advantages result in lower operational and investment costs [4,5]. As a consequence, all these aspects lead to a technology as attractive as PTC technology. LFR has a significant development potential that explains the current research efforts on this solar collector technology.

A LFR system consists of an array of slightly curved reflectors composing the optical solar concentrator and a receiver, which is fixed and collects the solar energy concentrated by the primary reflectors. These reflectors need to track the sun throughout daytime in order to reflect the incoming solar rays to the receiver position. The receiver may be composed of one single tube or a group of parallel tubes, located horizontally above the reflectors plane.

To ensure low investment costs, it is important to design as efficient components as possible. For example, a way to achieve this is using a less amount of material but maintaining the durability and performance requirements of the components. This work is focused on a manufacturing procedure of facets for LFR's primary reflectors. Each step of the procedure defined is detailed, altogether with different assembly possibilities that have been tested.

2. Background on LFR Mirrors Geometry

LFR works as an optical concentrator, which means that the mirror facets that compose the primary reflector, concentrate the solar rays in the receiver. As explained in Section 1, the primary reflector is the component where the concentration takes place. The shape of the mirrors that form the reflector is essential to the concentration rate that the collector reaches. The geometrical shape of the mirrors is defined from their transversal profile; the shapes addressed in this work are flat, parabolic and cylindrical (Figure 1).

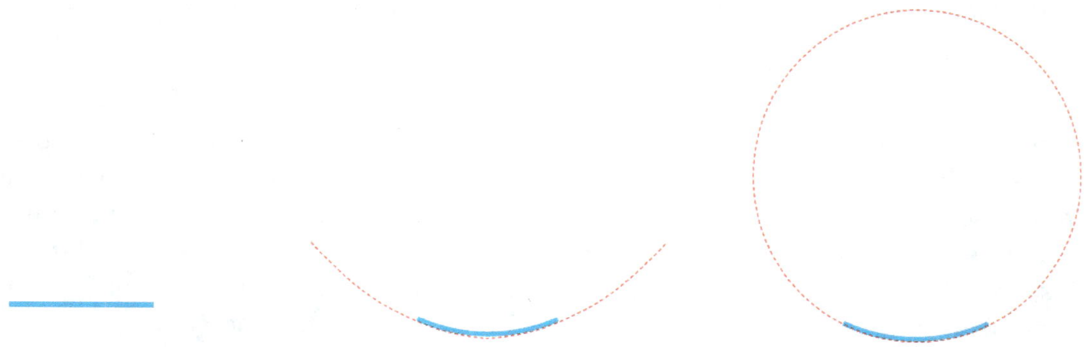

(a) (b) (c)

Figure 1. Scheme of the three mirror's shape addressed: (**a**) flat-shape mirrors, (**b**) parabolic-shaped mirrors, and (**c**) cylindrical-shaped mirrors.

2.1. Flat Mirrors

Flat-shaped mirrors (Figure 1a) are the simplest configuration; in this case, the transversal mirror profile is a straight line. From the optical point of view, the width of the image reflected by a flat-shaped mirror is practically the same as the mirror's actual width. The setback of flat-shaped mirrors is that one single mirror does not concentrate the radiation by itself, because the area of reflection is the same as the area of the reflected image. The concentration takes part when the reflected image of a group of mirrors is directed to the same place. Another disadvantage related to the use of flat-shaped mirrors is that the width of the mirrors must be virtually the same as the width of the receiver, because if the mirrors were wider, some of the reflected rays would not hit the receiver (overflow effect). For this reason, the width of flat-shaped mirrors and receivers are restricted to each other.

2.2. Parabolic Mirrors

Parabolic geometry (Figure 1b) perfectly fits with optical concentration technologies, because, by definition, the rays impinging perpendicularly to the aperture are reflected into the focal point of the parabola. In the case of solar collectors with parabolic reflectors, this condition is achieved when the reflectors track the movement of the sun during daytime. According to this geometrical principle, PTCs and LFRs have been using parabolic mirrors since the beginning of their operations [3,6].

Particularly in the LFR technology, from the construction point of view, to achieve a parabolic profile in the transversal section of the mirror is a difficult task to accomplish compared to other options. One option is to manufacture the mirror directly with the desired shape. However, this option is discarded because it limits the availability of suppliers and implies higher costs. An alternative option consists on assembling mechanically a straight mirror into a preformed pattern with the desired parabolic profile. In order to achieve this assemblage and to adopt the desired shape, the mirrors must be as thin as possible. This pattern could be a solid plate or a group of profiles named "ribs". Both possibilities require the material to be previously manufactured into the desired shape. The result is optimal optical behavior of the primary reflector. However, economically, the cost of manufacturing the pattern is higher compared to other geometries.

2.3. *Cylindrical Mirrors*

From the geometrical point of view, cylindrical mirrors (Figure 1c) are not as accurate as parabolic mirrors, but constructively, they are easier and cheaper to manufacture. With this geometry, the concentration of the rays is achieved when the distance from the mirror to the receiver is considerably larger than the width of the mirror [7].

In Figure 2, the scheme shows a cylindrical mirror, which is a segment of a circumference, and a ray, r, impinging perpendicularly to the aperture of the mirror in a point P. The ray is reflected and directed to f. The angles formed by the reflected ray and the incidence angle, θ, which are angles referring to the normal vector of the mirror, are equal according to specular reflection laws [8]. The point f is at a distance B from the center O of the circumference. The distance between the points O and M is the radius of the circumference that contains the cylindrical mirror.

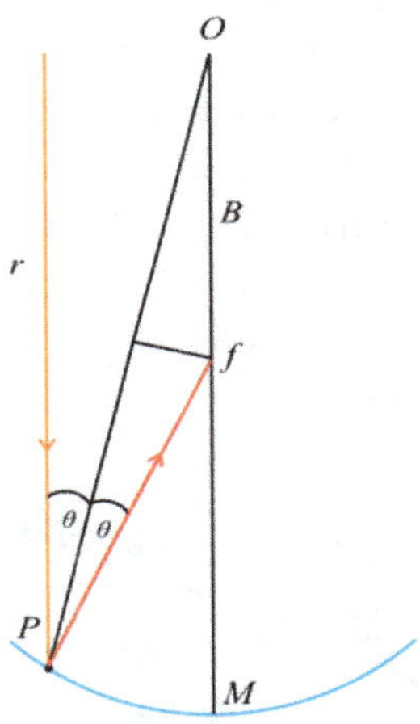

Figure 2. Scheme of the reflection in a cylindrical mirror.

The points P, f and O form an isosceles triangle, with θ in both equal angles. This triangle can be divided into two right triangles. The relation between the distances B, the radius R of the circumference that contains the mirror, and the angle of incidence/reflection θ is given by:

$$B = R/2 \cdot \cos\theta \tag{1}$$

As previously said, the condition to accomplish the concentration of the rays for a cylindrical geometry requires the width of the mirror to be smaller than the distance to the receiver. The former leads to a small value of the angle θ, then $\cos\theta \approx 1$. Therefore, the rays impinge into the middle of the circumference radius, $B = R/2$. This position is where the receiver of the LFR is placed.

As consequence, the mirror takes the form of a segment of a circumference with a radius two times larger than the focal distance, i.e., the distance from the mirror to the receiver.

From an optical point of view, every mirror has a singular focal distance to the receiver, which leads to a different radius of curvature for each mirror. Manufacturing different preformed patterns with every radius of curvature would increase the cost of the LFR.

3. Materials and Methods

3.1. Definition of Mirrors Geometry

From the different options presented in Section 2, it is initially stated that the cylindrical geometry can lead to an adequate solution in terms of cost and performance for a medium-size LFR and is comparable in performance to the parabolic shape.

For the definition of the mirror's geometry, an iteration process is performed using a previously created in-house ray trace code [9]. The goal of the process is to maximize the power impinging on the receiver while using one single radius of curvature for all mirrors (cylindrical shape). At each iteration, a specific radius of curvature is employed for all mirrors, which is changing from the focal distance of the farthest mirror from the receiver to the focal distance of the nearest one. As result of the application of this procedure, it is established that the radius of curvature of the farthest mirror delivers the best results. In addition, it is concluded that the optical performance when the farthest radius is used, is rather similar to the performance of a configuration that involves a particular and theoretician radius for every mirror (the average decrease estimated is 2%).

In order to endorse the optical behavior of cylindrical mirrors compared with parabolic mirrors, a new set of simulations are carried out. First, cylindrical mirrors are considered in the model and the power impinging on the receiver is obtained. Then, parabolic mirrors replace the cylindrical mirrors under the same conditions of the previous simulation. The results show that the optical behavior is rather similar for both configurations. As a consequence, it is stated that the optical handicap of using cylindrical mirrors instead of parabolic mirrors is marginal [9].

The LFR considered in this study is composed of 12 lines of reflectors, 10-m-long and 28-cm-width. Every 10 m line is divided into 10 facets, 1-m-long each (Figure 3). The procedure and analysis described from now on is related to a 1-m-long and 28-cm-width single facet.

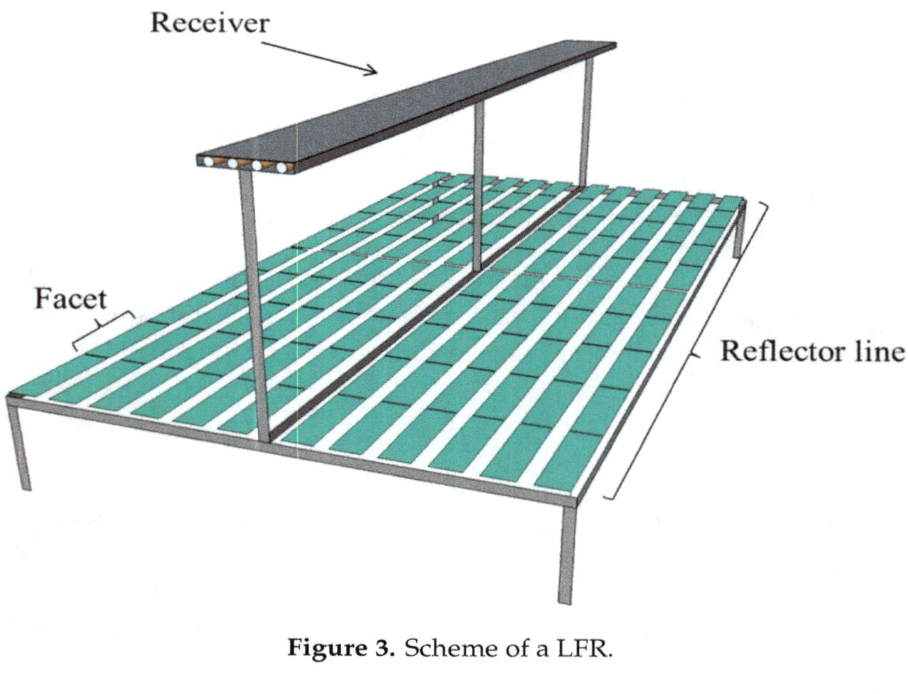

Figure 3. Scheme of a LFR.

The weight of the facets in the mechanical behavior of the collector is absolutely relevant. The goal is to accomplish a facet as light as possible, decreasing the load on the engine used for the movement of the whole reflectors' line to track the Sun. The maximum weight of one facet is set at 4 kg. The facets are assembled from two main components, the mirror and the structure that holds and shapes the mirror into the cylindrical geometry. Regardless of the type of structure and according to the weight restriction imposed, the material used is a 3-mm-thick metallic plate resistant to outdoor conditions. As explained, due to cost saving, the mirror does not have a pre-shaped geometry, and so the mirrors should be as thin as possible in order to properly adopt the geometry imposed by the structure. The mirror used is a 1-mm-thick highly-reflective glass-silvered mirror. Considering the characteristics of the materials selected, it is stablished that the maximum permissible geometrical deviation of the facets should be 0.5 mm from the sag parameter, which is defined in Section 4.

3.2. *Mirrors Structure Configuration*

Different configurations of the structure that support and shape the mirror into the facet are considered in this study, taking as premises the simplicity and the economy of the approaches without losing sight of the optical performance. The following subsections describe each configuration.

3.2.1. Naked Mirror over Preformed Ribs

The first approach to manufacture the facet's support structure is to mechanically bend the mirror in a cylindrical shape by clamping it between a set of preformed "ribs". Each rib is composed of two aluminum pieces (Figure 4a), the bottom piece that supports the mirror has a concave cylindrical geometry and the upper part, pushing down the mirror, has a convex cylindrical shape (Figure 4b). The bottom part of the ribs should also work as a union among three pairs of ribs that are used per individual facet, and to join to the others facets composing one of the 10-m-long reflectors' lines of the collector.

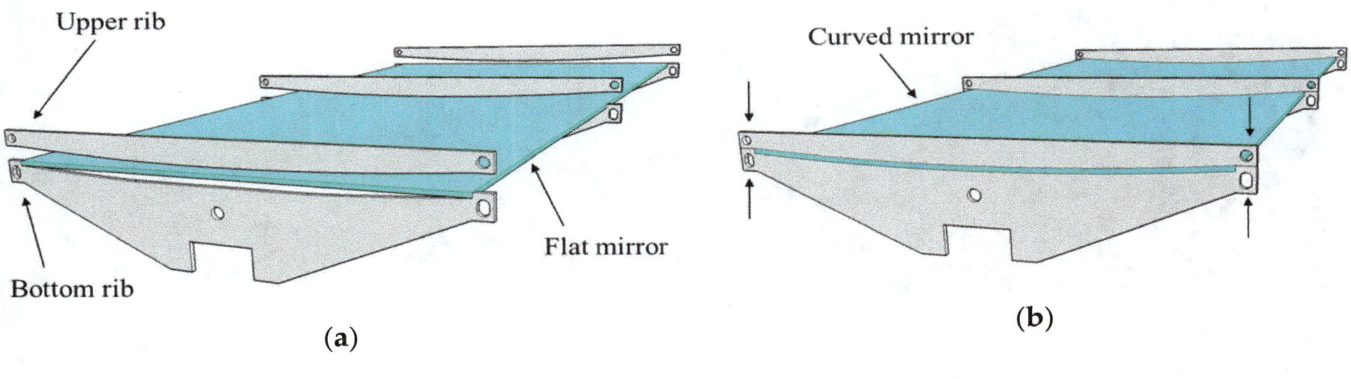

Figure 4. Scheme of the assembly for naked mirror over shaped ribs: (**a**) mirror before ribs clamping and (**b**) mirror after ribs clamping.

3.2.2. Naked Mirror Mechanically Bended by Metallic Bars Glued to it

This configuration uses a different approach to bend the mirror into a cylindrical shape. First of all, an aluminum bar is glued to the middle of the longitudinal axis of the mirror with structural silicone. Later, three transversal bars are fixed to the longitudinal bar (Figure 5a). In each extreme of the transversal bar, a pushing bolt (marked in red in Figure 5a) is installed. With this bolt, both sides of the mirror can be pushed and raised to achieve the desired shape, while the center of the mirror stays down, attached to the longitudinal bar. The result is a curved mirror created by two equal and opposite forces (See F_{push} and F_{hold} in Figure 5b).

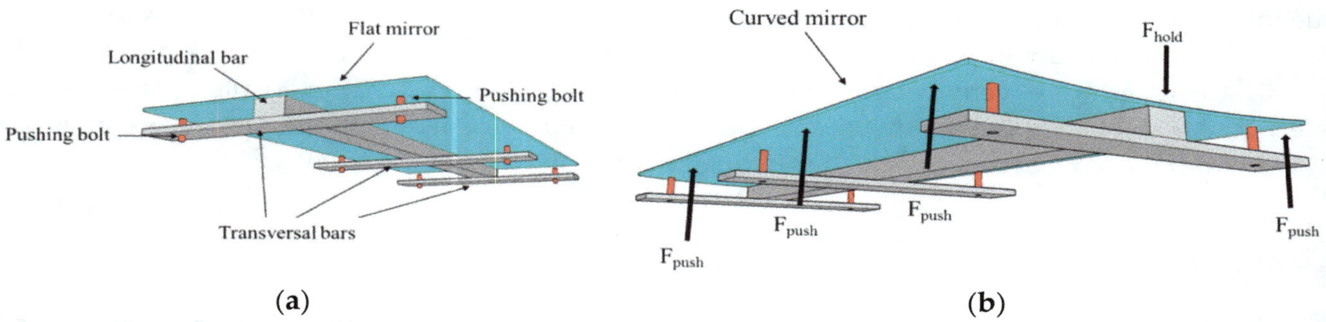

Figure 5. Scheme of the assembly for naked mirror mechanically bended by metallic bars glued to it: (**a**) mirror before being pushed by bolts and (**b**) mirror after being pushed by bolts.

3.2.3. Mirror Glued by Vacuum over a Curved Rectangular Plate Attached to a set of Preformed Ribs

Unlike the two former assemblies, the objective of this third configuration is to deliver a solid structure to support the mirror over its entire surface (Figure 6a). A pre-formed rectangular thin plate, made of 3-mm-thick stainless steel, is considered. This plate has the same dimension as the mirror and makes up the cylindrical shape of the facet. This curved plate is welded to a group of stainless steel ribs that provide the needed stiffness to keep the cylindrical shape. The ribs are aligned in a longitudinal direction by two metallic tubes welded to them (Figure 6b). The selection of steel as a constructive material instead of aluminum is due to the difficulty of welding the aluminum plate to the ribs. Due to the high thermal diffusivity of the aluminum, during the welding process, the whole piece could be heated, which along with the small thickness of the plate might entail the deformation of the plate when the material is chilling.

Figure 6. Assembly of a curved rectangular plate attached to a set of preformed ribs: (**a**) front view of the assembly and (**b**) lateral (longitudinal) view of the assembly.

In Figure 6, the configuration of the proposed assembly is shown. The flat mirror is located over the curved plate; previously, an adhesive is sprayed over the back face of the mirror and on the upper face of the metallic plate. Once the mirror and plate are joined, the assembly is introduced into a plastic bag to complete the sealing process. A vacuum pump is connected to an air outlet valve of the plastic bag to evacuate the air inside the bag, reducing the possibility of bubbles appearing between the mirror and the metal plate. This gluing process takes 15 minutes approximately. After this time, both the mirror and the metallic plate should be completely attached, with the mirror adopting the cylindrical shape of the plate.

3.2.4. Mirror Glued by Vacuum to a Curved Plate

The idea of gluing the mirror to a preformed plate used in the former configuration (Section 3.2.3) is considered as a useful solution to build facets for LFR. The mechanism of completing the gluing process in vacuum delivers optimal results. The challenge is to minimize the weight of the assembly described in Section 3.2.3; therefore, the first action is to replace the steel plate by another type of plate made of a material with lower density, e.g., aluminum.

The curvature of the plate must be achieved as accurately and repetitively as possible. The process selected to achieve the desired geometry consists of inserting the flat rectangular plate into a bending machine. The bending machine used in the experimental set up is made up of three longitudinal cylinders; first, the user adapts the position of two of the three cylinders to the thickness of the processed plate and, second, the third cylinder is moved up and down until the desired curvature of the structural plate is achieved. Once the positions of the cylinders are settled, the metallic plate is introduced and automatically curved.

The next step is to provide rigidity to the curved plate, which is done, for this specific facet size, by pleating 1.5 cm of the longitudinal extreme of the plate (Figure 7a). This pleat gives the plate enough stiffness in the longitudinal axis to avoid the plate losing linearity.

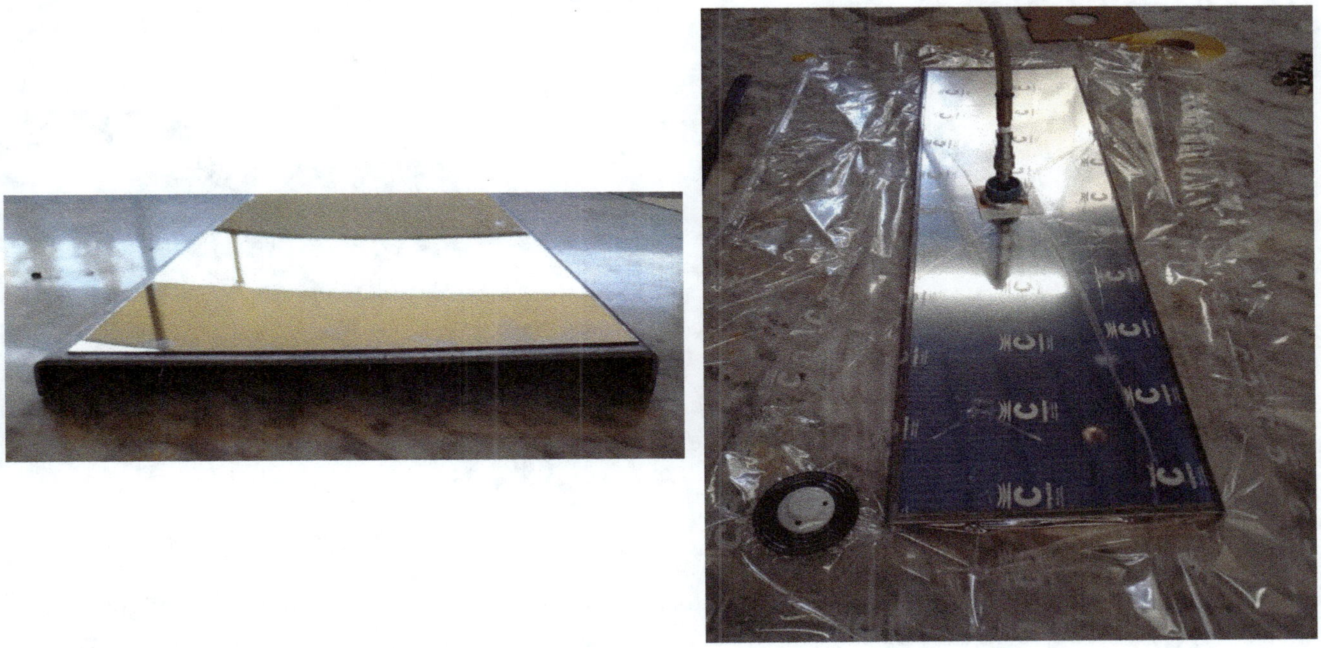

(a) (b)

Figure 7. Mirror glued by vacuum to a curved plate: (**a**) pleated mirror facet and (**b**) vacuum gluing process.

Once the plate is curved and pleated, the mirror is glued to the metallic plate following the same procedure explained in Section 3.2.3. The glue is sprayed onto the mirror and the plate, and then the set is put in a plastic bag, sealed and connected to a vacuum pump (Figure 7b).

At this point, the reflective part of the facet is completed, but as shown in Figure 7a, there are no structural bars or other elements in the aluminum plate to connect the facet neither to the LFR's main structure nor to the other facets. The welding of structural bars to the aluminum plate is not an adequate solution, as explained in Section 3.2.3. The solution proposed and designed is a suction cups system attached with silicone to the back face of the aluminum plate. These cups attach the facet to a structural bar, altogether with every one of the other nine facets conforming one 10-m-long line of 12 mirrors that compose the primary concentrator of the LFR collector.

The suction cups are composed by a perforated 3-mm-thick plate and two bolts welded to this plate. The perforations of the plate are designed to allow the silicone to overflow through it. The structural bar fit to the suction cup by two welded segments profiles type "L". These squads have a perforation in which the bolt fits and is adjusted with a pair of nuts (Figure 8e,f).

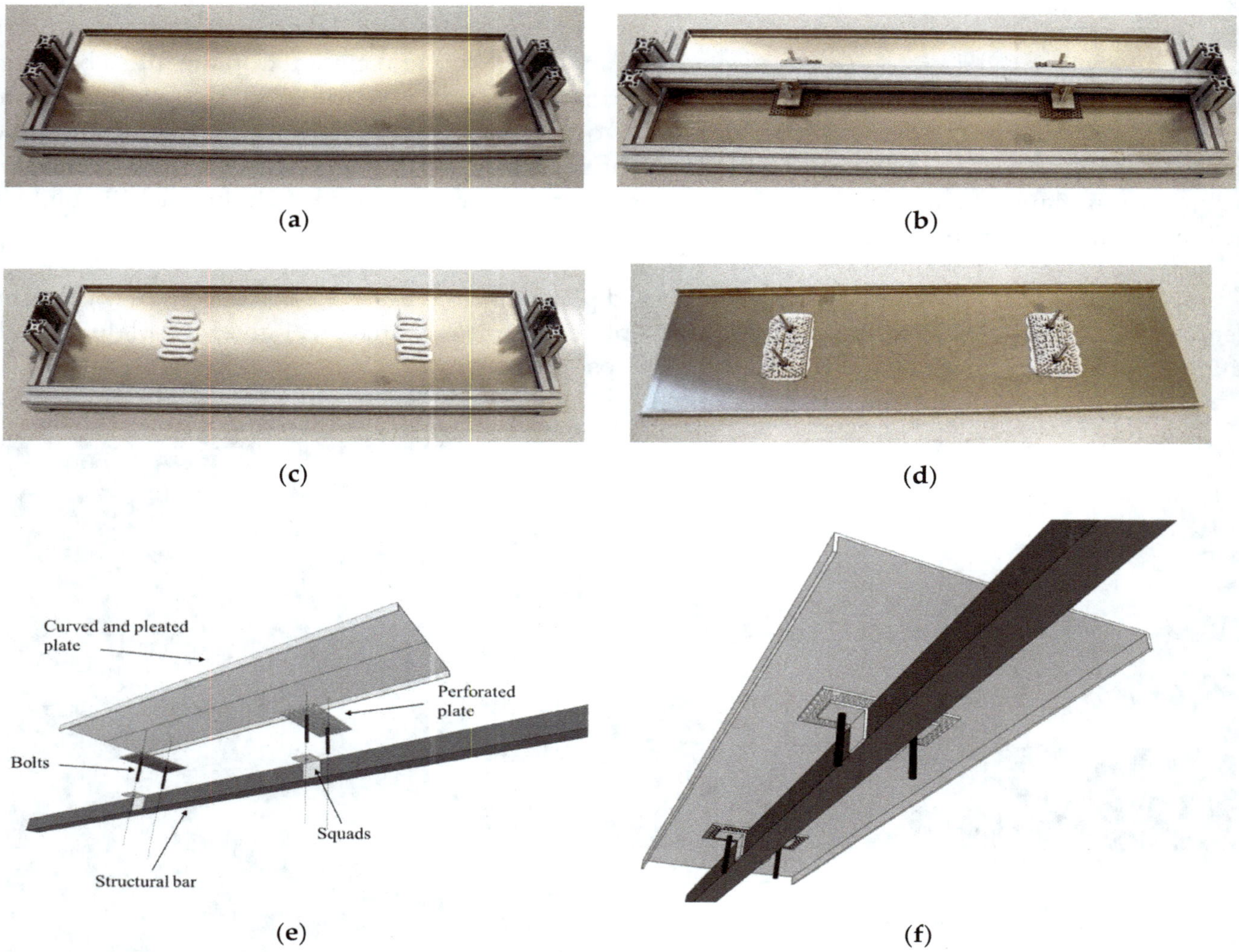

Figure 8. Assembly scheme of the suction cup into a mirror facet: (**a**) facet presented over the alignment device, (**b**) facet adjusted over the alignment device, (**c**) silicone cordon applied on the facet, (**d**) suction cups attached on the facet, (**e**) exploded view of the suction cup mounting process, and (**f**) suction cup mounted on the facet.

The complete assembly process of the suction cups into a mirror facet is shown in Figure 8. First, the facet (which has already been curved, pleated and the mirrors glued to it) is faced down into an alignment device (Figure 8a). This object is designed to line up the suction cups between them and with the facet. A longitudinal bar, which fits in the frame of the alignment device, is used to place the suction cups in the desired position on the facet (Figure 8b). The suction cups are held to the bar through a pair of squads. Once the position of the suction cup on the back side of the facet is defined, the bar is removed and a thick silicone cordon is applied on the back side of the plate (Figure 8c). Finally, the suction cups are placed again with the bar pressing down strongly until the silicone overflows through the perforations of the plate (Figure 8d).

4. Results

This section details the results obtained during the manufacturing process of facets assembled following the four procedures presented in Section 3 for the mirrors structure configuration. The geometrical quality of mirror facets manufactured following the manufacturing process that delivers the best results is also analyzed.

4.1. *Naked Mirror over Preformed Ribs*

The image reflected by this facet does not have the quality required to concentrate the rays on the receiver. Figure 9 shows how the image reflected is adjusted when each of the three ribs are clamped consecutively.

The first image (Figure 9a) corresponds to the mirror before clamping the ribs; a reflected rectangular image can be seen that perfectly matches the dimension of the mirror. Figure 9b is the image reflected when the center rib is clamped; it is clear how the image is adjusted to the cylindrical shape of the rib's range; a small area of concentrated heat flux in this zone emerges. When a second rib (one of the extremes) is clamped, another change in the reflected image is noticeable, and again a solid concentration region is created in the rib's range. However, in Figure 9c, it can be seen that the mirror does not adopt the desired curvature in the area between both ribs because a flat region persists in the reflected image, i.e., beam solar radiation is reflected but non-concentrated. Finally, the third rib (located in the opposite extreme) is clamped and the result is the same as the previous (Figure 9d), i.e., the mirror adopts the cylindrical geometry only in the rib's range. A similar configuration was tested adding two more ribs, but the results were the same; the mirror adopts the cylindrical shape exclusively in the ribs range and the flatness in other areas of the mirror facet persisted. This is due to the elastic property of the mirror used. It was concluded that this is not an acceptable solution to manufacture mirror facets for the LFR configuration and size considered.

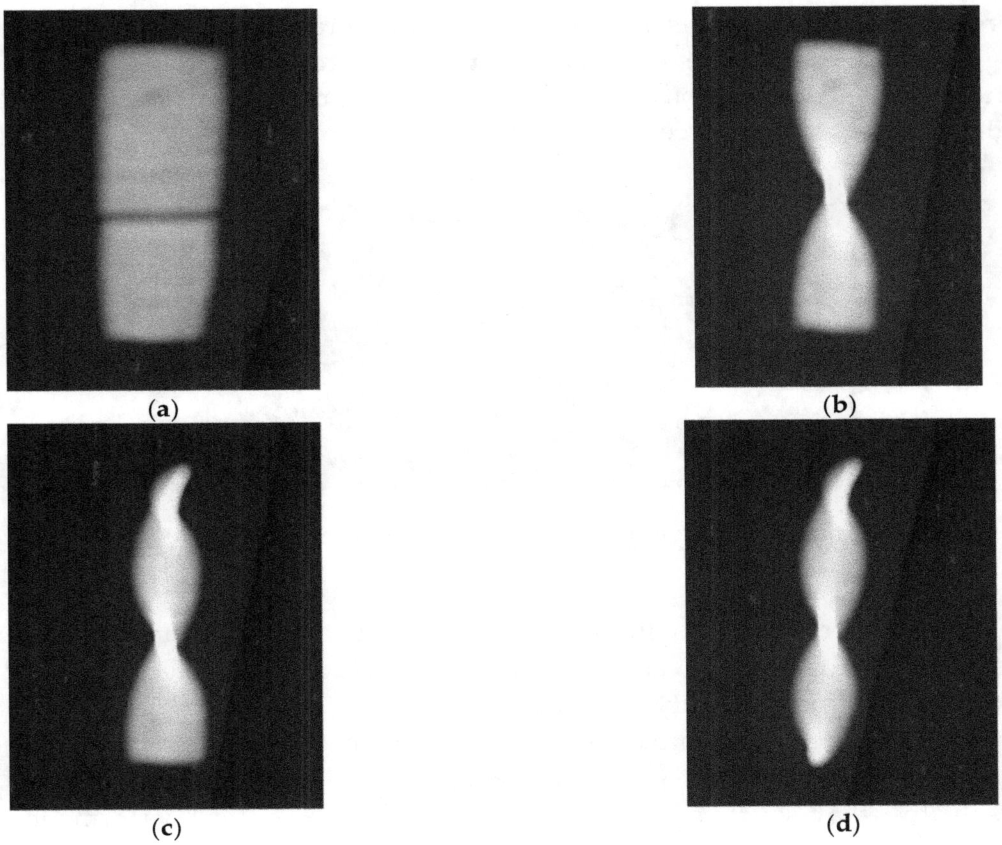

Figure 9. Reflected image for the assembly of naked mirror over shaped ribs facet configuration. (**a**) Non ribs clamped; (**b**) Middle ribs clamped; (**c**) Middle and one external ribs clamped; (**d**) Middle and both external ribs clamped.

4.2. *Naked Mirror Mechanically Bended by Metallic Bars Glued to it*

The application of this solution to configure and manufacture a mirror facet for the LFR proposed did not offer the targeted cylindrical geometry. The glued section attached to the longitudinal bar creates a flat zone in the center of the mirror. Then, the two free extremes of the mirror are pushed up, resulting in a particular geometry, as represented in Figure 10.

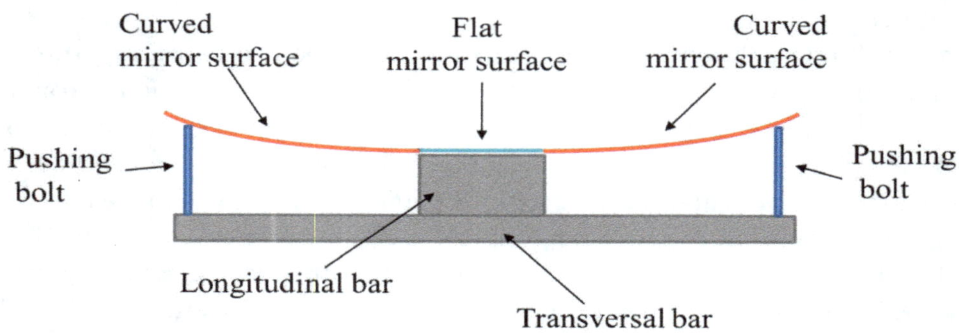

Figure 10. Scheme of transversal section of naked mirror mechanically bended by metallic bars glued to it.

The effect of this particular geometry onto the reflected image is an upgrade compared with the former configuration. Nevertheless, there are some diffused areas in the reflected image where the concentration of the solar radiation does not take place or is less noticeable than expected. Figure 11 shows the image reflected by a mirror facet manufactured considering this configuration. It can be seen that the reflected image presents a significant amount of rays concentrated onto the center line, creating a bright thick line. Also, the dispersion of other rays around the center line forming a lighter illuminated area can be seen. There is a discontinuation of the concentrated line. Presumably, this is caused by a glued malfunction between the bar and the mirror.

Figure 11. Reflected image for the assembly of naked mirror mechanically bended by metallic bars glued to it.

Besides the optical performance of this configuration, there is a significant setback in the assembly. To achieve the desired focus length, every facet must be individually adjusted to it. That means every one of the six bolts must be manually screwed up until the desired focus length is accomplished. The procedure is not viable taking into consideration that the solar collector unit considered in this study is composed of 120 facets. Moreover, screwing the bolt to the optimal position in which the mirror reaches the focus is a source of possible errors leading to an optical performance's uncertainty in the operation of the collector.

4.3. Mirror Glued by Vacuum over a Curved Rectangular Plate Attached to a set of Preformed Ribs

Regardless of the optical outcome, this configuration is discarded because of the excessive weight of the components. The use of steel, with density equal or higher than 7850 kg/m^3 (value depends on the type of steel used), leads to a weight higher than 6.5 kg per facet. This is contrary to the premise of getting lightweight facets.

4.4. Mirror Glued by Vacuum to a Single Curved Plate

The type of mirror facet built following this manufacturing process is composed of an aluminum curved plate with a thin glass-silvered mirror that is glued to its surface. The density of the aluminum is about 2700 kg/m^3, which makes lighter facets when compared to the use of steel for the support structure.

The result of this assembly (see Video S1) is considerably more satisfactory than the former. Figure 12 shows the concentrated solar radiation reflected by one facet on the surface of a Lambertian target, which was specially prepared with two longitudinal marks creating a 5 cm channel that allows appreciating the geometrical quality of the reflector.

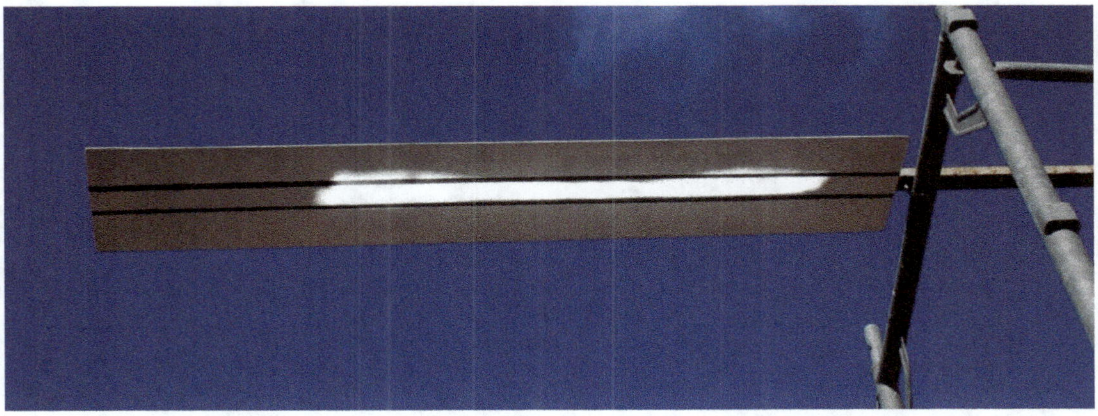

Figure 12. Concentrated radiation reflected into a Lambertian target.

The target is localized at the same height as the receiver, consequently at an equal distance to the focal length of the facet, which is 2.3 m for the LFR design considered in this study. Figure 12 shows how, in this preliminary test, a large amount of the concentrated solar radiation is within the 5 cm longitudinal channel. The image reflected is virtually a solid rectangle of 5 cm width and 1 m long, with almost non-dispersed radiation or discontinuity. Taking into consideration that the width of the mirror is 28 cm, the concentration ratio of the facet is 5.6.

After studying all four configurations and the results obtained from manufacturing the mirror facets for the considered LFR, the configuration based on a mirror glued by vacuum to a single curved aluminum plate is the one selected to manufacture 120 facets for the LFR collector prototype considered.

Geometrical Qualification of the Manufactured Facets

To complement and sustain the manufacturing process of the LFR facets, a verification of the geometry and the optic is carried out. The technique applied to measure the geometrical quality of the facets is close-range photogrammetry [10].

The photogrammetry process involves taking a set of well-planned photographs to the reflective surface of the facet. The facet's surface is previously covered with a target pattern, printed in an adhesive white vinyl sheet. The targets are black circles of 1 cm diameter arranged in rows and separated from each other by 2 cm. Finally, 323 target circles are distributed in the reflective surface as shown in Figure 13.

Once the target is settled, the reflective surface is photographed from different angles covering 360 degrees around the facet, positioning the camera in horizontal and vertical orientation to compensate for the optical aberration of the system.

The next step consists in post-processing the images. In Figure 14a, a representation of each point that compiles the reflective surface can be seen. Every blue circle represents a specific point of the facet's reflective surface and their spatial position is represented by their Cartesian coordinates.

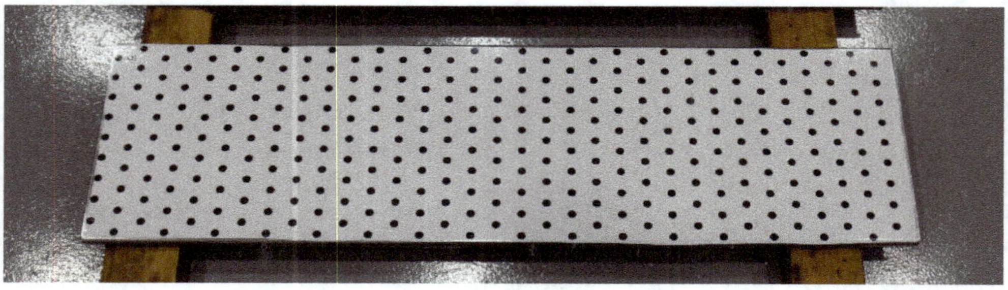

Figure 13. Target pattern over the reflector surface of the facet for the photogrammetry process.

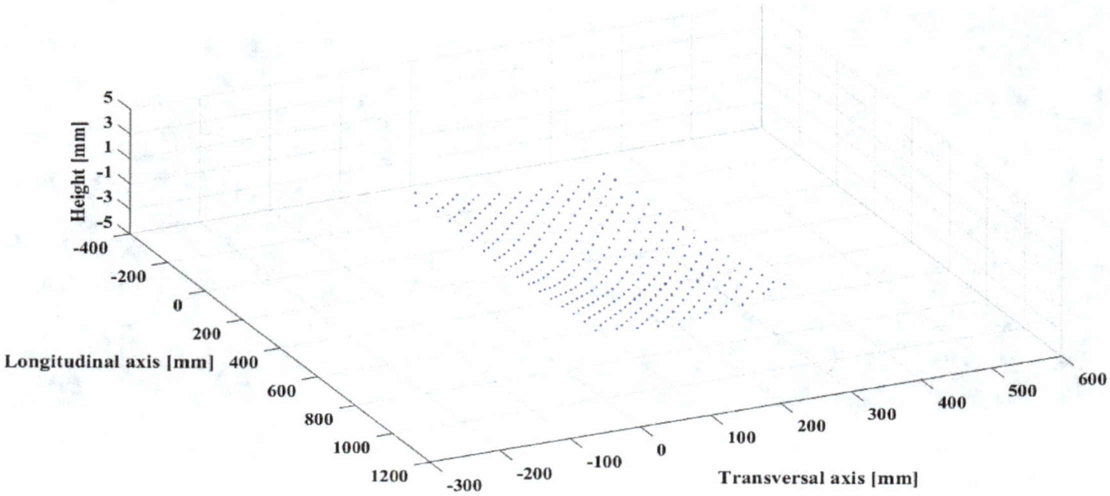

(**a**)

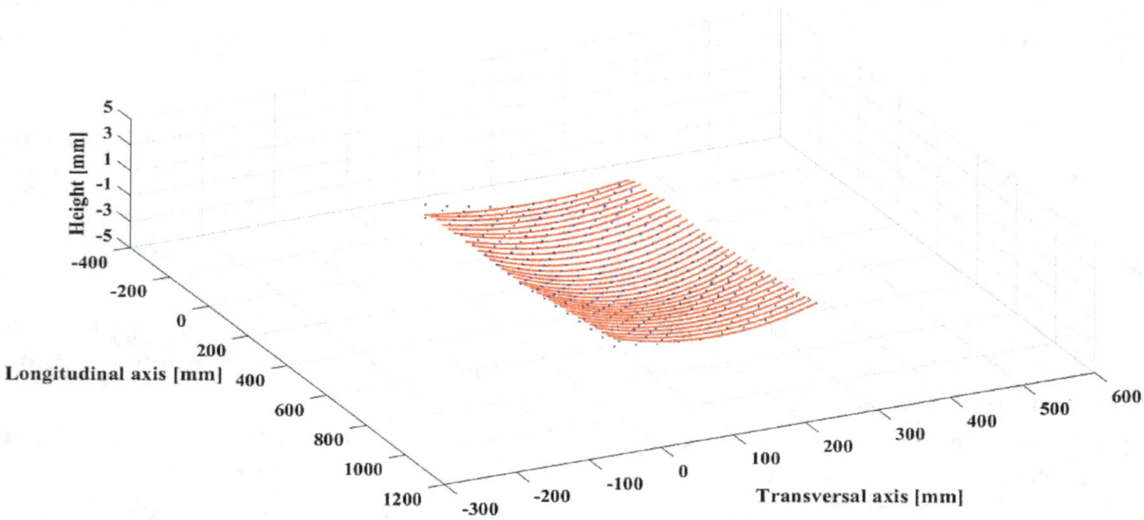

(**b**)

Figure 14. Photogrammetry's points and representative arcs spatially represented: (**a**) photogrammetry points and (**b**) representative arcs.

The sag of a circumference arc is selected as the parameter to evaluate the accuracy of the accomplished facet's geometry. The sag is defined as the height of the arc; it is the perpendicular distance between the midpoint of the arc's chord to the arc itself. This parameter is directly related to the radius of the facet's curvature. Every arc formed by the 34 groups of targets aligned in the transversal axis is built, and then they are compared among them. The facets are designed with a specific radius of curvature, which is 6.5 m in the design considered in this study, and the sag that corresponds to this radius is 1.51 mm. Figure 14b shows the arc that best fits to every transversal aligned target group.

The sag of each arc contained in the reflective surface depends on the optical behavior of the facet. The accuracy of the reflected image corresponds to a constant value of the sag along the facet. However, the manufacturing process entails different error sources that divert the final geometry from the defined one. The graph shown in Figure 15 represents the deviation of each calculated arc's sag in the longitudinal axis of a facet compared to the sag of the reference curvature radius.

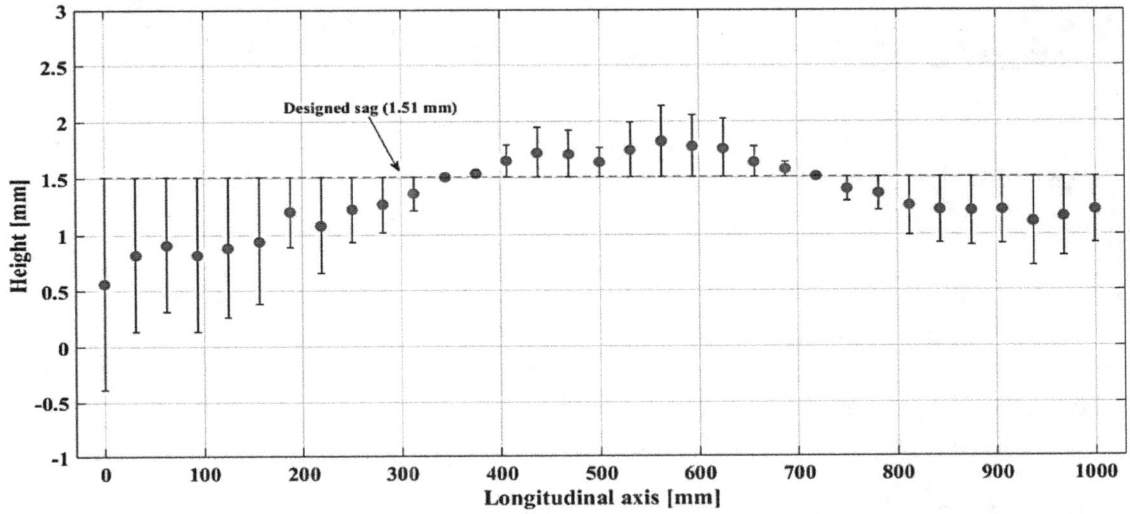

Figure 15. Deviation in millimeters of the calculated (measured) sag with respect to the sag of the designed geometry.

Each blue dot represents the sag of the corresponding circumference arc at a specific longitudinal position in the facet. The points are arranged according to the value of the designed sag represented in a dotted red line (1.51 mm). The magnitude of the deviation in the black lines across each point can be appreciated. Smaller values of the sag are linked to bigger curvature radius. Figure 15 shows a maximum deviation of 0.95 mm when the inferred value of the sag is 0.56 mm at the left side of the longitudinal axis. Then, the magnitude decreases until reaching the middle zone in which the sag is negligible and slightly higher than the designed value. Finally, the value of the sag decreases again with a smoother ratio.

As it was mentioned before, the variation of the sag along the longitudinal axis of the facet has a consequence in the reflected image. Figure 16 shows the same image as Figure 12, converted into a luminance map level using a RBG colormap, where 255 is the maximum value in the gray color scale representing the higher concentrated flux. The luminance distribution presents a perfect match with the deviation of the sag graph (Figure 15). The reflected image displays a thicker and dispersed pattern at the extremes of the longitudinal axis, which corresponds to a lower value of the sag. On the other hand, in the center of the facet, where the value is closer to the optimal, the image is consistently collimated. To address the geometrical errors in the manufacturing process, it is convenient to say that the bending machine used to shape the aluminum plates is designed to achieve pronounced or closed curves, and certainly, does not have the required optical precision to guarantee the desired consistency in the manufactured plates. To improve the quality of facets manufactured following the procedure presented in Section 3.2.4, an adequate selection of the bending machine is crucial to guarantee the optical quality of the mirror facets manufactured.

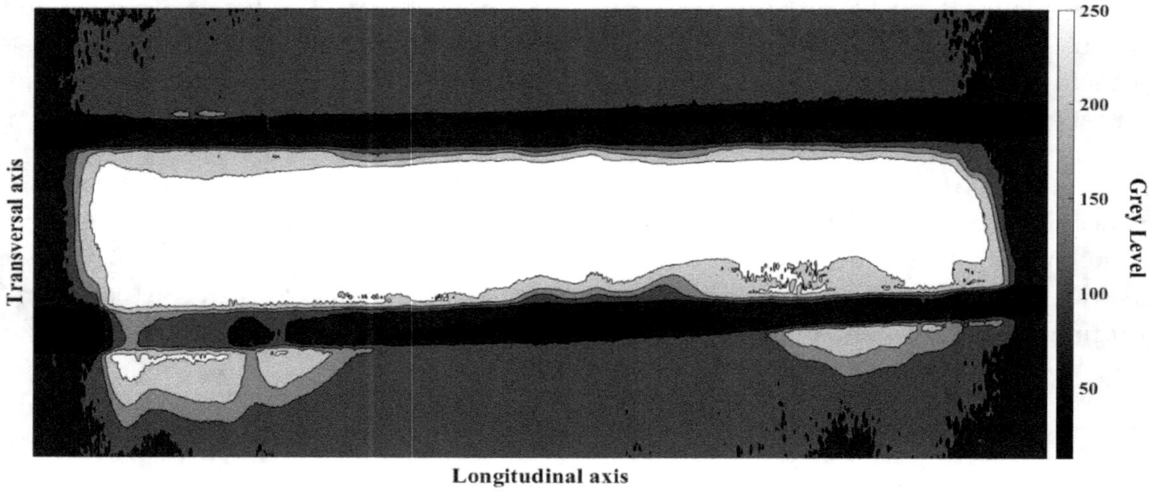

Figure 16. Luminance distribution of the concentrated radiation reflected by a prototype LFR mirror facet into a Lambertian target.

5. Conclusions

Different procedures to manufacture linear Fresnel reflector's mirror facets are detailed in this article. The first step before starting the manufacturing is the selection of the geometrical shape of the facet. Three mirror shapes are addressed: flat, cylindrical and parabolic. The one that leads to the simplest manufacturing process without compromising the optical behavior is the cylindrical-shaped mirror, which is defined as the proper configuration.

Four configurations of the structure that support and shape the mirror into the facet are described. It is found that the assembly composed by a mirror glued by vacuum to a single curved aluminum plate delivered the best results. This is based on the optical performance accomplished, altogether with technical aspects of the assembly process that provide specific and easy-to-apply steps to accomplish a larger manufacturing process. The sag of a circumference arc of the facets is determined as the geometrical parameter to characterize the accuracy of the curvature accomplished.

From the possible error sources found, one step is highlighted as the most susceptible to improvement. This is the bending process of the metallic plate used as support of the mirror. The use of a more precise bending machine would create a smaller deviation of the sag along the facet. Besides the errors, the image reflected is definitely optimum in terms of the amount of concentrated light into the target area. Also, the shape of the reflected image is within the expected boundaries.

Finally, this work aims to be a useful tool in the construction of linear Fresnel solar collectors, spreading the practical experience on the development of this concentrating solar technology development.

Author Contributions: Conceptualization, D.P.-I. and L.V.; methodology, D.P.-I. and L.V.; experimentation D.P.-I., J.G. and J.R.; validation, D.P.-I. and J.F.-R.; writing, D.P-I. and L.V; funding acquisition, L.V.

Acknowledgments: The authors would like to thank the Operation and Maintenance team of the Plataforma Solar de Almería for their continuous support in the whole manufacturing process.

Acronyms

LFR	linear Fresnel reflector
PTC	parabolic trough collector

Symbols

r	ray impinging in mirror's aperture
P	mirror point
f	focal distance of the cylindrical mirror (m)
θ	incidence angle (°)
B	distance from the center of circumference to the focal point (m)
O	center of the circumference
M	mirror point
F	force applied to the mirror

References

1. Vannoni, C.; Tobergte, D.R.; Curtis, S. Potential for Solar Hear in Industrial Processes. *J. Chem. Inf. Model.* **2013**, *53*, 1689–1699. [CrossRef]
2. Kalogirou, S.A. Solar thermal collectors and applications. *Prog. Energy Combust. Sci.* **2004**, *30*, 231–295. [CrossRef]
3. Fernandez-Garcia, A.; Zarza, E.; Valenzuela, L.; Perez, M. Parabolic-trough solar collectors and their applications. *Renew. Sustain. Energy Rev.* **2010**, *14*, 1695–1721. [CrossRef]
4. Sahoo, S.S.; Singh, S.; Banerjee, R. Analysis of heat losses from a trapezoidal cavity used for Linear Fresnel Reflector system. *Sol. Energy* **2012**, *86*, 1313–1322. [CrossRef]
5. Sait, H.H.; Martinez-Val, J.M.; Abbas, R.; Munoz-Anton, J. Fresnel-based modular solar fields for performance/cost optimization in solar thermal power plants: A comparison with parabolic trough collectors. *Appl. Energy* **2015**, *141*, 175–189. [CrossRef]
6. Feldhoff, J.F. *Linear Fresnel Collectors A Technology Overview of CSP Systems*; SFERA Summer School: Almería, Spain, 2012; pp. 1–30.
7. Abbas, R.; Muñoz, J.; Martínez-Val, J.M. Steady-state thermal analysis of an innovative receiver for linear Fresnel reflectors. *Appl. Energy* **2012**, *92*, 503–515. [CrossRef]
8. Kidger, M.J. *Fundamental Optical Design*; SPIE press: Bellingham, WA, USA, 2002.
9. Pulido-Iparraguirre, D.; Valenzuela, L.; Serrano-Aguilera, J.J.; Fernández-García, A. Optimized design of a Linear Fresnel reflector for solar process heat applications. *Renew. Energy* **2019**, *131*, 1089–1106. [CrossRef]
10. Fernández-Reche, J.; Valenzuela, L. Geometrical assessment of solar concentrators using close-range photogrammetry. *Energy Procedia* **2012**, *30*, 84–90. [CrossRef]

7

Research of the Energy Losses of Photovoltaic (PV) Modules after Hail Simulation using a Newly-Created Testbed

Kristina Kilikevičienė [1], Jonas Matijošius [2,*], Artūras Kilikevičius [2], Mindaugas Jurevičius [1], Vytautas Makarskas [1], Jacek Caban [3] and Andrzej Marczuk [3]

1 Department of Mechanical and Material Engineering, Vilnius Gediminas Technical University, J. Basanaviciaus St. 28, LT-03224 Vilnius, Lithuania; kristina.kilikeviciene@vgtu.lt (K.K.); mindaugas.jurevicius@vgtu.lt (M.J.); vytautas.makarskas@vgtu.lt (V.M.)

2 Institute of Mechanical Science, Vilnius Gediminas Technical University, J. Basanaviciaus St. 28, LT-03224 Vilnius, Lithuania; arturas.kilikevicius@vgtu.lt

3 Department of Agricultural, Forestry and Transport Machines, University of Life Sciences in Lublin, 28 Głęboka Street, 20-612 Lublin, Poland; jacek.caban@up.lublin.pl (J.C.); andrzej.marczuk@up.lublin.pl (A.M.)

* Correspondence: jonas.matijosius@vgtu.lt.

Abstract: The impact of hail ice cubes on composite structures (such as solar cells) causes actual defects. This article presents a series of tests, in which solar cell modules were exposed to hail simulation testbed balls, allowing to assess the following: the impact energy, which causes the major defects in solar cells; the formed micro-cracks in the structure of solar cells, resulting in the loss of power generated by a solar cell; and the solar cell parameters necessary for modelling. In addition, this article presents a digital analysis of hail simulation. Information received from the digital analysis was used to optimize the structure of solar cells in order to improve its resistance properties. The aim of this study was to present a simple method for experimental hail simulation. The proposed hail impact estimation method can be successfully applied to study the influence of the mechanical–dynamic impact of photovoltaic (PV) modules of different structures on the technical characteristics of these modules (structural stability, power generation, etc.). The study showed that PV modules are subjected to an irreversible effect of the excitation force (i.e., micro-cracking) and it can reduce the generated power by 2.33% to 4.83%.

Keywords: photovoltaic (PV) module; hail simulation; resistance of photovoltaic modules; energy losses; newly-created testbed

---

1. Introduction

The operation and maintenance of photovoltaic (PV) power-generating modules have become increasingly more important in recent years, as the number of such modules is being raised to ensure maximum power generation throughout their life cycle. All types of degradations and failures adversely affect the reliability and lifetime of PV module installations. Analysis of the impact of these factors on technical and financial problems is becoming more important for operators and investors [1]. PV electricity generation has become an important factor in the electricity mix in many countries and is expected to become even more important in the future [2]. PV failures have a significant impact on the safety, reliability, and energy balance of PV devices [3]. Similarly, malfunctions directly result in loss of power, which reduces the return on investment. The manufacturing process of PV modules from silicon solar cells is conventionally performed by soldering copper strips to the solder and metallizing the contact area. Contact shooting and screen printing are used for this purpose [4].

PV modules are either composites or laminates made up of different layers with very different mechanical properties and functions. Some layers of photovoltaic modules are designed to generate electricity, while others protect the modules from environmental effects. Two types of photovoltaic modules are usually used [5]: (1) rigid plates with glass layers on the outside to ensure the rigidity and protection of the modules from mechanical loads; (2) semi-flexible plates that must be fixed to flat or curved surfaces with a protective polymer layer, which allows the PV module to have some flexibility. Since PV modules are installed outdoors, thermo-hygrometric cycles, wind gusts, snow, and hail are the main sources of damage and degradation over the estimated 25 years of operation. To determine whether the solar cell and mechanical stresses exceed the permissible stress levels, several studies used the Finite Element Method (FEM) [4]. Beinert et al. analyzed mechanical stresses in frame-based and non-composite PV modules, where mechanical stresses were induced [6]. Other studies focused on determining the influence of joints [7] and their coating, mechanical properties [8], and geometry [4] on the mechanical stresses of PV modules. Mechanical impacts are among the factors most affecting the reliability and rigidity of PV modules [8–10]. Their impact on durability and electrical operation of PV modules were reviewed [11–18]. When it comes to mechanical impacts, the impact of hail on PV modules is significant, as it may result in the breakdown of silicon, which in turn may lead to significant electricity loss. Hail tests are usually carried out using pneumatic equipment (the diameter of an ice cube is 25 mm and its speed at impact reaches 23 m/s) and by analyzing the curves of electrical current losses from the current voltage (I-V), as well as the electroluminescence pictures that allow clear visual identification of micro-cracks and their impact on electrical response [19–21].

All newly-manufactured PV modules in Europe, regardless of their type, must meet the requirements of the International Electrotechnical Commission (IEC) 61215 standards [22]. The IEC also defines hail impact simulation tests which must be performed by firing a formed ice ball on a PV module, using a pneumatic actuator [23]. The standard IEC 61215 defines an ice ball diameter of 25 mm and an impact velocity of 23 m/s. However, during a hail, the parameters (the shape and size) and impact velocity of the ice balls vary, therefore the test aims to assess whether the PV module itself can withstand the hail.

It is very important that the correlation between impact velocity and the diameter of the ice ball is determined based on the requirements of IEC 61215. On the other hand, the repeatability of the tests is important in the test process too. This is very difficult to ensure because the ice balls used must have the correct shape and no cracks or other defects that will affect the test results [24,25]. Thus, the preliminary series of shots we used allowed us to identify and select the most appropriate impact velocity and use it during the tests.

The research methodology applied in this work allowed to simplify the test procedure and to ensure its repeatability by using a 40 mm diameter sphere instead of an ice ball (the material of the sphere was also selected similar in density to the ice ball). Therefore, the requirements of the IEC standards did not need to be applied, since the conversion of the linear velocity of the ice ball to obtain the same impulse was not applied.

Carrying out the theoretical research of the problem is important in order to supplement experimental research. The problem of impact and contact mechanics in the dynamics has been extensively studied from both the experimental and theoretical perspectives, but most studies addressed the problems concerning the aerospace and defense fields [26,27]. In the aforementioned research, composite cellular plates [28,29] dominated as the main subject. This article analyzes the possibilities and challenges of the newly-created testbed for hail simulation, for which the market demand is very high—especially in small and medium-sized enterprises—and which in turn determines the cost of new technologies [30,31]. The newly-developed testbed allows achieving the parameters used in similar research (ball speed and the impact force) [11]. Most studies simulate the effects of hail using pneumatic equipment that provides the energy needed to move a piece of ice [2,4,32–36]. The hail test requires expensive, time-consuming equipment. Thus, one of the main aims of this study was to provide a simple method for experimental hail simulation. The proposed hail impact estimation methodology can be successfully applied to study the influence of the mechanical–dynamic impact of PV modules of different structures on the technical characteristics of these modules (structural stability, power generation, etc.) [37–53]. This study employed a newly-developed hail simulation bench, which allows the value of the impact force applied to the PV module to be varied and the angle of the action of that force to be adjusted. The ability to change and adjust the action angle brings the experiment closer to the real conditions of the PV module since the PV modules are almost always angled. The paper investigates the potential of the newly-developed bench and assesses the impact of different forces acting on the PV module on the technical characteristics of that module.

2. Hail Simulation Testbed and Equipment for Measuring Dynamic Parameters

The subject of this research was the newly-created hail simulation testbed (see Figure 1). The testbed consisted of three parts: the hail simulation part, the part for measuring acceleration and force, and the part for detecting defects of solar cells (electroluminescence). The hail simulation part consisted of a motor with a frequency converter, hail simulation components, and a solar module moving on rails. The part for measuring acceleration and force parameters comprised force and acceleration sensors and equipment for processing and storing measurement results. The part for detecting the defects of the PV module consisted of a camera and equipment for processing and storing results. The specifics of the PV module chosen for this research: cell size 37 × 156 mm; number of cells 10; front-side 3.2 mm hardened solar glass; dimensions (L × W) 322 × 204 mm; weight 625 g. This type of a PV module is commonly used as a roof tile. The PV module consisted of: 3.2-mm non-iron glass/EVA film (0.4 mm)/The matrix of PV modules/EVA film/PPE plastic.

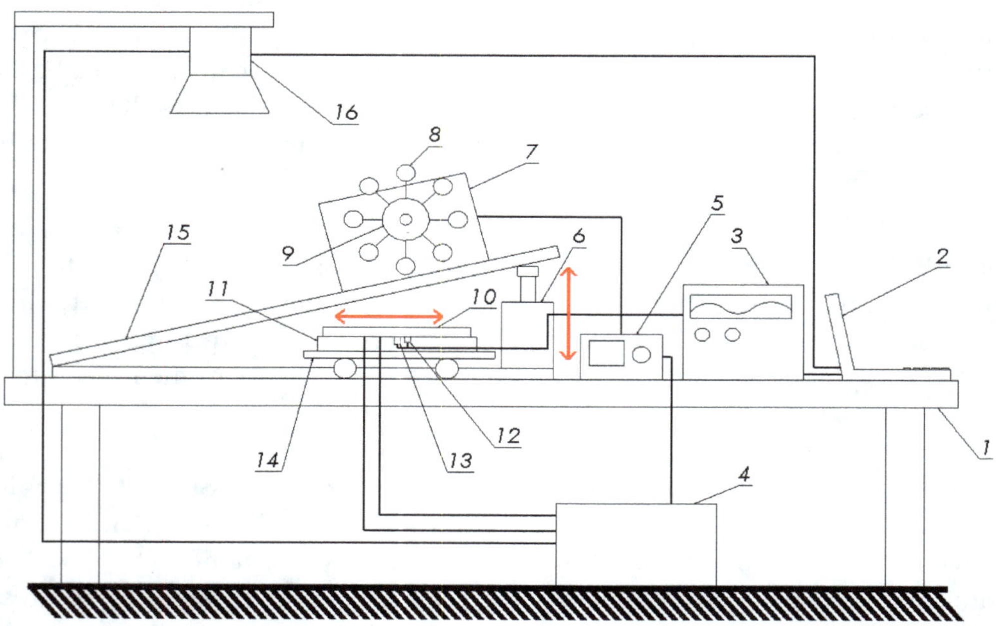

(a)

Figure 1. *Cont.*

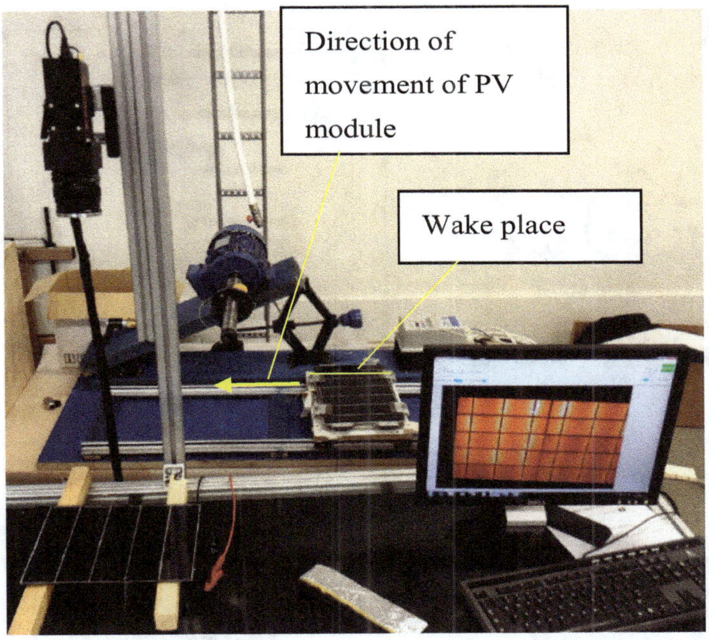

(b)

Figure 1. Scheme of the photovoltaic (PV) module testbed (**a**) and general view (**b**). In (**a**): where: 1–a table, 2—a computer, 3—data processing and storage equipment, 4—a generator, 5—a control unit, 6—lifting equipment, 7—a motor, 8—balls (material—PPE plastic), 9—a bushing, 10—a PV module, 11—a frame, 12—a force transducer, 13—an accelerometer, 14—a carriage, 15—a portable base, 16—an electroluminescent stand (Greateyes GE 1024 1024 BI MID CCD camera; NIR (near-infrared) spectrum: 780-1,100 nm). The Brüel & Kjær meters were used to measure oscillations and force parameters (see Figure 2): (1) portable equipment for processing, storing and managing measurement results Machine Diagnostics Toolbox—Type 9727; (2) a single-axis accelerometer 8341; (3) a force transducer 8230.

The aims of the study were to find out the possibilities of the newly-created testbed, to perform hail simulation tests, and to determine the impact of dynamic effects on the structure and performance of PV modules. During the experiment, the rotational speed was changed to simulate dynamic mechanical loads that correspond to different speeds of ice cubes during hail. The PV module was firmly attached to the carriage that moved on the rail. The motor pushed the mechanism with fastening balls, which excited PV modules. The motor was connected to the control unit, which controlled frequencies (in the 5–60 Hz range) and revolutions. The accelerometer was affixed to the centre of the PV module, at the bottom, in order to measure the values and nature of the incoming signal (acceleration and displacement). The force transducer designated for measuring ball impacts was affixed next to the accelerometer. The accelerometer and the force transducer were connected to data storage and processing equipment. The data storage and processing equipment were connected to the computer, which graphically displayed the results. Before and after the hail simulation, PV modules were photographed (using an electroluminescence camera) on the electroluminescent stand with an electric current (the current depended on the size of the PV modules; it can reach up to 4–5 A) supplied to the PV modules. Electroluminescent imaging is usually used to check for defects [54–56] because the intensity of radiation is proportional to the current density. The pictures taken helped detect any defects in the PV modules.

Figure 2 illustrates the block diagram of the testbed.

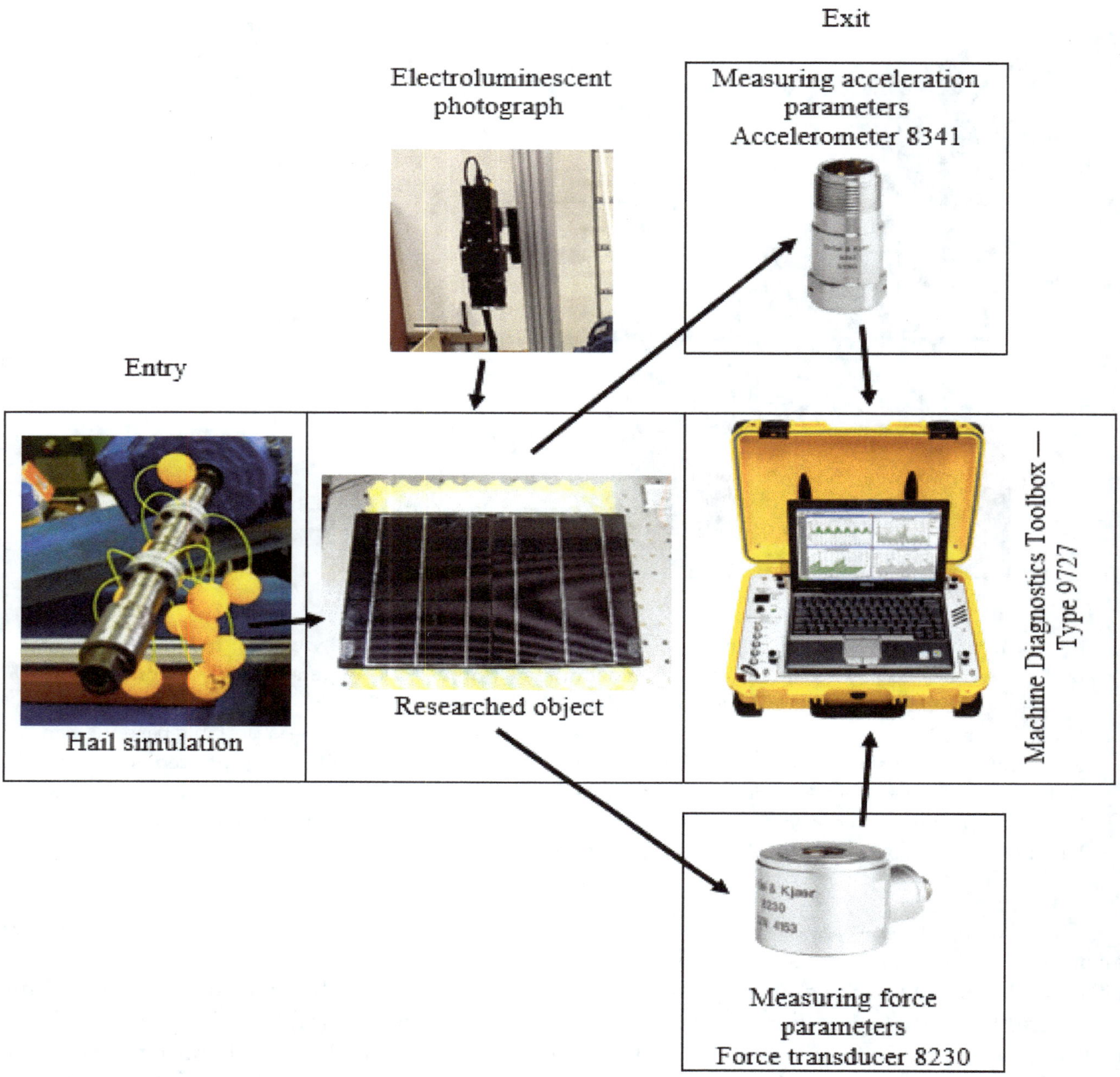

Figure 2. Block diagram of the dynamic mechanical load-generating facility.

3. Experimental and Theoretical Research Results

The first stage of the experimental research was aimed at finding out the necessary force that creates cracks in PV modules. The excitation force affecting PV modules was gradually increased (see the first peak in Figure 3) during the first stage, which was done by changing the speed of engine rotations (see position 7, Figure 1a). In the second stage, after the force required to create cracks was determined, a number of tests was carried out to assess the regularities of the cracks. The force and acceleration were measured during the tests. Figure 3 shows the acceleration values. The aim of the tests was to determine the frequency characteristics of the PV module.

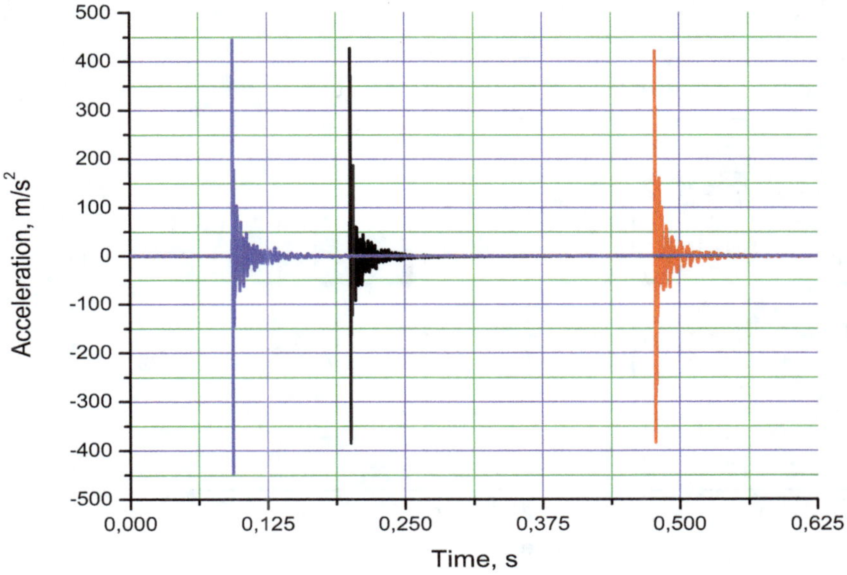

(**a**)

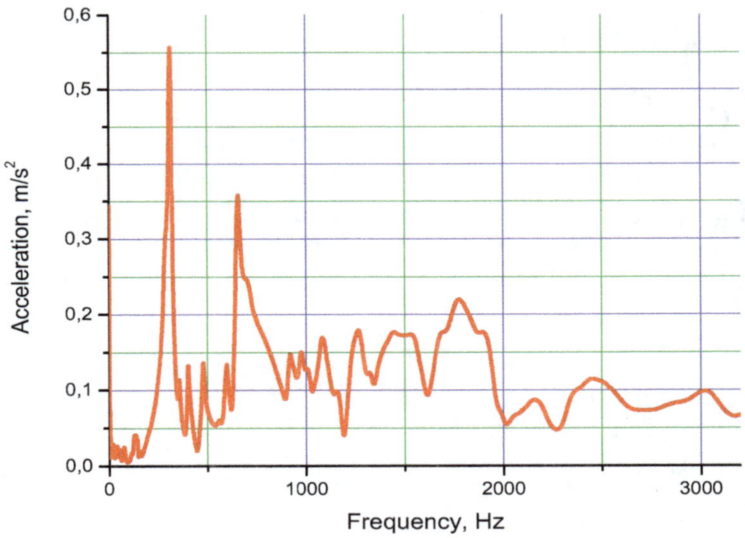

(**b**)

Figure 3. PV module acceleration (**a**) and its spectral density (**b**) (results of three tests and the average spectral density curve).

The analysis of the first-stage results (see Figure 4) revealed that internal cracks in the PV module formed at the force of 582.06 N, and glass cracks appeared at the force of 1129 N. After the interaction, a visual PV module inspection was performed to check for any cracks. The tests of the second stage aimed to maintain similar forces.

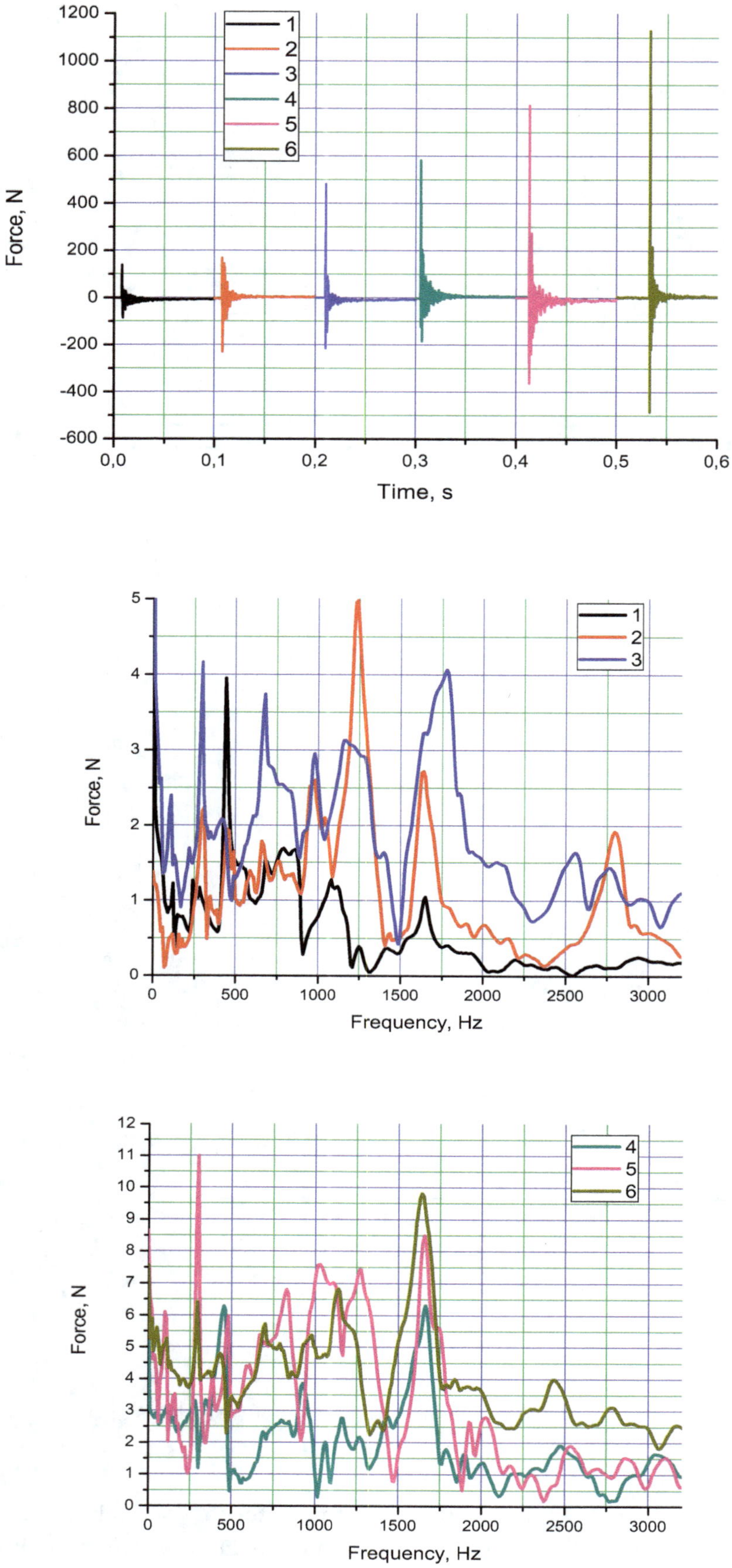

(a)

(b)

Figure 4. The force which the PV module is exposed to (**a**) and its spectral density (**b**) (black—137.92 N; red—166.07 N; blue—480.81 N; green—582.06 N; pink—811.9 N; dark yellow—1129 N).

4. Research of the Resistance of Photovoltaic Modules

Solar cell cracks and their regularities were analyzed in the second stage of the experimental tests. The purpose of these tests was to find and identify defects (e.g., cracks) in the PV modules, in their glass and their interior structure. In order to compare the main parameters of the PV module after the impact of the hail testbed, the PV module was photographed with the electroluminescent stand at the beginning of the study and measured using a voltaic characteristic of the testing stand. First damage, marked in yellow, was observed at the force of hail simulation balls of 582.06 N (The middle image of PV module 1 in Table 1). An electroluminescent photo of the test at the force of 1129 N can be seen on the right side of Table 1. The electroluminescent photo of the test at the force of 1129 N was presented accordingly; the majority of defects, which are marked in yellow, formed during this test. More findings have been presented in previous studies [14]. A uniform excitation force was maintained during the test; the force was selected in accordance with a previous study which found that the first cracks in the PV module occurred at a force of 582.06 N. The weight of the hail imaging ball used in the experiment was 30 g; diameter 40 mm; movement speed 27.5 m/s; and energy 11.1 J (these parameters corresponded to the values recommended in IEC 61215). More features of the stand are described in the article [11]. During the experiment, one ball was used and the PV module was uniformly moved from right to left (see Figure 1b); the PV modules were excited one by one in 4 strokes. The PV module was excited along the middle section (see Figure 1b).

Table 1 was compiled out of the two drafted voltaic data (before the tests and after the sixth test). It compares the three key parameters: the generated power (W), the maximum voltage (V), and the maximum current (A). Table 1 shows the characteristics of the PV module 1 after the test at 1129 N. In Table 1, the characteristics of PV modules were obtained using I–V (Voltage Amps) measuring equipment: Pasan Highlight 3, class AAA+, measuring time 10 ± 1 ms. Measured under standard conditions: 1000 ± 30 W/m^2, solar spectrum AM 1.5 g, 25 °C [57–60].

Table 1. Changes in performance characteristics of PV cells after the hail simulation test.

	Before Hail Simulation	**After Hail Simulation**	%
		The PV module 1	
	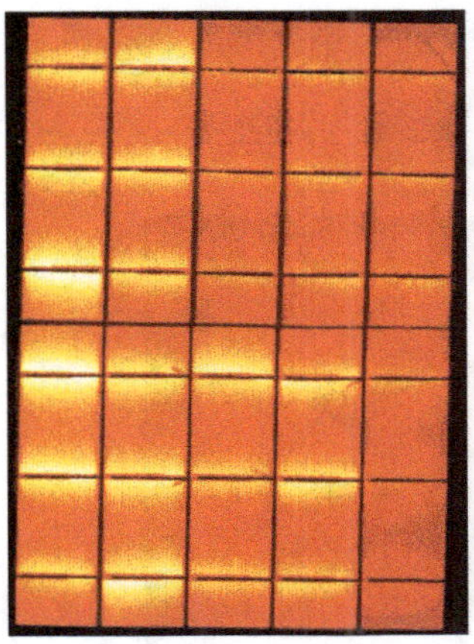	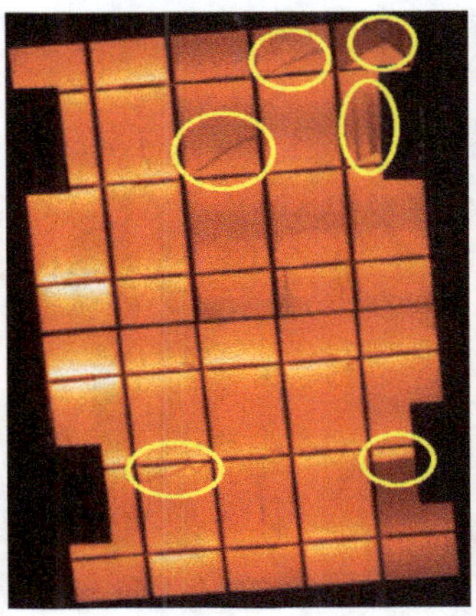	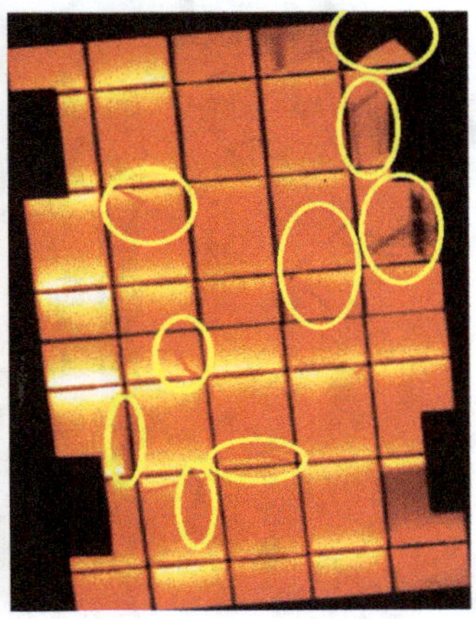
Power (W)	9.720	8.914	8.29%
Voltage (V)	5.227	5.257	0.57%
Current (A)	1.860	1.695	8.87%
		The PV module 2	

Table 1. *Cont.*

	Before Hail Simulation	After Hail Simulation	%
	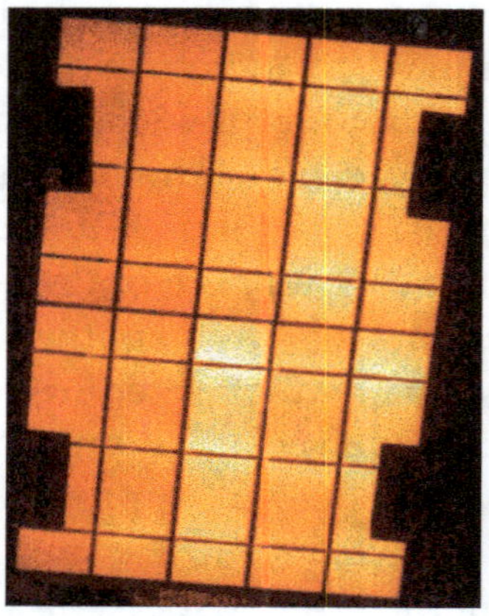	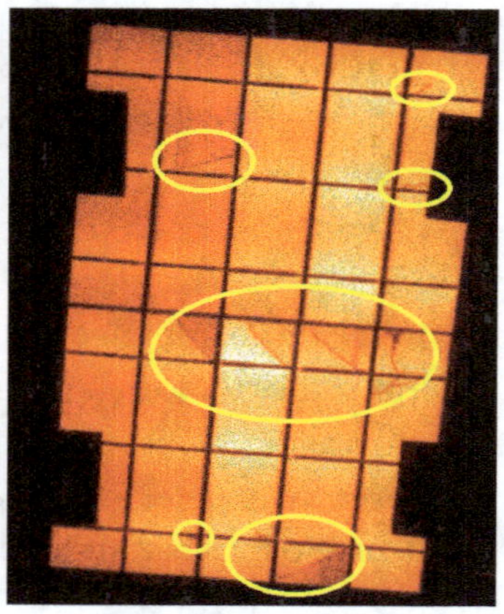	
Power (W)	9.315	8.987	3.52%
Voltage (V)	5.013	4.952	1.22%
Current (A)	1.858	1.815	2.31%
	The PV module 3		
		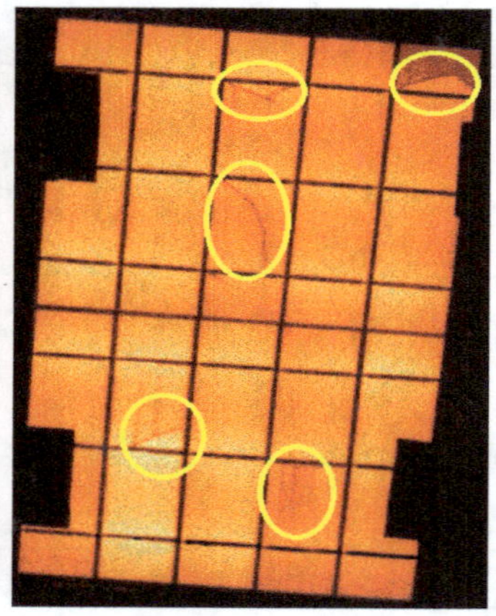	
Power (W)	9.457	9.237	2.33%
Voltage (V)	5.080	5.009	1.40%
Current (A)	1.861	1.844	0.91%
	The PV module 4		
		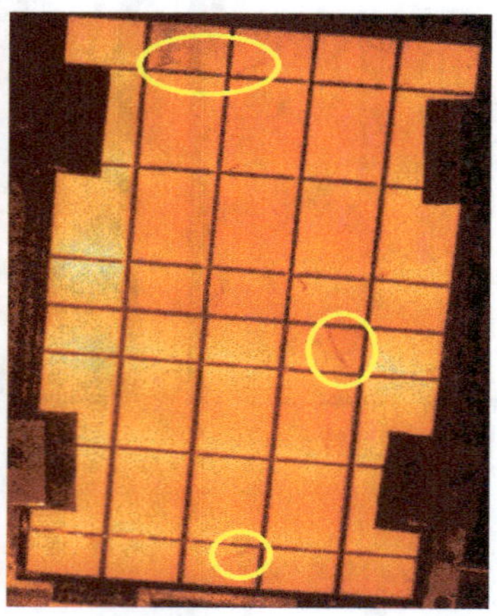	

Table 1. *Cont.*

	Before Hail Simulation	After Hail Simulation	%
Power (W)	9.461	9.125	3.55%
Voltage (V)	5.064	4.966	1.94%
Current (A)	1.868	1.837	1.66%
The PV module 5			
		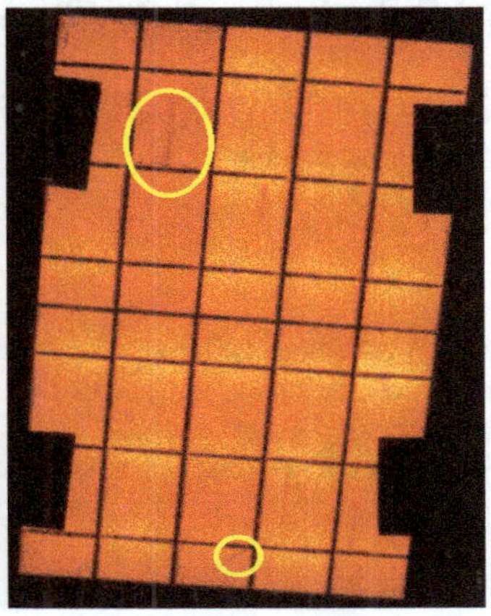	
Power (W)	9.413	9.089	3.44%
Voltage (V)	5.070	4.958	2.21%
Current (A)	1.857	1.833	1.29%
The PV module 6			
		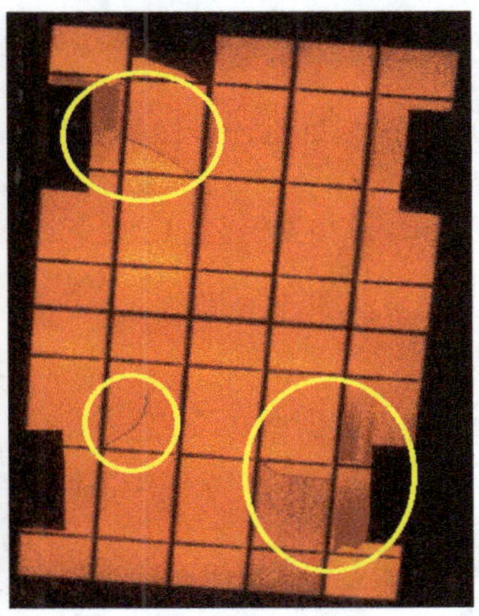	
Power (W)	9.345	8.894	4.83%
Voltage (V)	5.012	4.893	2.37%
Current (A)	1.865	1.818	2.52%

The analysis of the results received (Table 1) revealed that micro-cracks mainly formed at the point of contact between the solar cell and the ball simulating an ice cube, and some micro-cracks appeared at the fastening points on the edges of the PV module. When it comes to power losses, the analysis of the results of Table 1 showed a loss of power ranging from 2.33% to 4.83% when the impact of the force-induced micro-cracking (i.e., the magnitude of the force) was assessed. Accordingly, the assessment of the impact of the maximum force (1129 N) used in the research showed that the loss of power was 8.29% and that a force, having increased twice, leads to a loss of power of 1.7 to 3.56 times. In summary, the researched solar modules suffered significant damage after the hail simulation test. The structure of the PV modules was mechanically damaged (cracks formed), resulting in a loss of power.

5. Conclusions

The present study examined a newly-created hail simulation testbed aimed at identifying the necessary conditions for hail simulation and assessing the performance parameters of PV modules before and after the dynamic impact.

After the hail simulation test, the tested PV modules suffered severe damage, resulting in cracks in their crystalline structure and the loss of generated power.

The conducted tests revealed that micro-cracks in the crystalline structure of the PV module formed when the module was exposed to the force. When the PV module was exposed to the said force, the generated power losses ranged from 2.33% to 4.83%. After increasing the force impacting the PV module, power loss increased to 8.29%.

Author Contributions: For research articles with several authors, a short paragraph specifying their individual contributions must be provided. The following statements should be used "conceptualization, K.K. and J.M.; methodology, A.K.; software, M.J.; validation, V.M., J.C. and A.M.; formal analysis, K.K.; investigation, J.M.; resources, A.K.; data curation, M.J.; writing—original draft preparation, V.M.; writing—review and editing, J.C.; visualization, A.M.; supervision, A.K.; project administration, J.M.

References

1. Buerhop, C.; Wirsching, S.; Bemm, A.; Pickel, T.; Hohmann, P.; Nieß, M.; Vodermayer, C.; Huber, A.; Glück, B.; Mergheim, J.; et al. Evolution of cell cracks in PV-modules under field and laboratory conditions. *Prog. Photovolt. Res. Appl.* **2018**, *26*, 261–272. [CrossRef]
2. Camus, C.; Offermann, P.; Weissmann, M.; Buerhop, C.; Hauch, J.; Brabec, C.J. Site-specific assessment of mechanical loads on photovoltaic modules from meteorological reanalysis data. *Sol. Energy* **2019**, *188*, 1134–1145. [CrossRef]
3. Lillo-Bravo, I.; González-Martínez, P.; Larrañeta, M.; Guasumba-Codena, J. Impact of Energy Losses Due to Failures on Photovoltaic Plant Energy Balance. *Energies* **2018**, *11*, 363. [CrossRef]
4. Rendler, L.C.; Romer, P.; Beinert, A.J.; Walter, J.; Stecklum, S.; Kraft, A.; Eitner, U.; Wiese, S. Thermomechanical stress in solar cells: Contact pad modeling and reliability analysis. *Sol. Energy Mater. Sol. Cells* **2019**, *196*, 167–177. [CrossRef]
5. Corrado, M.; Infuso, A.; Paggi, M. Simulated hail impacts on flexible photovoltaic laminates: Testing and modelling. *Meccanica* **2017**, *52*, 1425–1439. [CrossRef]
6. Beinert, A.J.; Ebert, M.; Eitner, U.; Aktaa, J. Influence of Photovoltaic Module Mounting Systems on the Thermo-Mechanical Stresses in Solar Cells by FEM Modelling. In Proceedings of the 32nd European Photovoltaic Solar Energy Conference and Exhibition, Munich, Germany, 20–24 June 2016; pp. 1833–1836.
7. Tippabhotla, S.K.; Radchenko, I.; Song, W.J.R.; Illya, G.; Handara, V.; Kunz, M.; Tamura, N.; Tay, A.A.O.; Budiman, A.S. From cells to laminate: Probing and modeling residual stress evolution in thin silicon photovoltaic modules using synchrotron X-ray micro-diffraction experiments and finite element simulations: Probing and modeling residual stress evolution. *Prog. Photovolt. Res. Appl.* **2017**, *25*, 791–809. [CrossRef]
8. Višniakov, N.; Kilikevičius, A.; Novickij, J.; Grainys, A.; Novickij, V. Low-cost experimental facility for evaluation of the effect of dynamic mechanical loads on photovoltaic modules. *Eksploat. Niezawodn. Maint. Reliab.* **2015**, *17*, 334–337. [CrossRef]
9. Kilikevičius, A.; Čereška, A.; Kilikevičienė, K. Analysis of external dynamic loads influence to photovoltaic module structural performance. *Eng. Fail. Anal.* **2016**, *66*, 445–454. [CrossRef]
10. Martins, A.C.; Chapuis, V.; Virtuani, A.; Li, H.-Y.; Perret-Aebi, L.-E.; Ballif, C. Thermo-mechanical stability of lightweight glass-free photovoltaic modules based on a composite substrate. *Sol. Energy Mater. Sol. Cells* **2018**, *187*, 82–90. [CrossRef]
11. Kilikevičienė, K.; Matijošius, J.; Fursenko, A.; Kilikevičius, A. Tests of hail simulation and research of the resulting impact on the structural reliability of solar cells. *Ekspolatacja Niezawodn. Maint. Reliab.* **2019**, *21*, 275–281. [CrossRef]

12. Paggi, M.; Corrado, M.; Berardone, I. A global/local approach for the prediction of the electric response of cracked solar cells in photovoltaic modules under the action of mechanical loads. *Eng. Fract. Mech.* **2016**, *168*, 40–57. [CrossRef]
13. Di Giacomo, F.; Fakharuddin, A.; Jose, R.; Brown, T.M. Progress, challenges and perspectives in flexible perovskite solar cells. *Energy Environ. Sci.* **2016**, *9*, 3007–3035. [CrossRef]
14. Said, S.A.M.; Hassan, G.; Walwil, H.M.; Al-Aqeeli, N. The effect of environmental factors and dust accumulation on photovoltaic modules and dust-accumulation mitigation strategies. *Renew. Sustain. Energy Rev.* **2018**, *82*, 743–760. [CrossRef]
15. Eisenträger, J.; Naumenko, K.; Altenbach, H.; Köppe, H. Application of the first-order shear deformation theory to the analysis of laminated glasses and photovoltaic panels. *Int. J. Mech. Sci.* **2015**, *96*, 163–171. [CrossRef]
16. Bouraiou, A.; Hamouda, M.; Chaker, A.; Mostefaoui, M.; Lachtar, S.; Sadok, M.; Boutasseta, N.; Othmani, M.; Issam, A. Analysis and evaluation of the impact of climatic conditions on the photovoltaic modules performance in the desert environment. *Energy Convers. Manag.* **2015**, *106*, 1345–1355. [CrossRef]
17. Ndiaye, A.; Kébé, C.M.F.; Charki, A.; Ndiaye, P.A.; Sambou, V.; Kobi, A. Degradation evaluation of crystalline-silicon photovoltaic modules after a few operation years in a tropical environment. *Sol. Energy* **2014**, *103*, 70–77. [CrossRef]
18. Kichou, S.; Silvestre, S.; Nofuentes, G.; Torres-Ramírez, M.; Chouder, A.; Guasch, D. Characterization of degradation and evaluation of model parameters of amorphous silicon photovoltaic modules under outdoor long term exposure. *Energy* **2016**, *96*, 231–241. [CrossRef]
19. Punge, H.J.; Kunz, M. Hail observations and hailstorm characteristics in Europe: A review. *Atmospheric Res.* **2016**, *176–177*, 159–184. [CrossRef]
20. Guo, B.; Javed, W.; Pett, C.; Wu, C.-Y.; Scheffe, J.R. Electrodynamic dust shield performance under simulated operating conditions for solar energy applications. *Sol. Energy Mater. Sol. Cells* **2018**, *185*, 80–85. [CrossRef]
21. Du, B.; Yang, R.; He, Y.; Wang, F.; Huang, S. Nondestructive inspection, testing and evaluation for Si-based, thin film and multi-junction solar cells: An overview. *Renew. Sustain. Energy Rev.* **2017**, *78*, 1117–1151. [CrossRef]
22. International Electrotechnical Commission. Technical Committee 82: Solar Photovoltaic Energy Systems. In *Modules photovoltaïques (PV) au silicium cristallin pour application terrestre: Qualification de la conception et homologation = Crystalline Silicon Terrestrial Photovoltaic (PV) Modules: Design Qualification and Type Approval*; International Electrotechnical Commission: Geneva, Switzerland, 2005; ISBN 978-2-8318-7963-5.
23. Kraemer, F.; Wiese, S.; Peter, E.; Seib, J. Mechanical problems of novel back contact solar modules. *Microelectron. Reliab.* **2013**, *53*, 1095–1100. [CrossRef]
24. Nikoletatos, J.; Halambalakis, G. Standards, Calibration, and Testing of PV Modules and Solar Cells. In *McEvoy's Handbook of Photovoltaics*; Elsevier: Chennai, India, 2018; pp. 1155–1182; ISBN 978-0-12-809921-6.
25. Osterwald, C.R. Standards, Calibration, and Testing of PV Modules and Solar Cells. In *Practical Handbook of Photovoltaics*; Elsevier: Waltham, MA, USA, 2012; pp. 1045–1069; ISBN 978-0-12-385934-1.
26. Perera, S.; Lam, N.; Pathirana, M.; Zhang, L.; Ruan, D.; Gad, E. Use of static tests for predicting damage to cladding panels caused by storm debris. *J. Build. Eng.* **2017**, *12*, 109–117. [CrossRef]
27. Pernas-Sánchez, J.; Artero-Guerrero, J.A.; López-Puente, J.; Varas, D. Numerical methodology to analyze the ice impact threat: Application to composite structures. *Mater. Des.* **2018**, *141*, 350–360. [CrossRef]
28. Sohel, K.M.A.; Richard Liew, J.Y.; Koh, C.G. Numerical modelling of lightweight Steel-Concrete-Steel sandwich composite beams subjected to impact. *Thin Walled Struct.* **2015**, *94*, 135–146. [CrossRef]
29. Zhang, D.; Fei, Q.; Zhang, P. Drop-weight impact behavior of honeycomb sandwich panels under a spherical impactor. *Compos. Struct.* **2017**, *168*, 633–645. [CrossRef]
30. Selech, J.; Joachimiak-Lechman, K.; Klos, Z.; Kulczycka, J.; Kurczewski, P. Life cycle thinking in small and medium enterprises: The results of research on the implementation of life cycle tools in Polish SMEs—Part 3: LCC-related aspects. *Int. J. Life Cycle Assess.* **2014**, *19*, 1119–1128. [CrossRef]
31. Andrzejczak, K.; Młyńczak, M.; Selech, J. Poisson-distributed failures in the predicting of the cost of corrective maintenance. *Eksploat. Niezawodn.Maint. Reliab.* **2018**, *20*, 602–609. [CrossRef]
32. Roy, S.; Kumar, S.; Gupta, R. Investigation and analysis of finger breakages in commercial crystalline silicon photovoltaic modules under standard thermal cycling test. *Eng. Fail. Anal.* **2019**, *101*, 309–319. [CrossRef]

33. Ceran, B.; Orłowska, A. The Impact of Power Source Performance Decrease in a PV/WT/FC Hybrid Power Generation System on the Result of a Multi-Criteria Analysis of Load Distribution. *Energies* **2019**, *12*, 3453. [CrossRef]
34. Song, Z.; Le, J.; Whisler, D.; Kim, H. Skin-stringer interface failure investigation of stringer-stiffened curved composite panels under hail ice impact. *Int. J. Impact Eng.* **2018**, *122*, 439–450. [CrossRef]
35. Cha, H.; Bhang, B.; Park, S.; Choi, J.; Ahn, H. Power Prediction of Bifacial Si PV Module with Different Reflection Conditions on Rooftop. *Appl. Sci.* **2018**, *8*, 1752. [CrossRef]
36. Beinert, A.J.; Büchler, A.; Romer, P.; Haueisen, V.; Rendler, L.C.; Schubert, M.C.; Heinrich, M.; Aktaa, J.; Eitner, U. Enabling the measurement of thermomechanical stress in solar cells and PV modules by confocal micro-Raman spectroscopy. *Sol. Energy Mater. Sol. Cells* **2019**, *193*, 351–360. [CrossRef]
37. Boumaaraf, B.; Boumaaraf, H.; Slimani, M.E.-A.; Tchoketch_Kebir, S.; Ait-cheikh, M.S.; Touafek, K. Performance evaluation of a locally modified PV module to a PV/T solar collector under climatic conditions of semi-arid region. *Math. Comput. Simul.* **2020**, *167*, 135–154. [CrossRef]
38. Ozden, T.; Carr, A.J.; Geerligs, B.L.; Turan, R.; Akinoglu, B.G. One-year performance evaluation of two newly developed back-contact solar modules in two different climates. *Renew. Energy* **2020**, *145*, 557–568. [CrossRef]
39. Devadhas, G.G. Design and development of new control technique for standalone PV system. *Microprocess. Microsyst.* **2020**, *72*, 102888.
40. Orioli, A. An accurate one-diode model suited to represent the current-voltage characteristics of crystalline and thin-film photovoltaic modules. *Renew. Energy* **2020**, *145*, 725–743. [CrossRef]
41. Thormeyer, C.; Sasse, J.-P.; Trutnevyte, E. Spatially-explicit models should consider real-world diffusion of renewable electricity: Solar PV example in Switzerland. *Renew. Energy* **2020**, *145*, 363–374. [CrossRef]
42. Fathabadi, H. Novel solar-powered photovoltaic/thermoelectric hybrid power source. *Renew. Energy* **2020**, *146*, 426–434. [CrossRef]
43. Jendoubi, A.; Tlili, F.; Bacha, F. Sliding mode control for a grid connected PV-system using interpolation polynomial MPPT approach. *Math. Comput. Simul.* **2020**, *167*, 202–218. [CrossRef]
44. Gürtürk, M.; Benli, H.; Ertürk, N.K. Determination of the effects of temperature changes on solar glass used in photovoltaic modules. *Renew. Energy* **2020**, *145*, 711–724. [CrossRef]
45. Liu, D.; Sun, Y.; Wilson, R.; Wu, Y. Comprehensive evaluation of window-integrated semi-transparent PV for building daylight performance. *Renew. Energy* **2020**, *145*, 1399–1411. [CrossRef]
46. Mundaca, L.; Samahita, M. What drives home solar PV uptake? Subsidies, peer effects and visibility in Sweden. *Energy Res. Soc. Sci.* **2020**, *60*, 101319. [CrossRef]
47. Klugmann-Radziemska, E.; Kuczyńska-Łażewska, A. The use of recycled semiconductor material in crystalline silicon photovoltaic modules production—A life cycle assessment of environmental impacts. *Sol. Energy Mater. Sol. Cells* **2020**, *205*, 110259. [CrossRef]
48. Castro, L.M.; Rodríguez-Rodríguez, J.R.; Martin-del-Campo, C. Modelling of PV systems as distributed energy resources for steady-state power flow studies. *Int. J. Electr. Power Energy Syst.* **2020**, *115*, 105505. [CrossRef]
49. Manohar, M.; Koley, E.; Ghosh, S.; Mohanta, D.K.; Bansal, R.C. Spatio-temporal information based protection scheme for PV integrated microgrid under solar irradiance intermittency using deep convolutional neural network. *Int. J. Electr. Power Energy Syst.* **2020**, *116*, 105576. [CrossRef]
50. Tsai, C.-Y.; Tsai, C.-Y. See-through, light-through, and color modules for large-area tandem amorphous/microcrystalline silicon thin-film solar modules: Technology development and practical considerations for building-integrated photovoltaic applications. *Renew. Energy* **2020**, *145*, 2637–2646. [CrossRef]
51. Obiwulu, A.U.; Chendo, M.A.C.; Erusiafe, N.; Nwokolo, S.C. Implicit meteorological parameter-based empirical models for estimating back temperature solar modules under varying tilt-angles in Lagos, Nigeria. *Renew. Energy* **2020**, *145*, 442–457. [CrossRef]
52. Hassan, A.; Wahab, A.; Qasim, M.A.; Janjua, M.M.; Ali, M.A.; Ali, H.M.; Jadoon, T.R.; Ali, E.; Raza, A.; Javaid, N. Thermal management and uniform temperature regulation of photovoltaic modules using hybrid phase change materials-nanofluids system. *Renew. Energy* **2020**, *145*, 282–293. [CrossRef]
53. Al Garni, H.Z.; Awasthi, A. A Monte Carlo approach applied to sensitivity analysis of criteria impacts on solar PV site selection. In *Handbook of Probabilistic Models*; Elsevier: Oxford, UK, 2020; pp. 489–504; ISBN 978-0-12-816514-0.

54. Singh, P.; Rajesh, B.; Bishnoi, S.; Swati, G.; Jaiswal, V.V.; Shanker, V.; Haranath, D. Optimization of processing parameters for designing an efficient AC driven powder electroluminescent device. *Ceram. Int.* **2016**, *42*, 17016–17022. [CrossRef]
55. Lin, K.; Lee, Y.-H.; Huang, W.-Y.; Chen, G.; Kuo, Y.-W.; Wang, L.-K.; Yang, S.-Y. Detection of soldering induced damages on crystalline silicon solar modules fabricated by hot-air soldering method. *Renew. Energy* **2015**, *83*, 749–758. [CrossRef]
56. Theis, D.; Oppolzer, H.; Ebbinghaus, G.; Schild, S. Cross-sectional transmission electron microscopy of electroluminescent thin films fabricated by various deposition methods. *J. Cryst. Growth* **1983**, *63*, 47–57. [CrossRef]
57. Lee, Y.; Tay, A.A.O. Stress Analysis of Silicon Wafer-Based Photovoltaic Modules Under IEC 61215 Mechanical Load Test. *Energy Procedia* **2013**, *33*, 265–271. [CrossRef]
58. Hülsmann, P.; Weiss, K.-A. Simulation of water ingress into PV-modules: IEC-testing versus outdoor exposure. *Sol. Energy* **2015**, *115*, 347–353. [CrossRef]
59. Hsu, S.-T.; Lin, W.-Y.; Wu, S.-J. Environmental Factors for Non-uniform Dynamic Mechanical Load Test due to Wind Actions on Photovoltaic Modules. *Energy Procedia* **2018**, *150*, 50–57. [CrossRef]
60. Fowler, S. Accelerated Environmental Chambers and Testing of PV Modules. In *Durability and Reliability of Polymers and Other Materials in Photovoltaic Modules*; Elsevier: Oxford, UK, 2019; pp. 257–277; ISBN 978-0-12-811545-9.

8

Influences of Optical Factors on the Performance of the Solar Furnace

Zhiying Cui [1,2,3,4], Fengwu Bai [1,2,3,4,*], Zhifeng Wang [1,2,3,4] and Fuqiang Wang [5]

1 Key Laboratory of Solar Thermal Energy and Photovoltaic System, Chinese Academy of Sciences, Beijing 100190, China; cuizhiying@mail.iee.ac.cn (Z.C.); zhifeng@vip.sina.com (Z.W.)
2 Institute of Electrical Engineering, Chinese Academy of Sciences, Beijing 100190, China
3 University of Chinese Academy of Sciences, Beijing 100049, China
4 Beijing Engineering Research Center of Solar Thermal Power, Beijing 100190, China
5 Harbin Institute of Technology, Weihai 264209, Shandong, China; wangfuqiang@hitwh.edu.cn
* Correspondence: baifw@mail.iee.ac.cn

Abstract: In this paper, an optical structure design for a solar furnace is described. Based on this configuration, Monte Carlo ray tracing simulations are carried out to analyze the influences of four optical factors on the concentrated solar heat flux distribution. According to the practical mirror shape adjustment approach, the curved surface of concentrator facet is obtained by using the finite element method. Due to the faceted reflector structure, the gaps between the adjacent mirror arrays and the orientations of facets are also considered in the simulation model. It gives the allowable error ranges or restrictions corresponding to the optical factors which individually effect the system in Beijing: The tilt error of heliostat should be less than 4 mrad; the tilt error of the concentrator in the orthogonal directions should be both less than 2 mrad; the concentrator facets with the shape most approaching paraboloid would greatly resolve slope error and layout errors arising in the concentrator. Besides, by comparing the experimentally measured irradiance with the simulated results, the optical performance of the facility is evaluated to investigate their comprehensive influence. The results are useful to help constructors have a better understanding of the solar furnace's optical behavior under conditions of multiple manufacture restrictions.

Keywords: solar furnace; Monte Carlo ray tracing; finite element method; factor influence

---

1. Introduction

In solar thermal power applications, the solar furnace has globally served as an ideal test-bed to develop key technologies required for high temperatures (up to 3500 K), and thus is used for its capability of concentrating solar radiation to factors of more than thousands of suns [1]. Solar furnaces are used for the investigation of high-temperature action on materials and equipment in many fields, such as thermochemistry [2] and hydrogen production [3]. During more than half of a century evolvement, largescale solar furnaces and solar furnaces with various optical structure designs have been constructed to allow for experiments in various fields such as thermochemistry. In Odeillo, France, the CNRS has been operating a 1000 kW thermal power solar furnace since 1968. It is known as one of the biggest solar furnaces in the world, consisting of 63 heliostats and a parabolic concentrator with an 1830 m^2 aperture area [4]. There is another facility with the same capacity that was commissioned in 1987 in Parkent, Uzbekistan [5]. The solar furnace in PSA [6], named SF40, has a similar optical structure design to the one in KIER [7]. They are composed of a flat heliostat with a no-concentration effect and a revolution paraboloid dish, and achieve very high concentrated solar heat flux and thermal gradients. Zhang et al. [8] proposed a line-focused solar concentrating system to extend the possible experimental support. The shape of the mirror arrays, also called facets, are chosen to be hexagonal for good fill in the concentrator area for the solar furnace in Mexico [9]. In addition, the solar furnace in DLR [10] takes the advantage of an off-axis solution to efficiently avoid the shadow from objects installed in the focal region onto the concentrator.

Typically, solar furnaces of the small and medium sizes usually adopt two sets of mirror devices, with one being a heliostat and one being a concentrator. As the decisive component responsible for concentrating sunlight, the concentrator's optical characteristics directly determine the operating performance of the solar furnace [6]. Despite a continuous support for the revolution paraboloid as the ideal design for the concentrator, the considerable cost in its manufacture and its time-consuming testing are crucial impediments for its commercial application, in addition to its extremely high demands in precision. Thus, a practical and economical alternative using a faceted structure covered with identical curved mirror arrays is more common practice. The study [11] has concluded that point-focused collectors have an acceptable performance that strongly depends on facet sizes and optical errors, although the final result differs from that of an equivalent ideal paraboloid. However, its optical characteristics in the solar furnace remain unknown in regards to the faceted concentrator receiving sunlight from the heliostat instead of the sun. Besides, there are few studies on how other optical factors effect the working performance of the solar furnace. In terms of an ideal high-temperature solar thermal facility capable of a reaching a substantial market, this study seeks to figure out the individual and coupling impacts of basic optical factors on this specific concentrating system, thereby providing better guidance in fabrication and assembling to achieve the optimal working conditions.

Concentrated solar heat flux distribution provides a basic measure to evaluate the optical performance of a concentrating system, including the peak heat flux and the spot size. Its value represents the highest attainable concentrated solar irradiance for a blackbody. In general, concentrated heat flux can be measured experimentally, which relies on numbers of precise and fast responsive microsensors [12]. Additionally, it can also be estimated by means of simulation. The Monte Carlo ray tracing (MCRT) method is very flexible and has been proved to have a great accuracy: Li et al. performed the MCRT simulation on a Xe-arc lamp, with results of radiation power 0.32% higher than the experiment measurement [13]; Zhao et al. set up the MCRT model for the parabolic trough collector, and it had good agreement with the theoretical formula [14]. Additionally, the MCRT method is effectively used to design and optimize the parameters for a solar concentrating system [15,16]. The principle of the MCRT is as follows: A large amount of solar rays' behavior in a mirror field such as reflection, intersection and escape are traced, and the final concentrated heat flux distribution is derived by statistically counting the number of rays reaching the surface of the focusing target. Moreover, if the representative variables that account for influencing factors are considered in the model, by the MCRT simulations, it equally allows an assessment of the concentration accuracy of the concentrating system [7]. Basically, factors mainly influencing the final concentrated heat flux distribution can be grouped in to two major categories: Optical errors and geometrical errors [17]. Among them, factors such as specular reflectance errors [18,19] and tracking errors [20] can now be controlled well in comparatively ideal conditions, due to the increasingly sophisticated knowledge and technology available. As a result, the optical factors which directly effect the reflective surface normal vector have become the main sources deteriorating the optical performance of the concentrating system. Their existence would even double the offset in sunlight reflection. Therefore, here in this work, the approach of adding error models in MCRT is chosen to evaluate the performance of the solar furnace and analyze influences of optical factors.

In this paper, an optical structure of the solar furnace is described. The solar furnace with this design, chosen as the study case, was constructed in the solar thermal power plant in Yanqing, Beijing [21,22]. Four types of optical factors that impact on the concentrated heat flux distribution are analyzed by MCRT simulations: The tilt error of the heliostat, which originates from heliostat facets deviating from the ideal plane; the slope error of the concentrator, referring to the irregular deformation of reflecting curves of the concentrator facets by the surface adjustment approach; the layout error of the concentrator, caused by the same or similar shape of facets attached to the concentrator frame;

and the tilt error of the concentrator, which arises from concentrator facets tilting. All of them strongly effect on the sunlight path in mirror field of the solar furnace optical system. In particular, results differ from the available references which take the integral concentrator mirror surface or facets of it as an idealized paraboloidal, whereas here we have adopted the finite element method (FEM) to better represent the practical reflecting surface. A high-order linear fitting curve synthesizes the concentrator facet surface by fitting discrete points obtained from the FEM. The constraints are the displacements of adjusting bolts in the rear of mirror arrays and position the connecting pad located. The validated MCRT model is mixed-programed in C++ and MATLAB. By analyzing the simulation results, the study gives the allowable error ranges when each factor effects individually. In the final part, comparisons between the results from simulations and the results obtained by experimental measures provide comprehensive influences of four factors, and indicate some possible causes leading to the differences.

2. Structure Design

On-axis configuration is applied in present case, for the consideration of minimizing the axis aberration effect. The main components of the optical system and receiver are placed on the same line, joining the center of the concentrator and the center of the heliostat. Due to the relatively high latitude of Beijing, it is decided to use a north-south arrangement, that locates the heliostat north of the concentrator and with the receiver placed at the focal area. As shown in Figure 1, the working principle is as follows: The heliostat uses a two-axis tracking strategy to track the sun and reflects solar rays horizontally and in parallel to the optical global axis of the concentrator. The concentrator concentrates incoming rays from the heliostat onto its focus. The receiving surface placed with testing material or equipment is eventually irradiated with a highly-concentrated solar radiation.

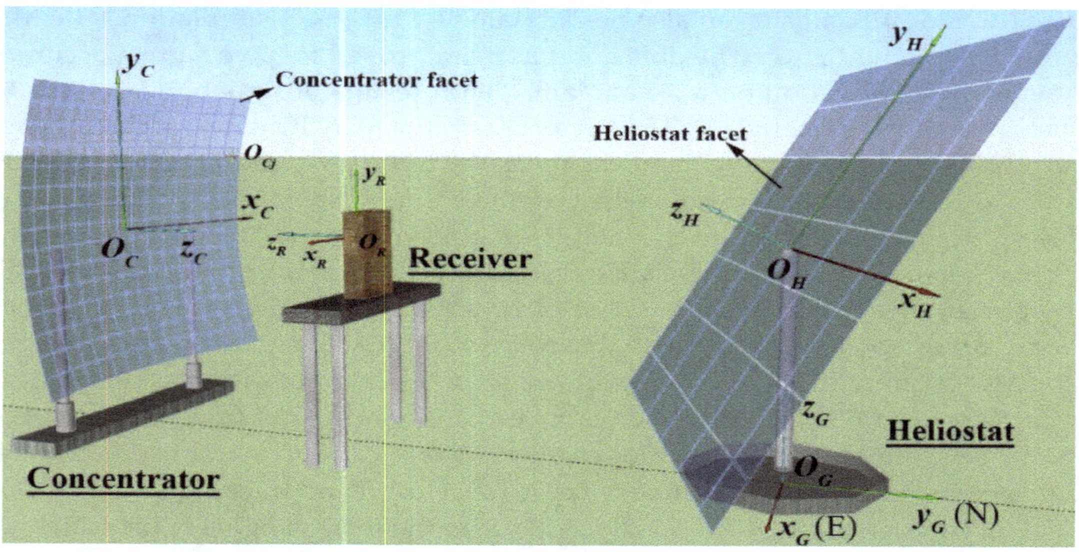

Figure 1. Schematic diagram of the solar furnace and coordinate systems.

The heliostat consists of flat reflective facets and rotating arms. It differs from those in the concentrating solar power plant, as the heliostat in the solar furnace has no concentration effect. Reflected rays from the heliostat are constantly kept in the expected direction, horizontal and parallel to the concentrator's optical axis. All facets are attached to the plane frame of the heliostat with connecting pieces. Also, they are aligned tightly in two perpendicular directions. Narrow gaps between adjacent facets are required for the leveling process. In addition, the two-axis rotating motion is continuously driven by an azimuth-evaluation tracking strategy in the computer with the PLC module.

The concentrator is the key component of the solar furnace. It is kept stationary on the ground to concentrate the sunlight onto the static position. Basically, it is the first priority to find the balance between the optical performance of the solar furnace and the cost primarily originated from the concentrator. Thus, it is the common practice to use the faceted structure covered with identical approximate paraboloid facets. Before being installed to the supporting frame, each facet's center is fixed to its bracket by a connecting pad, contacted with a finite number of adjusting bolts (eight in this paper) in the rear. By the method of regulating the lengths of the adjusting bolts, the flat mirror can be deformed to a desired curved surface which meets the requirement. Those facets with similar curvature are subsequently attached to the concentrator frame with an individual optical axis which is normal to their respective positions. Certainly, in such way, it may reduce the performance in concentrating solar power to some extent. However, in return, this approach greatly lowers the price and difficulty in fabrication and shortens the commissioning time. Besides, the focusing effect in the optimal conditions is totally acceptable compared with the ideal situation, which will be fully discussed in the following sections. As similar to the heliostat, facets of the concentrator must be carefully aligned. In addition, the receiver platform is placed between two mirror devices at the focal point of the concentrator. In this way, the entrance area could be exposed to the highest attainable solar irradiance.

3. Methods

For using the MCRT method, the mirror field is first set up based on working conditions of the concentrating system, including the physical placement of the main mirror devices, and orientation of the reflecting surface. According to the working principle of the solar furnace, the solar power concentrating process is divided into four parts: The solar position tracking, primary reflection by the heliostat, secondary concentration by the concentrator, and heat flux calculation on the receiver surface. Figure 1 shows five right-handed cartesian systems adopted in modelling. They are, respectively, the ground co-ordinate system O_G, heliostat co-ordinate system O_H, concentrator co-ordinate system O_C, concentrator facet co-ordinate system O_{c_j} and receiver co-ordinate system O_R. Here, the methodology is based on the following assumptions:

1. Incidence sunlight is treated as an optic cone.
2. Heliostat facets are perfect flat surfaces.
3. Concentrator facets are perfect curved surfaces.
4. The center of the facets on the faceted concentrator is located at the point on a continuous revolution paraboloid with the equivalent focal distance.
5. All traced solar rays have equal solar energy regardless of the angles of incidence.

3.1. Sun Position Tracking

To achieve the tracking strategy, it requires to acquire the sun position before determining heliostat's rotating motion. Here, classical simplified equations of the sun position are used [23], in O_G:

$$\begin{aligned} \alpha_s &= \text{asin}(\sin\phi\sin\delta + \cos\phi\cos\delta\cos\omega) \\ \gamma_s &= \tfrac{3\pi}{2} - \text{sign}\omega\left|\text{acos}\left(\tfrac{\sin\alpha_{sun}\sin\phi - \sin\delta}{\cos\alpha_{sun}\cos\delta}\right)\right| \end{aligned} \tag{1}$$

where α_s is the solar altitude angle, positive above the horizontal plane; γ_s is the solar azimuth angle, positive counterclockwise from the east; ϕ is the local latitude; δ is the declination; ω is the hour angle. The latter three variables can be solved, as in the reference [23]. Note that, in this paper, the Greek letter α refers to the altitude angle of a vector, and the letter γ refers to the azimuth angle.

The heliostat adopts an azimuth-elevation biaxial tracking strategy. The first rotating axis is perpendicular to the ground, and the heliostat rotates along this axis with value γ_H, defined as the azimuth angle of the heliostat's global axis. The second axis is parallel to the ground and perpendicular to the first axis, and the heliostat does rotation along this axis with value α_H, defined as the altitude angle of the heliostat's global axis. Both of two terms determine the tracking state of the heliostat and make sure the incoming rays from the sun can be reflected in the expected direction on the concentrator.

In O_G, the solar incidence optic cone center vector $\vec{s_s}$ and the primary reflected optic cone center $\vec{t_s}$ are given as Equation (2). $\vec{s_s}$ can be derived from Equation (1) and the azimuth angle and altitude angle in $\vec{t_s}$, respectively γ_{h2t} and α_{h2t}, are known from the relative position between the center of the integral heliostat mirror surface and the tracking target on the concentrator. Normally, in on-axis configuration, the tracking point is the center of the integral concentrator mirror surface if there is no geographic height difference between the two devices. On the basis of the law of reflection, it is easy to get the values of γ_H and α_H by solving Equation (3), where $\vec{n_H}$ is the unit vector of the heliostat global optical axis, and θ_s is the solar incidence angle. Noted that R_s is the solar rotation matrix and all the matrixes concerning rotation and coordinate transformation are listed in the Appendix A.

$$\begin{aligned} \vec{s_s} &= [0,0,1]R_s = [\cos\alpha_s \cos\gamma_s, \cos\alpha_s \sin\gamma_s, \sin\alpha_s] \\ \vec{t_s} &= [\cos\alpha_{h2t} \cos\gamma_{h2t}, \cos\alpha_{h2t} \sin\gamma_{h2t}, \sin\alpha_{h2t}] \end{aligned} \tag{2}$$

$$\begin{aligned} \vec{s_s} + \vec{t_s} &= 2 \cdot \vec{n_H} \cdot \cos\theta_s \\ \cos 2\theta_s &= \vec{s_s} \cdot \vec{t_s} \\ \vec{n_H} &= [\cos\alpha_H \cos\gamma_H, \cos\alpha_H \sin\gamma_H, \sin\alpha_H] \end{aligned} \tag{3}$$

3.2. Optical System Models

In co-ordinate O_H, a random solar incidence ray is represented as Equation (4). The radial angle φ and circumferential angle ψ for a random solar ray are expressed as Equation (5) [14]:

$$\vec{s_1} = -[\sin\varphi \cos\psi, \sin\varphi \sin\psi, \cos\varphi] \cdot R_s \cdot M_{G2H} \tag{4}$$

$$\begin{aligned} \varphi &= \operatorname{atan}\left(\sqrt{\varepsilon_1} \tan\varphi_{\max}\right) \\ \psi &= 2\pi\varepsilon_2 \end{aligned} \tag{5}$$

where M_{G2H} is co-ordinate transform matrix from O_G to O_H, ε_1 and ε_2 are random numbers between 0 and 1, $\varphi_{\max}$ represents the radial angle of the incident solar optic cone.

To set up a mirror field, it starts with the heliostat optical model. The integral heliostat surface is segmented into many pieces of flat mirrors. It is necessary to number facets to find which one of them intersects with and reflects the random solar ray. When considering error models of the heliostat, the error resulted from facets rotating in the heliostat mirror plane is neglected, due to the confinement of very narrow gaps between the adjacent facets. Furthermore, engineering measures can now better control a facet's vertical offset, within a few millimeters. Its impact on the final result is minuscule, thus, which is also neglected. Based on assumption 2, the optical model for the heliostat can be further simplified. Here, it is assumed that a heliostat facet is kept in the ideal position and only the local optical axis is deflected because of facet's tilting. The area changes in the integral sunlight receiving surface is the sum of the cosine values of facets between the local optical axis and the ideal surface normal vector (which is parallel to the z-axis of O_H). As illustrated in Figure 2, the local optical axis with tilting error of the i-th heliostat facet $\vec{n_{h_i}}$ is expressed as Equation (6). The area changes in the integral sunlight receiving surface of the heliostat, η_1, is given as Equation (7).

$$\begin{aligned} \vec{n_{h_i}} &= \vec{n_H} \cdot M_{G2H} \cdot E_{h_i} = [0,0,1] \cdot E_{h_i} \\ &= \left[\sin\varphi_{h_i} \cos\psi_{h_i}, \sin\varphi_{h_i} \sin\psi_{h_i}, \cos\varphi_{h_i}\right] \end{aligned} \quad i = 1, 2, \ldots, r_h \cdot c_h \tag{6}$$

$$\eta_1 = \frac{\sum_{i=1}^{r_h \cdot c_h} \vec{n_{h_i}} \cdot \left(\vec{n_H} \cdot M_{G2H}\right) \cdot S_{h_i}}{\sum_{i=1}^{r_h \cdot c_h} S_{h_i}} = \frac{\sum_{i=1}^{r_h \cdot c_h} \vec{n_{h_i}}(z) \cdot S_{h_i}}{S_H} \tag{7}$$

where the radial angle φ_{h_i} and circumferential angle ψ_{h_i} introduced into the ideal normal vector represent the i-th heliostat facet tilting error; r_h and c_h are the row and column number of heliostat facets; E_{h_i} is defined as the i-th heliostat facet's tilting error matrix; S_{h_i} is the area of the i-th facet; and S_H is the neat area of the heliostat.

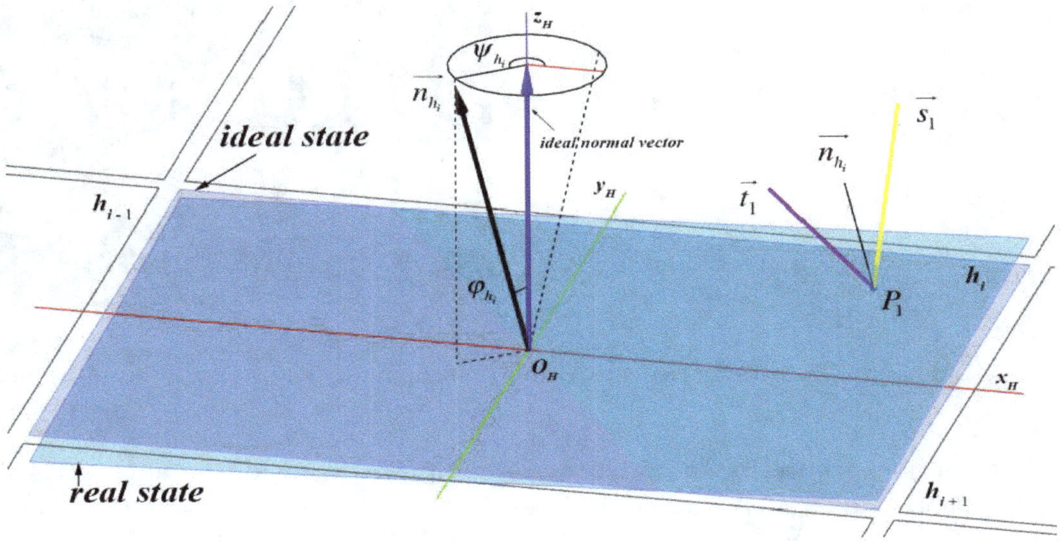

Figure 2. Schematic diagram of a heliostat facet.

Based on the heliostat optical model described above, the primary reflection modeling is as follows: A random solar incidence ray $\vec{s_1}$ intersects with the i-th facet of the heliostat on point $P_1(x_{P_1}, y_{P_1}, z_{P_1})$ where the local optical axis is $\vec{n_{h_i}}$, then the primary reflected ray $\vec{t_1}$ can be solved as Equation (8), where H_{h_i} is the reflection matrix of the i-th heliostat facet.

$$\vec{t_1} = \vec{s_1} \cdot H_{h_i} = \vec{s_1} \cdot \left(\vec{e}\,(3) - 2 \cdot \vec{n_{h_i}}^{T} \cdot \vec{n_{h_i}} \right) \tag{8}$$

Similarly, the integral concentrator concentrating surface is faceted, and also the gaps between the mirrors are considered and facets are numbered when modeling. According to the aforementioned surface adjustment approach, the chosen structure with eight adjusting bolts and one connecting pad is shown as Figure 3. By this approach, the facet is deformed to a non-ideal complex curved surface, which causes the slope error. It could not be expressed with a simple formula. Thus, the FEM is applied to first obtain discrete points on the facet with the constraints of displacements of adjusting bolts and the fixed position of connecting pad. The fourth-order polynomial fitting is then used to obtain the facet equation for a higher fitting accuracy. All facets attached to the concentrator share the same curved surface equation under the same constraints. Take the j-th concentrator facet as an example, in O_{c_j}, its surface equation is expressed as Equation (9):

$$z = f_{c_j}\left(x^{(4)}, y^{(4)}\right) \quad j = 1, 2, \ldots, r_c \cdot c_c \tag{9}$$

where r_c and c_c are the row and column number of concentrator facets, respectively.

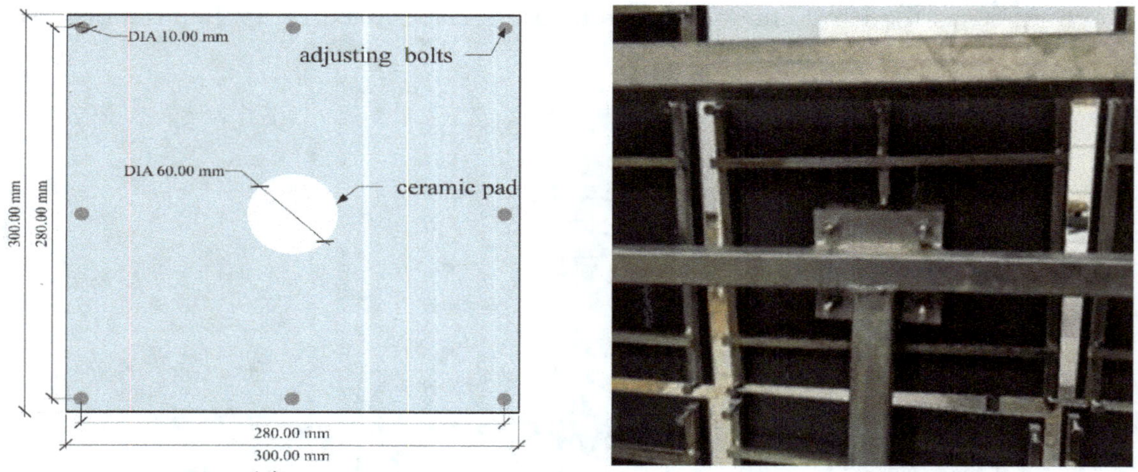

(a) (b)

Figure 3. Backside of a concentrator facet structure: (**a**) Schematic diagram of a facet; (**b**) Picture of a facet.

The tilting error of concentrator should be added in three dimensions, that involves the rotating around the x_{c_j}, y_{c_j}, and z_{c_j} axis as illustrated in Figure 4. Considering the less than tens of millimeters gap size between facets, the rotation around the z-axis is neglected here. The rest of them take major parts in the deviation of the surface normal vector of each reflected point from its ideal direction. It is defined as the tilting error matrix of the j-th concentrator facet as E_{c_j}, in which the rotating angle around the positive direction of x_{c_j} is φ_{c_j}, and that of y_{c_j} is ψ_{c_j}.

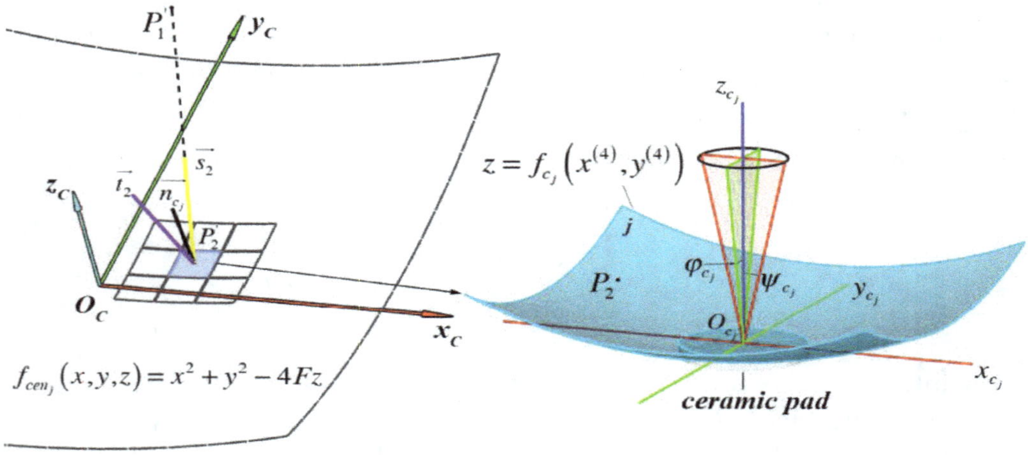

Figure 4. Schematic of a concentrator facet.

Moreover, by the same surface adjustment method, mirror arrays with the same or similar curvature cover the revolution paraboloid concentrator frame, which inevitably reduces off the concentrating performance, here defined as layout error. Facets placed on the respective positions have their own individual local optical axis normal to their center. Based on assumption 4, in O_C, the ideal paraboloid where facets center locate is given as Equation (10) with a focal length F. The azimuth and altitude angles of the global optical axis are γ_C and α_C, respectively, obtained from the concentrator's stationary orientation. With those two variables, the co-ordinate transform matrix M_{G2C} can be calculated. The j-th facet's optical axis is as Equation (11). The x-axis rotation angle γ_{c_j} and y-axis rotation angle α_{c_j} in the transform matrix M_{C2c_j} are calculated as Equation (12).

$$f_{cen_j}(x, y, z) = x^2 + y^2 - 4Fz \tag{10}$$

$$n_{cen_j} = \frac{-\left[\frac{\partial fcen_j}{\partial x}, \frac{\partial fcen_j}{\partial y}, \frac{\partial fcen_j}{\partial z}\right]}{\left[\frac{\partial fcen_j}{\partial x}, \frac{\partial fcen_j}{\partial y}, \frac{\partial fcen_j}{\partial z}\right]_2} \tag{11}$$

$$\begin{aligned} \gamma_{c_j} &= \mathrm{atan}\left(n_{cen_j}(x)/n_{cen_j}(z)\right) \\ \alpha_{c_j} &= -\mathrm{asin}\left(n_{cen_j}(y)\right) \end{aligned} \tag{12}$$

To sum up, the concentrator secondary concentration modeling is as follows: In O_C, an incidence ray $\vec{s_2}$ from the point P_1' on the heliostat intersects with the j-th concentrator facet on point P_2'. The point P_2' is derived from the point $P_2\left(x_{p_2}, y_{p_2}, z_{p_2}\right)$ in O_{c_j} by co-ordinates conversion between co-ordinate systems. As knowing $\vec{s_2}$ and the normal vector $\vec{n_{c_j}}\big|_{P_2'}$ given as Equation (14) on P_2', the secondary reflected ray $\vec{t_2}$ then can be solved from Equation (15). Here, H_{c_j} is the reflection matrix of the j-th concentrator facet. In O_G, $T_{G2H} = \overrightarrow{O_G O_H}$, $T_{G2C} = \overrightarrow{O_G O_C}$.

$$\begin{aligned} \vec{s_2} &= \vec{t_1} \cdot M_{G2H}^{-1} \cdot M_{G2C} \\ P_1' &= \left(P_1 \cdot M_{G2H}^{-1} + T_{G2H} - T_{G2C}\right) \cdot M_{G2C} \\ P_2' &= P_2 \cdot M_{C2c_j}^{-1} + \left[x_{cen_j}, y_{cen_j}, z_{cen_j}\right] \end{aligned} \tag{13}$$

$$\vec{n_{c_j}}\Big|_{P_2'} = \frac{\left[-\frac{\partial f_{c_j}}{\partial x_{p_2}}, -\frac{\partial f_{c_j}}{\partial y_{p_2}}, 1\right]}{\left\|\left[-\frac{\partial f_{c_j}}{\partial x_{p_2}}, -\frac{\partial f_{c_j}}{\partial y_{p_2}}, 1\right]\right\|_2} \cdot E_{c_j}^{-1} \cdot M_{C2c_j}^{-1} \tag{14}$$

$$\vec{t_2} = \vec{s_2} \cdot H_{c_j}\Big|_{P_2'} = \vec{s_2} \cdot \left(\vec{e}(3) - 2 \cdot \vec{n_{c_j}}\Big|_{P_2'}^{T} \cdot \vec{n_{c_j}}\Big|_{P_2'}\right) \tag{15}$$

3.3. Heat Flux Calculation

The solar furnace mainly uses the direct normal irradiance (DNI), and Equation (22) is adopted for the simulated value because it is suitable for the conditions in Beijing [24]:

$$\mathrm{DNI} = \frac{1376\left(1 + 0.033\cos\frac{2\pi n}{365}\sin\alpha_s\right)}{\sin a_s + \eta_a} \tag{16}$$

where n is the date order, and in sunny day $\eta_a = 0.33$.

The azimuth angle γ_R and the altitude angle α_R of the global axis of the receiver surface can be known from its settled orientation, which is as same as the concentrator. Furthermore, the co-ordinate transform matrix M_{G2R} is derived on the basis of those two variables. In O_R, given the secodary reflected ray $\vec{s_3}$, incidence ray from point P_2'' on the concentrator, the intersection point with the receiver surface can be solved. Based on the MCRT method, incidence solar rays with the number of N_t are traced and every single ray repeats the concentration procedure described above. To obtain the final concentrated heat flux distribution, the receiver surface is meshed into small rectangular area with the number of $r_r \cdot c_r$. The result is calculated as Equation (19).

$$\vec{s_3} = \vec{t_2} \cdot M_{G2C}^{-1} \cdot M_{G2R} \tag{17}$$

$$P_2'' = \left(P_2' \cdot M_{G2C}^{-1} + T_{G2C} - T_{G2R}\right) \cdot M_{G2R} \tag{18}$$

$$I_k = \frac{N_k \cdot I_0}{S_k} = \frac{N_k}{S_k} \cdot \frac{\mathrm{DNI} \cdot S_H \cdot \eta_1 \cdot \eta_2 \cdot \eta_3 \cdot \eta_4 \cdot \eta_5}{N_t} \quad k = 1, 2, \ldots, r_r \cdot c_r \tag{19}$$

where I_k is the heat flux of the k-th mesh; N_k is the number of rays fall in the k-th mesh; S_k is the k-th mesh area; I_0 is the solar energy of one traced ray; η_2 is the cosine efficiency; η_3 is the reflectivity of heliostat; η_4 is the reflectivity of concentrator; and η_5 represents the degree of shading area.

4. Results and Discussion

4.1. Validation for the MCRT Model

The MCRT codes are programmed in mixed programing languages, which should be validated before the further study. This MCRT model is validated against the results available in the reference [11]. The parameters used for validation are summarized in Table 1 and the comparisons are presented in Figure 5.

Table 1. Parameters used in the validation [11].

Name	Values
φ_{max} (mrad)	4.7
Heliostat (m × m)	12 × 10
Radius of the Circular of the Paraboloid Concentrator (m)	4
Focal Length (m)	15.512
DNI (W/m^2)	970
η_3	0.9
η_4	0.9

Generally, the accuracy of the MCRT method strongly depends on the numbers of tracing rays. Here, the number of 1×10^8 rays are chosen. Besides, the distribution of the final concentrated heat flux is related to the both mesh size on the receiver surface and the solar model [25]. In the validation, it is decided to use square meshes of 10 × 10 mm for reducing computing time on heat flux calculation, and the same uniform distribution solar model as in the reference. From the Figure 5b, it agrees well with results in the available reference, which can verify the accuracy of the MCRT model used in the current paper.

Since it has to take the local fabrication capability into consideration, the practical parameters of the solar furnace in Yanqing, Beijing are listed in Table 2, and the main components are presented as Figure 6. In the following cases analyzing optical factors influence, it adopts 12:00 PM on 21st March as the time, and Yanqing district in the north of Beijing (40.3833 N, 115.9367 E) as the location for the simulations, DNI 958 W/m^2, sun position $\gamma_s = 4.66$ rad, $\alpha_s = 0.86$ rad, radial angle $\varphi_{max} = 4.65$ mrad, and the reflectivity of heliostat and concentrator are both set as 0.8.

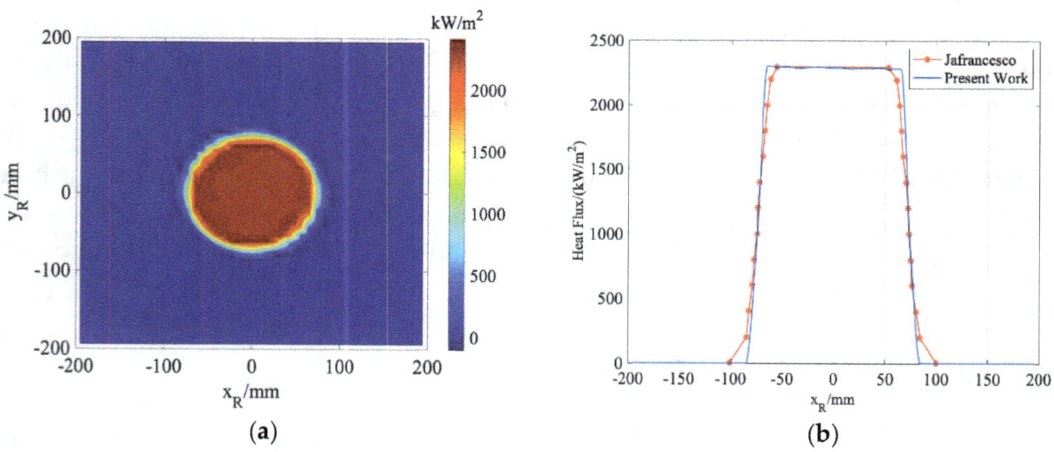

Figure 5. Validation of the Monte Carlo ray tracing MCRT model: (**a**) Simulated concentrated heat flux; (**b**) Comparisons between the present work and the reference.

Table 2. Parameters of the solar furnace in Yanqing, Beijing.

Name	Size (m × m)	Row & Column	Gap (mm)	Focal Length (m)
Heliostat	6.250 × 5.975	7 × 6	30	3.8
Concentrator	4.572 × 4.572 (Aperture)	15 × 15	10	

To discern each optical factor effect, the simulations in the following four sections are based on an ideal reference solution of the solar furnace: The mirror arrays of the heliostat are perfectly flat and fixed on the same plane, and the facets of the concentrator are the perfect paraboloid curved with the focal of 3.8 m, and the local optical axis of each mirror units are oriented to the ideal direction normal to their respective positions.

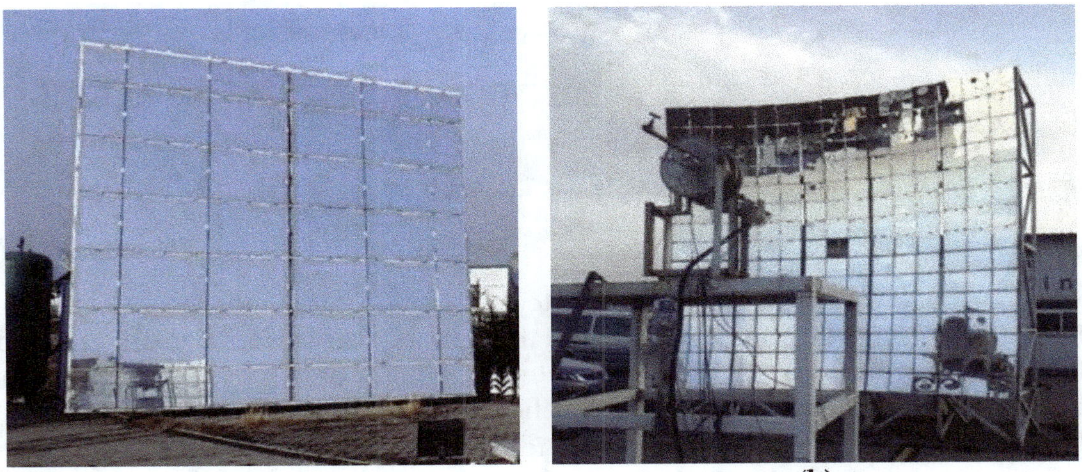

(a) (b)

Figure 6. Solar furnace in Yanqing, Beijing: (**a**) Heliostat; (**b**) Concentrator and receiver.

4.2. Tilt Error of the Heliostat

Once facets are attached to the frame structure, their respective tilting angles are fixed. However, it is impossible to install or assemble the facets facing to the exactly ideal direction in an engineering practice, which also means the tilt error of facets has great randomness and uncertainty. To better represent the tilt error in the heliostat model, parameters φ_{h_i} and ψ_{h_i} of facet tilt error are given by Equation (20), which is borrowed from the solar model representing uniform randomness.

$$\begin{aligned} \varphi_{h_i} &= \operatorname{atan}\left(\sqrt{\varepsilon_3}\tan\varphi_{h\max}\right) \\ \psi_{h_i} &= 2\pi\varepsilon_4 \end{aligned} \tag{20}$$

where ε_3 and ε_4 are random numbers between 0 and 1, with the tilt error range of heliostat defined as $e_h = 2 \cdot \varphi_{h\max}$.

Figure 7 shows how the center concentrated heat flux distribution varies with e_h. As is expected, the concentrated solar irradiance is irregularly scattering around as the error increases. Apparently, the error of the facets tilting away from ideal position deviates the local optical axis of the mirror pieces. In consequence, reflected rays from heliostat are no longer parallel to the global optical axis of concentrator. From the simulation cases, it can be seen that: When $e_h \le 8$ mrad, the heat flux distribution still maintains a good-shape Gaussian distribution, in particular, when $e_h = 8$ mrad, the peak heat flux decreases to 84.3% of that in ideal case; when $e_h = 4$ mrad, the heat flux distribution almost coincides with that in ideal conditions; as $e_h \ge 12$ mrad, the divergence of the concentrated solar power are serious, which means the optical performance of the solar furnace drops sharply. Nonetheless, even in the worst situation where facets tilt error is intolerable, the change in the sunlight receiving area of the heliostat is very small, specifically: $\eta_1 = 0.999975$, as $e_h = 24$ mrad, which provides a reference for further simplifying the heliostat model in the solar furnace.

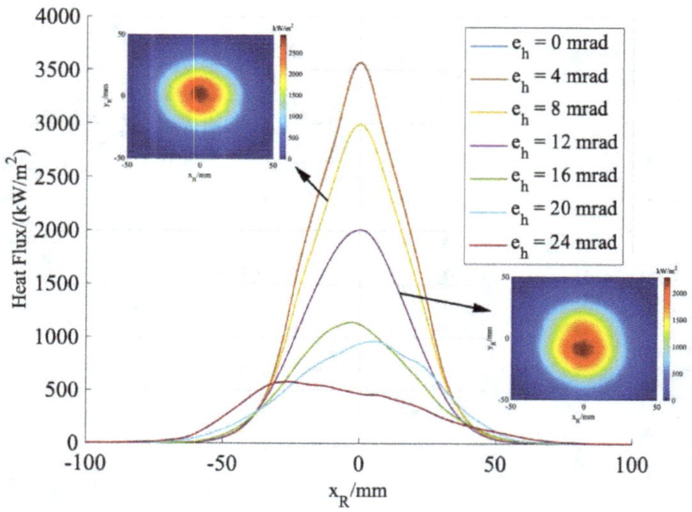

Figure 7. Influence of the tilt error of the heliostat.

4.3. Slope Error of the Concentrator

The properties of concentrator facets are listed in Table 3. The concentrator facet is a square mirror with a side length of 300 mm. As illustrated in Figure 3a, the surface adjustment is achieved by pushing eight bolts numbered from 1 to 8, contacted in the rear of the mirror. Table 4 provides five different facet shapes simulated in this section. The connecting pad fixed at the center of the facet are all with the diameter of 60 mm. It should be noted that if the facet is placed at the center of the concentrator, displacements of the adjusting bolts in facet *a* equal to sagitta lengths in an equivalent revolution paraboloid. Among those facets, facet *a* shows the closest curvature as the paraboloidal with the same focal. The displacements in facet *e* is measured data in practical.

Table 3. Properties of the concentrator facet.

Name	Values
Diameter of connecting pad (mm)	600
Facet mirror thickness (mm)	4
Density (kg/m^3)	2800
Young's modulus (GPa)	75
Poisson ratio	0.25

Table 4. Parameters of concentrator facets in simulation cases.

Name	Displacements of the Adjusting Bolts (mm)							
	1	2	3	4	5	6	7	8
Facet *a*	2.58	1.29	2.58	1.29	2.58	1.29	2.58	1.29
Facet *b*	3.00	1.50	3.00	1.50	3.00	1.50	3.00	1.50
Facet *c*	2.60	1.20	2.60	1.20	2.60	1.20	2.60	1.20
Facet *d*	2.80	1.80	2.50	1.00	3.50	0.80	2.20	1.20
Facet *e*	An ideal revolution paraboloid with a focal distance of 3.8 m							

Figure 8 gives the curved surface of facet *a* obtained by the FEM. From the result, the plane area is occurred at the center due to the displacement constraint of the connecting pieces. Along the direction from the center to the periphery, the circumferential curvature near the center changes similarly, and the gradient in curvature becomes larger near the outer circle. In the solar furnace, five forms of the facets' ability to concentrate solar power show obvious differences in Figure 9. Facet *e* results in a cylindrical irradiance distribution. Its peak heat flux is 30.41 kW/m^2 and the spot is around 10 mm radius (where 90% of the solar power are collected). Concentrated heat flux results for facet *a* and facet *c* present the Gaussian distribution, normally seen in the practical cases. Their peak heat fluxes are 22.84 kW/m^2 and 19.91 kW/m^2, respectively. Under the same thermal power, those values decrease by 24.9% and 34.5% compared with that of facet *e*. When compared mutually, facet *a*'s ability to concentrate solar power is better than that of facet *c*, in terms of smaller spot size and higher peak heat flux. With the surface adjustment approach mentioned above, the curved surface of facet *a* is the most ideal one can be achieved in practical. From the rest simulation cases, adjusting bolts near four vertices of facet *b* push the mirror so hard that its surface is excessively deformed, and in result, the concentrated heat flux shows a multi-peak central symmetrical distribution with a minimum value at the center. As to facet *d*, its peak heat flux position slightly shifts from the center of the receiver surface with the value of 12.95 kW/m^2. The distribution is scattering and irregular in shape. In general, the displacements of adjusting bolts need to be controlled within the accuracy of millimeters, otherwise, the slope error resulted from the surface deformation would obviously reduce the facet's concentrating ability.

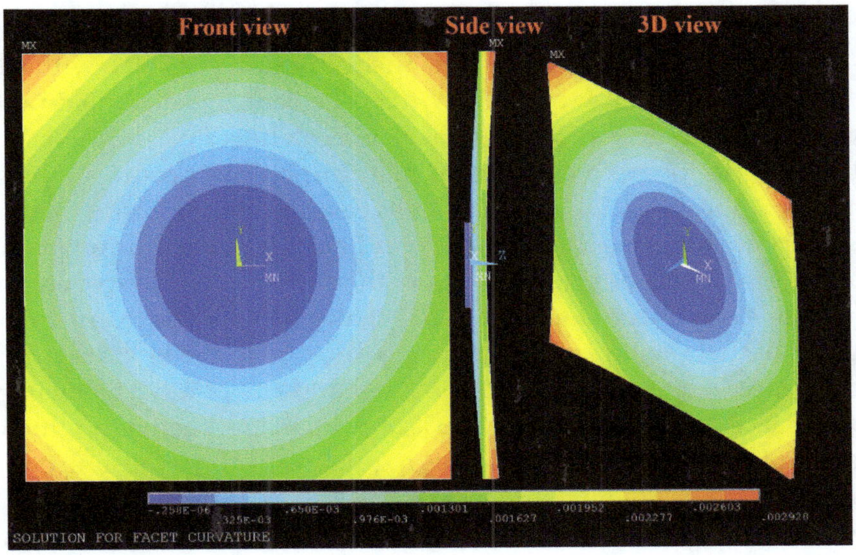

Figure 8. *z*-axis displacements color mapping of facet *a*.

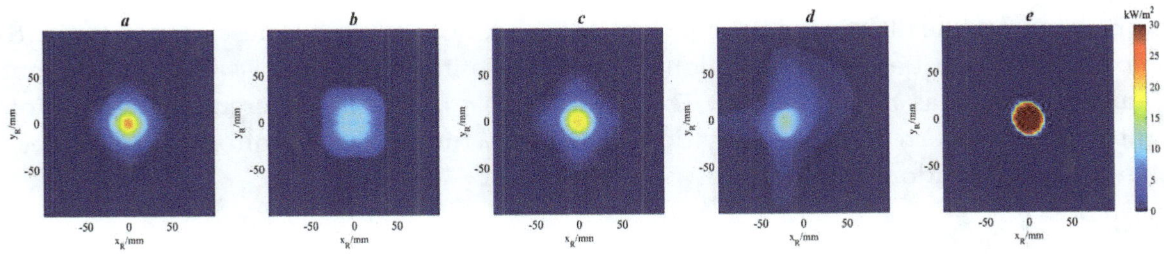

Figure 9. Comparisons of concentrator facets on the concentrating ability.

4.4. Layout Error of the Concentrator

We have selected five kinds of concentrator layouts to analyze the influence of the layout error. The mirror arrays attached to four of them are, respectively, facet *a*, *b*, *c*, and *e* in Table 3. The last one is a continuous revolution paraboloid concentrator, taken as a reference in the theoretically ideal condition. The simulation results are presented in Figure 10.

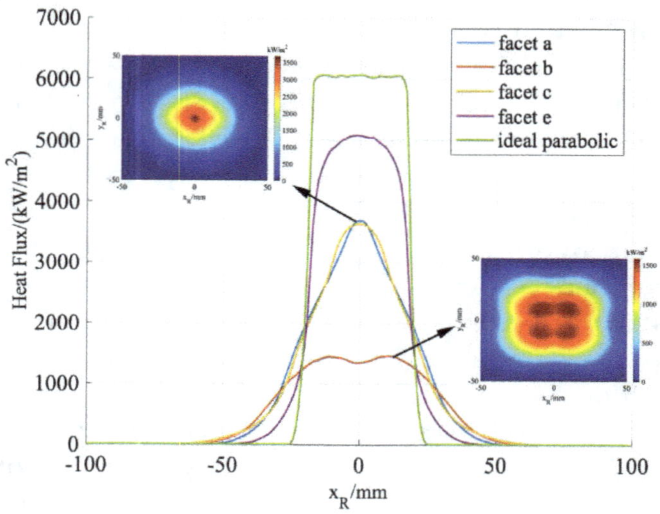

Figure 10. Comparisons of center heat flux distribution of different concentrator layouts.

The thermal power of the five cases on the receiver surface is equal. The green line synthetizes the reference case. As is intuitively expected, it has the minimum spot size with the diameter 50 mm, and the maximum peak heat flux is up to 6071 kW/m^2. The concentrator filled with facet *e* shows the second-best performance. Compared to the reference, its peak heat flux is decreased by 16.2%, and the concentrated irradiance is no longer a cylindrical distribution but occurs as the gradient decreases from the center to the periphery. However, those two cases might not be achievable in the practical situation. The most approaching result is a Gaussian heat flux distribution, as illustrated with the blue and yellow lines. Combined with conclusions in the previous section, even the focusing effect of single-piece facet in Table 3 has the obvious difference that the overall performance of the concentrator does not reflect the sum of their slope errors. That is to say, due to the layout error, the error influence caused by concentrator facets deformation is diminished at some point. The peak heat flux of layouts covered with facet *a* and facet *c* are reduced by 39.3% and 40.3%, respectively, which is still tantamount to more than 3000 suns of solar energy. It indicates a good optical performance for the solar furnace. Nonetheless, for the concentrator layout using facet *e*, the flux distribution keeps in a multi-peak shape and also the overall value is low, which further emphasizes the importance to ensure the accuracy of the displacements of adjusting bolts within 1~2 mm.

4.5. Tilt Error of Concentrator

As differing from the tilt error in the heliostat, for the concentrator, it is described in two orthogonal directions. The vertical title error range along y_{c_j}-axis is defined as $e_{cx} = 2 \cdot \varphi_{c\max}$, and the horizontal along x_{c_j}-axis is defined as $e_{cy} = 2 \cdot \psi_{c\max}$. The radial angle φ_{c_j} and circumferential angle ψ_{c_j} of the *j*-th concentrator facet's tilt error are given by Equation (21), which is also a uniform randomness model. Here, ε_5, ε_6 ε_7 ε_8 are random numbers between 0 and 1.

$$\begin{aligned} \varphi_{c_j} &= \text{sign}(\varepsilon_5 - 0.5) \cdot \text{atan}\left(\sqrt{\varepsilon_6} \tan \varphi_{c\max} \right) \\ \psi_{c_j} &= \text{sign}(\varepsilon_7 - 0.5) \cdot \text{atan}\left(\sqrt{\varepsilon_8} \tan \psi_{c\max} \right) \end{aligned} \tag{21}$$

It is observed from Figure 11 that as e_{cx} increases, the concentrated heat flux gradually diverges along the y_R-axis, and the position of the peak heat flux slightly shifts vertically, and distribution along the x_R-axis maintains a good Gaussian shape, but values decreases. Likewise, the factor e_{cy} has the same effect but in opposite direction. When tilt errors are less than 2 mrad, the performance of solar furnace is closed to that in the ideal; as tilt errors are less than 4 mrad, the concentrated heat flux distribution can still maintain a good Gaussian shape.

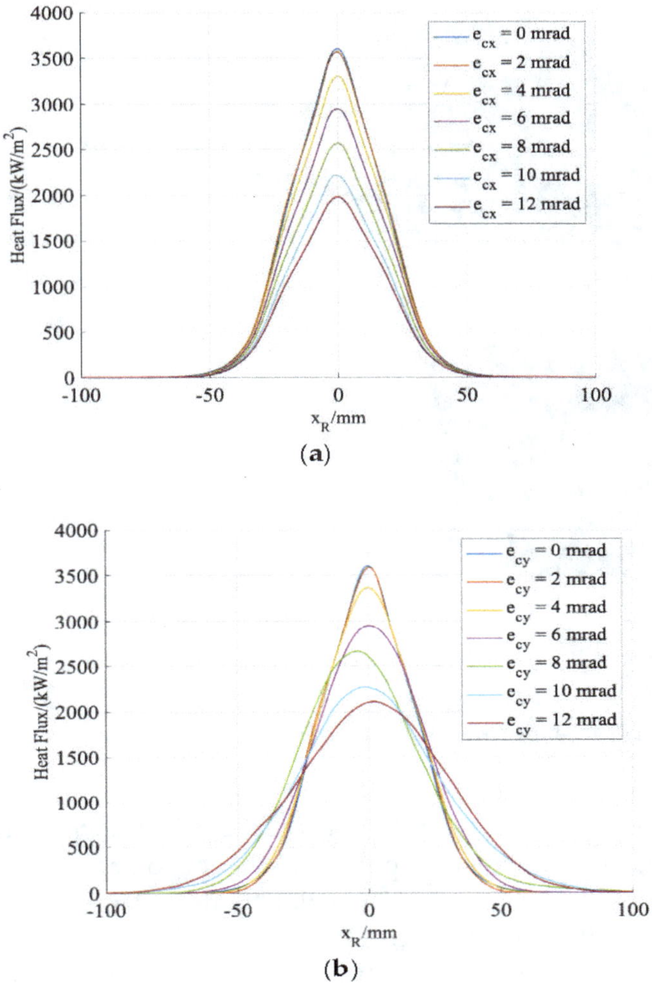

Figure 11. Influences of the tilt error of concentrator: (**a**) Along the vertical direction; (**b**) Along the horizontal direction.

4.6. Comprehensive Factor Influence

Take an experimentally measured concentrated heat flux to analyze the comprehensive influence of four optical factors. The selected data was measured at 13:38 on 18th June 2015. At this moment, the measured DNI is 851 W/m^2, with the measured reflectance as $\eta_3 = 0.80$, $\eta_4 = 0.83$. From the simulation, it shows that gap loss in the heliostat is 5.3%, the sum of incoming rays' overflowing and the gap loss in concentrator is 24.1%. The total number of traced rays is 1×10^8, and 7.19×10^7 of them reach the receiver surface. The cosine efficiency is $\eta_2 = 0.63$, and the shadow is $\eta_3 = 0.90$. From the discussions above, if factors in the optical system effect individually, it could attain the comparatively ideal working performance of the solar furnace when $e_h \leq 4$ mrad or using facet *a* or $e_{cx} \leq 2$ mrad, $e_{cy} \leq 2$ mrad, which coincides with the conclusion in reference [9] that the upper boundary of 4 mrad for optical errors can achieve the desirable structure design. Three forms of solar furnace used to study the comprehensive influence of errors are listed in Table 5. Comparisons between the simulations and the experiment is given in Figure 12.

Table 5. Parameters of solar furnace with different errors.

Name	Tilt Error in Heliostat	Tilt Error in Concentrator	Concentrator Facet
Solar furnace *a*	$e_h = 0$ mrad	$e_{cx} = 0$ mrad, $e_{cy} = 0$ mrad	Facet *a*
Solar furnace *b*	$e_h = 4$ mrad	$e_{cx} = 0$ mrad, $e_{cy} = 2$ mrad	Facet *a*
Solar furnace *c*	$e_h = 12$ mrad	$e_{cx} = 3$ mrad, $e_{cy} = 8$ mrad	Facet *a*

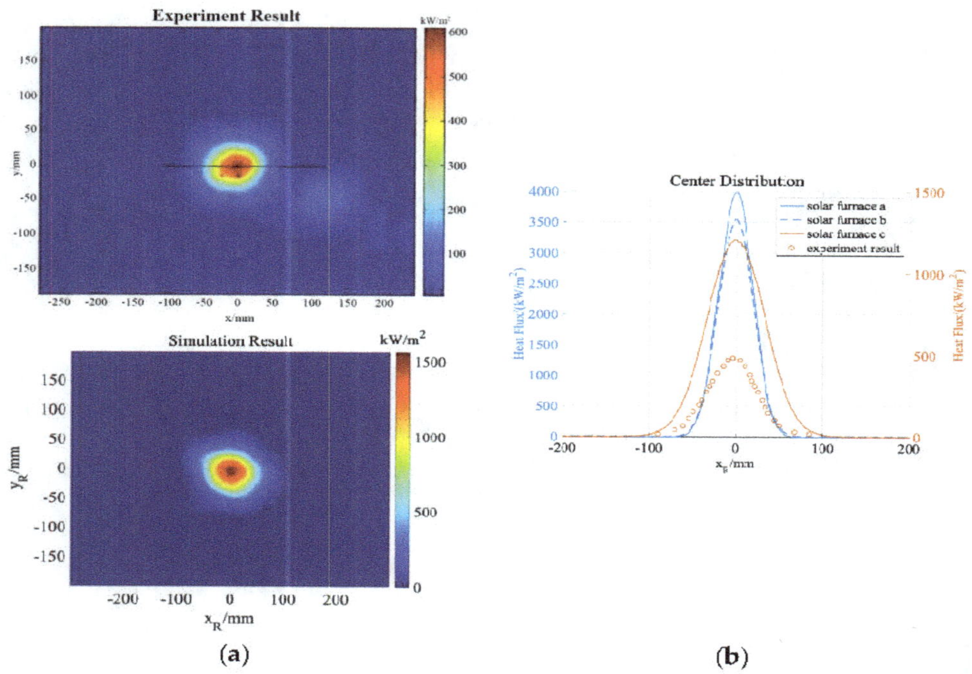

Figure 12. Evaluation of the solar furnace in Yanqing, Beijing: (**a**) Comparison of concentrated heat flux distribution between the experiment and solar furnace *c*; (**b**) Comparisons of the center heat flux distribution.

Solar furnace *a* represents the ideally optimal optical performance in the engineering practice for the present structure design. From the results, peak heat flux of solar furnace *a*, *b*, and *c* are 4036 kW/m^2, 2586 kW/m^2, and 1290 kW/m^2, respectively. The solar furnace *b*'s is 11.2% lower than that value of solar furnace *a*. It indicates that even each error can be well controlled, for the reason of the optical system demanding extremely high precision, the comprehensive influence of factors can obviously decrease the facility's performance. For the solar furnace *c* in Figure 12a, the integral of the heat flux equals to the thermal power of 8.28 kW, which agrees with the value calculated by Equation (22). It can further verify the accuracy of the model.

$$P = \text{DNI} \cdot S_H \cdot \frac{\sum N_k}{N_t} \cdot \eta_2 \cdot \eta_3 \cdot \eta_4 \cdot \eta_5 = 8.29 \text{ kW} \tag{22}$$

As shown in Figure 12a, the flux distribution in the simulation shows Gaussian shape, as the measured result presents. Their trend of gradient change and the shape and size of the spot are approximately similar. However, the overall heat flux values in the simulation are much higher than the experiment. In particular, the measured peak flux is 484 kW/m^2 and the thermal power is around 2.5 kW, which is 30% of that in the simulation. Apparently, there are lots of uncertainties unexpectedly involved in the practical system, more than the four factors focused on in the present work. Based on the comparisons in Figure 12b, there are several possible explanations given, as follows:

1. In terms of simulation, the MCRT model mainly consider optical factors caused by facets, not including the heliostat tracking error, the pillar tilting, environmental factors, etc. The actual situation is impossible to be exactly duplicated. Therefore, there must be differences between the simulation and the experiment.
2. The experimental measurement procedure is as follows: A CCD camera captures the spot on the Lambertian intercepted surface and outputs the greyscale image. The heat flux result is then exported by comparing to the dependence of brightness and grey value on the standard point where a microsensor real-time measuring. The accuracy of the measurement strongly depends on

the microsensor's calibration. Thus, the accuracy of the experimental results may also be open to question.

3. The difference in results between the experiment and the simulation, in the other aspect, also indicates that the comprehensive influence from various factors has an obvious impact on the overall operating performance of solar furnace. The key to a higher optical behavior is controlling each error well within the allowable range.

5. Conclusions

This paper introduces both the optical structure design and the MCRT model for the solar furnace in detail. The optical structure design adopts one two-axis tracking heliostat and one faceted concentrator using the identical facets. In the concentrator, all curved facets are deformed to an approximate revolution paraboloid by adjusting bolts in the rear, and then attached to the frame orienting normal to their respective position. The result using the MCRT model agrees well with the available references, which could validate its accuracy. By analyzing the simulations, influences of four optical factors on the optical performance are discussed:

1. The tilt error of heliostat affects the non-parallelism degree of the reflected sunlight. As the error increases, both the concentrated heat flux distribution and spot are diverging.

2. The FEM is used to obtain the curved surface of concentrator facet. The slope error of the concentrator causes the mirror unit to become no longer an ideal surface, but rather complex and irregular under constraints. The error in the 1~2 mm could greatly reduce the concentration ability.

3. The layout error occurs when facets attached to the concentrator frame share a similar curvature. Its existence will not enlarge the influence of the slope error.

4. The tilt error of concentrator facets directly impacts on the focusing effect. As the tilt error increases in a certain direction, heat flux along the direction is scattering, while the orthogonal direction keeps in the Gaussian shape, but overall value decreases.

Based on the facility in Yanqing, Beijing, allowable ranges for corresponding errors effect individually are given: The tilt error of heliostat $e_h \leq 4$ mrad; concentrator facets adopts facet a, in which displacements of adjusting bolts equal to sagitta lengths in the equivalent paraboloid; tilt error of concentrator $e_{cx} \leq 2$ mrad and $e_{cy} \leq 2$ mrad. Besides, the comprehensive influence of the four factors has been studied. The study shows that even when controlling well for each optical factor within the allowable range, the optical performance of the solar furnace will drop significantly because of their coexistence. Compared with the experimentally measured concentrated heat flux, it indicates that there are still considerable uncertainties remained to be discussed in the further study. Besides, it is also worth to figure out how the optical factors influence the thermal behavior on the testing materials. Results in the present work are useful for the developers to have a better understanding of the influences of factors on the final result. It is helpful in the process of manufacture and assembling to attain the optimal performance of the solar furnace.

Author Contributions: Conceptualization, Z.C. and F.B.; methodology, Z.C.; software, Z.C.; validation, Z.C.; formal analysis, Z.C. and F.B.; investigation, Z.C. and F.B.; writing—original draft preparation, Z.C., F.B., Z.W., and F.W.; writing—review and editing, Z.C., F.B., Z.W. and F.W.; visualization, Z.C.; supervision, F.B. and Z.W.; project administration, F.W., Z.W. and F.W.; funding acquisition, F.W.

Acknowledgments: This work was supported by the National Natural Science Foundation of China (No. 51676061).

Appendix A

Solar rotation matrix

$$R_s = \begin{bmatrix} \cos\left(\frac{\pi}{2} - \alpha_s\right) & 0 & -\sin\left(\frac{\pi}{2} - \alpha_s\right) \\ 0 & 1 & 0 \\ \sin\left(\frac{\pi}{2} - \alpha_s\right) & 0 & \cos\left(\frac{\pi}{2} - \alpha_s\right) \end{bmatrix} \cdot \begin{bmatrix} \cos\gamma_s & \sin\gamma_s & 0 \\ -\sin\gamma_s & \cos\gamma_s & 0 \\ 0 & 0 & 1 \end{bmatrix} \tag{A1}$$

The i-th heliostat facet's deflection error matrix

$$E_{h_i} = \begin{bmatrix} \cos\varphi_{h_i} & 0 & -\sin\varphi_{h_i} \\ 0 & 1 & 0 \\ \sin\varphi_{h_i} & 0 & \cos\varphi_{h_i} \end{bmatrix} \cdot \begin{bmatrix} \cos\psi_{h_i} & \sin\psi_{h_i} & 0 \\ -\sin\psi_{h_i} & \cos\psi_{h_i} & 0 \\ 0 & 0 & 1 \end{bmatrix} \tag{A2}$$

The j-th concentrator facet's deflection error matrix

$$E_{c_j} = \begin{bmatrix} 1 & 0 & 0 \\ 0 & \cos\varphi_{c_j} & -\sin\varphi_{c_j} \\ 0 & \sin\varphi_{c_j} & \cos\varphi_{c_j} \end{bmatrix} \cdot \begin{bmatrix} \cos\psi_{c_j} & 0 & \sin\psi_{c_j} \\ 0 & 1 & 0 \\ -\sin\psi_{c_j} & 0 & \cos\psi_{c_j} \end{bmatrix} \tag{A3}$$

Coordinate transform matrix from O_G to O_H

$$M_{G2H} = \begin{bmatrix} \cos\left(\frac{\pi}{2} + \gamma_h\right) & -\sin\left(\frac{\pi}{2} + \gamma_h\right) & 0 \\ \sin\left(\frac{\pi}{2} + \gamma_h\right) & \cos\left(\frac{\pi}{2} + \gamma_h\right) & 0 \\ 0 & 0 & 1 \end{bmatrix} \cdot \begin{bmatrix} 1 & 0 & 0 \\ 0 & \cos\left(\frac{\pi}{2} - \alpha_h\right) & -\sin\left(\frac{\pi}{2} - \alpha_h\right) \\ 0 & \sin\left(\frac{\pi}{2} - \alpha_h\right) & \cos\left(\frac{\pi}{2} - \alpha_h\right) \end{bmatrix} \tag{A4}$$

Coordinate transform matrix from O_G to O_C

$$M_{G2C} = \begin{bmatrix} \cos\left(\frac{\pi}{2} + \gamma_s\right) & -\sin\left(\frac{\pi}{2} + \gamma_s\right) & 0 \\ \sin\left(\frac{\pi}{2} + \gamma_s\right) & \cos\left(\frac{\pi}{2} + \gamma_s\right) & 0 \\ 0 & 0 & 1 \end{bmatrix} \cdot \begin{bmatrix} 1 & 0 & 0 \\ 0 & \cos\left(\frac{\pi}{2} - \alpha_s\right) & -\sin\left(\frac{\pi}{2} - \alpha_s\right) \\ 0 & \sin\left(\frac{\pi}{2} - \alpha_s\right) & \cos\left(\frac{\pi}{2} - \alpha_s\right) \end{bmatrix} \tag{A5}$$

Coordinate transform matrix from O_C to O_{c_j}

$$M_{C2c_j} = \begin{bmatrix} \cos\gamma_{c_j} & 0 & \sin\gamma_{c_j} \\ 0 & 1 & 0 \\ -\sin\gamma_{c_j} & 0 & \cos\gamma_{c_j} \end{bmatrix} \cdot \begin{bmatrix} 1 & 0 & 0 \\ 0 & \cos\alpha_{c_j} & -\sin\alpha_{c_j} \\ 0 & \sin\alpha_{c_j} & \cos\alpha_{c_j} \end{bmatrix} \tag{A6}$$

References

1. Trefilov, V.; Schur, D.; Pishuk, V.; Zaginaichenko, S.; Choba, A.; Nagornaya, N. The solar furnaces for scientific and technological investigation. *Renew. Energy* **1999**, *16*, 757–760. [CrossRef]
2. Oliveira, F.A.C.; Fernandes, J.C.; Badie, J.-M.; Granier, B.; Rosa, L.G.; Shohoji, N. High meta-stability of tungsten sub-carbide W2C formed from tungsten/carbon powder mixture during eruptive heating in a solar furnace. *Int. J. Refract. Met. Hard Mater.* **2007**, *25*, 101–106. [CrossRef]
3. Villafán-Vidales, H.; Arancibia-Bulnes, C.; Riveros-Rosas, D.; Romero-Paredes, H.; Estrada, C. An overview of the solar thermochemical processes for hydrogen and syngas production: Reactors, and facilities. *Renew. Sustain. Energy Rev.* **2017**, *75*, 894–908. [CrossRef]
4. Guillot, E.; Rodriguez, R.; Boullet, N.; Sans, J.-L. Some details about the third rejuvenation of the 1000 kWth solar furnace in Odeillo: Extreme performance heliostats. *AIP Conf. Proc.* **2018**, *2033*, 040016.
5. Abdurakhmanov, A.A.; Zainutdinova, K.K.; Mamatkosimov, M.A.; Paizullakhanov, M.S.; Saragoza, G. Solar technologies in Uzbekistan: State, priorities, and perspectives of development. *Appl. Sol. Energy* **2012**, *48*, 84–91. [CrossRef]

6. Rodriguez, J.; Cañadas, I.; Zarza, E. New PSA high concentration solar furnace SF40. *AIP Conf. Proc.* **2016**, *1734*, 070028.
7. Lee, H.; Chai, K.; Kim, J.; Lee, S.; Yoon, H.; Yu, C.; Kang, Y. Optical performance evaluation of a solar furnace by measuring the highly concentrated solar flux. *Energy* **2014**, *66*, 63–69. [CrossRef]
8. Zhang, X.; Cui, Z.; Zhang, J.; Bai, F.; Wang, Z. Optical performance analysis of an innovative linear focus secondary trough solar concentrating system. *Front. Energy* **2018**, *13*, 590–596. [CrossRef]
9. Riveros-Rosas, D.; Herrera-Vázquez, J.; Pérez-Rábago, C.; Bulnes, C.A.A.; Vázquez-Montiel, S.; Sanchez-Gonzalez, M.; Granados-Agustín, F.; Jaramillo, O.; Estrada, C.; Jaramillo, O. Optical design of a high radiative flux solar furnace for Mexico. *Sol. Energy* **2010**, *84*, 792–800. [CrossRef]
10. Neumann, A.; Groer, U. Experimenting with concentrated sunlight using the DLR solar furnace. *Sol. Energy* **1996**, *58*, 181–190. [CrossRef]
11. Jafrancesco, D.; Sansoni, P.; Francini, F.; Contento, G.; Cancro, C.; Privato, C.; Graditi, G.; Ferruzzi, D.; Mercatelli, L.; Sani, E.; et al. Mirrors array for a solar furnace: Optical analysis and simulation results. *Renew. Energy* **2014**, *63*, 263–271. [CrossRef]
12. Ballestrín, J.; Monterreal, R. Hybrid heat flux measurement system for solar central receiver evaluation. *Energy* **2004**, *29*, 915–924. [CrossRef]
13. Li, Z.; Tang, D.; Du, J.; Li, T. Study on the radiation flux and temperature distributions of the concentrator—Receiver system in a solar dish/Stirling power facility. *Appl. Therm. Eng.* **2011**, *31*, 1780–1789. [CrossRef]
14. Zhao, D.; Xu, E.; Wang, Z.; Yu, Q.; Xu, L.; Zhu, L. Influences of installation and tracking errors on the optical performance of a solar parabolic trough collector. *Renew. Energy* **2016**, *94*, 197–212. [CrossRef]
15. Shuai, Y.; Xia, X.-L.; Tan, H.-P. Radiation performance of dish solar concentrator/cavity receiver systems. *Sol. Energy* **2008**, *82*, 13–21. [CrossRef]
16. Zou, B.; Dong, J.; Yao, Y.; Jiang, Y. A detailed study on the optical performance of parabolic trough solar collectors with Monte Carlo Ray Tracing method based on theoretical analysis. *Sol. Energy* **2017**, *147*, 189–201. [CrossRef]
17. Bonanos, A.M. Error analysis for concentrated solar collectors. *J. Renew. Sustain. Energy* **2012**, *4*, 063125. [CrossRef]
18. Meyen, S.; Lupfert, E.; DLR Gernam Aerospace Center; Ciemat, A.F.-G.; Nrel, C.K. *Standardization of Solar Mirror Reflectance Measurements—Round Robin Test*; National Renewable Energy Lab. (NREL): Golden, CO, USA, 2010.
19. Meyen, S.; Lüpfert, E.; Pernpeintner, J.; Fend, T.; Schiricke, B. Optical Characterization of Reflector Material for Concentrating Solar Power Technology. *In SolarPaces Conference*. 2009. Available online: https://www.google.com/url?sa=t&rct=j&q=&esrc=s&source=web&cd=1&ved=2ahUKEwir9tCr55_lAhVnzIsBHV8eD48QFjAAegQIAhAC&url=http%3A%2F%2Felib.dlr.de%2F61684%2F1%2FSolarPaces2009_Mirror-Characterisation.png&usg=AOvVaw29cvWfKwbxWpYCBZeAaa_f (accessed on 15 October 2019).
20. Guo, M.; Wang, Z.; Zhang, J.; Sun, F.; Zhang, X. Accurate altitude—Azimuth tracking angle formulas for a heliostat with mirror—Pivot offset and other fixed geometrical errors. *Sol. Energy* **2011**, *85*, 1091–1100. [CrossRef]
21. Wang, F.; Bai, F.; Wang, T.; Li, Q.; Wang, Z. Experimental study of a single quartz tube solid particle air receiver. *Sol. Energy* **2016**, *123*, 185–205. [CrossRef]
22. Ma, L.; Wang, Z.; Lei, D.; Xu, L. Establishment, Validation, and Application of a Comprehensive Thermal Hydraulic Model for a Parabolic Trough Solar Field. *Energies* **2019**, *12*, 3161. [CrossRef]
23. Duffie, J.A.; Beckman, W.A. *Solar Engineering of Thermal Processes*; Wiley: Hoboken, NJ, USA, 2013; p. 23.
24. Xu, Y.; Cui, K.; Liu, D. The development of a software for solar radiation and its verification by the measurement results on the spot. *Energy Technol.* **2002**, *23*, 237–239.
25. Evans, D. On the performance of cylindrical parabolic solar concentrators with flat absorbers. *Sol. Energy* **1977**, *19*, 379–385. [CrossRef]

9

Methodology for an Opto-Geometric Optimization of a Linear Fresnel Reflector for Direct Steam Generation

Eduardo González-Mora * and Ma. Dolores Durán García

Ingeniería en Sistemas Energéticos Sustentables, Facultad de Ingeniería, Universidad Autónoma del Estado de México, Cerro de Coatepec S/N, Toluca 50100, Mexico; mddg_2210@hotmail.com

* Correspondence: egonzalezm022@profesor.uaemex.mx

Abstract: A methodology for an optical optimization of the intercept factor concerning a linear Fresnel reflector is described to increase the amount of solar irradiation that will be delivered in the absorber for Agua Prieta, Sonora; taking the FRESDEMO's Fresnel field as the reference design. For the performed optimization, the intercept factor is determined as a function of the receiver's height, establishing a simple criterion for the optimization. The FRESDEMO's field description is determined and briefly discussed, next compared with the proposed optimization. The compound parabolic concentrator (CPC) Winston function for a circular absorber is modified to relocate the cusp of the reflector and the absorber. This modified CPC will redirect all the reflected rays that do not hit directly the absorber, as in the FRESDEMO field, so all of them are captured by the absorber. Through ray-tracing, the graphic flux distribution in the receiver aperture is conceived. This flux distribution is compared with the FRESDEMO field and with a PTC with a flat absorber, ensuring an adequate optimization regarding the intercept factor. The result of the opto-geometric optimization is compared between the FRESDEMO and the optimized field for a specified thermal process, addressing a considerable reduction in the length of the loops.

Keywords: LFR; IAM; intercept factor; CPC; opto-geometric optimization; ray-tracing

1. Introduction

Currently, concentrated solar power plants (CSP), are an alternative to produce electricity against non-renewable energy sources such as oil, coal, and nuclear energy. In these systems, different designs and geometries of concentrators are used. CSP plants concentrate the solar rays to heat a working fluid that runs a thermodynamic power cycle, usually a steam Rankine cycle [1]. Most of these solar plants use solar fields with parabolic trough collectors (PTC) [2]. Notwithstanding, in recent years, Fresnel reflectors are among the most promising technologies to produce energy from concentrated solar radiation [3,4]; therefore, there has been an increasing interest in adopting Fresnel reflectors due to the low operating costs [5].

This interest has led to different studies discussing the opto-geometric parameters to be determined and optimized. Abbas and Martínez-Val [6] analyzed the optical design of Fresnel Reflectors width and shifts of the mirrors, Boito and Grena [7,8] developed a model to find the optimal focal length of the primary mirrors and also an optical optimization relating the plant cost and the concentrated solar radiation. Qiu et al. [9] presented a strategy to homogenize the heat flux in the receiver area employing multi-objective genetic algorithms and Monte Carlo ray-tracing. However, there are no optimizations in the intercept factor for the typical fields to be analyzed (FRESDEMO, PUERTO ERRADO, etc.), this is developed in the present work considering the FRESDEMO field as the starting point for optimization since it is a field that has proven to be technically operational for the DSG.

Concentrating the sun's rays allows a working fluid (HTF, heat transfer fluid) to reach high temperatures that ensure a good conversion of heat-electricity efficiency. The HTF can be synthetic oil, molten-salt or water-steam; the latter known as Direct Steam Generation (DSG). Commercially, the DSG is not applied yet because, in PTCs, the thermal stress problems produce several problems in the receiver integrity [10,11]. However, Fresnel reflectors do not present this disadvantage since the heat flux is distributed more uniformly over the absorber and also because the receiver is fixed, so, one of the trends is towards this direction since it also avoids the use of heat exchangers [12,13]. With these advantages, the thermal efficiency is increased by simplifying the configuration of the entire system [14–16].

The linear Fresnel reflector (LFR, which will be referred hereafter only as Fresnel reflectors), simplifies the design of the PTCs. The LFR resembles the parabolic shape with a collection of flat (or nearly flat) mirrors positioned at ground level [2]. These mirrors move on a single axis to track and reflect the Sun's rays toward the receiver. Unlike PTCs, the receiver of the LFRs is fixed in space and the reflectors rotate simultaneously to keep the focus continuously on the receiver, as shown in Figure 1 [1,3].

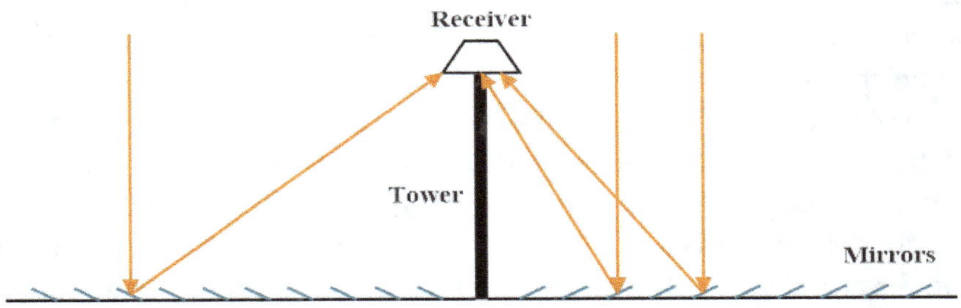

Figure 1. Linear Fresnel reflector with a central receiver. Adapted from reference [17].

All rows of the LFR match the same tracking as the apparent movement of the Sun. However, each row is tilted at different angles at any time, since they have different positions regarding the receiver. This allows a very large primary unit composed of relatively small reflectors, with a long focal length, which indeed leads to the possibility to use almost flat and low-cost glass mirror elements [4,5].

One difficulty with LFRs technology is that avoiding shading and blocking between adjacent reflectors leads to a greater spacing between the reflectors leading to a greater use of land. Blocking and shading between mirrors can be reduced significantly by increasing the height of the receivers, but this increases the cost of the system.

The main advantages of LFR over PTC are:

- fixed receiver.
- low wind load in the mirrors.
- largest collection area for each receiver (which reduces the cost of receivers and simplifies the management of fluid circulation).
- small moving parts (the mirrors are much smaller than the only one mirror of a PTC).
- lower costs of the optical components (the mirrors are almost flat, and their construction is simpler).
- the receiver is always illuminated from below, but the addition of a CPC allows a more uniform heat flux distribution over the absorber tube, regardless of the angle of the Sun, minimizing thermal stresses.
- an easier system maintenance facility.

The principal disadvantages are [18–21]:

- reduced optical efficiency, especially when the Sun is far from the focal plane.
- greater sensitivity to optical and tracking errors, due to the greater distances between the mirrors and the receiver.

Taking into consideration the above, and evaluating the advantages and disadvantages of the LFR compared to the PTC, it is of great interest the implementation of a solar field that could use linear Fresnel reflectors for DSG; since it implies that solar power plants would be able to start using a simpler technology that could be competitive and commercial in the near future.

In addition, if the solar field is optically optimized for a certain locality, the decrease in the length of the field becomes evident, which would reduce the cost of the installation; however, all the simulations based on the DSG with LFR are using the FRESDEMO configuration, therefore, it is interesting to optimize the solar field to analyze the behavior of the system and confirm the reduction of the length of the entire field.

2. Methodology

The LFRs, like the PTCs, follow the Sun while moving the mirrors in prescribed paths minimizing the angle of incidence of the radiation on the surfaces and thus, maximize the radiation collected of the incident rays. The aforementioned is done through calculated rotations on a single axis (that may have any orientation, although, in practice, it is usually horizontal from East to West, or from North to South). All the linear concentrators used for power generation are designed with an East-West tracking, so the concentration system is oriented in a North-South direction; this is done to maximize the use of irradiance on the concentrator.

The design of the solar field is the most important component to be determined in a solar plant for the generation of electrical energy. A poor design can trigger low operating times, an increment in the thermal energy storage capacity, low thermal performance and, even lead to the inoperability of the power block since the concentrated solar energy density is not enough to be useful; so the optimization of the opto-geometric description of a linear Fresnel reflector for direct steam generation is essential to ensure an optimum optical performance. For the reasons exposed, in the present work, a methodology for the opto-geometric optimization was developed; this methodology is schematized in Figure 2.

First, for the optimization process, the location of the power plant must be defined. This allowed calculating the solar angles to define the incidence and direction of the irradiation (Section 2.1). In a second step, the location of the primary mirrors and the intercept factor is determined (Section 2.2). Third, the intercept factor is computed as a function of the receiver's height and a criterion for the optimum height is established; once the criterion is fulfilled, the incident angle modifier (IAM) is computed, and the CPC geometry is defined (Section 3). Fourth, a ray-tracing is performed to compare the flux distribution between a PTC, FRESDEMO and the optimized field (Section 4). Afterward, a thermal process is computed to compare the results of the optimized field against the FRESDEMO field.

2.1. Solar Angles in A LFR

In order to define the layout and the design of the LFR, the solar angles must be determined to ensure that the reflected solar radiation is redirected in the best way to the receiver. For a solar field oriented in the direction N-S (solar fields use this orientation to maximize the radiation capture at solar midday so that more irradiance can be used annually.) (as seen in Figure 3a); the transverse solar altitude angle α_T can be determined by the solar altitude angle α_s and the azimuthal angle γ_s [14,16,22]; the celestial coordinates are determined by:

$$\cos\theta = \left(\cos^2\theta_z + \cos^2\delta_s \sin^2\omega\right)^{\frac{1}{2}} \tag{1}$$

$$\tan\beta_s = \tan\theta_z \left|\cos(\gamma_s - \gamma)\right| \tag{2}$$

$$\gamma = \begin{cases} 90^\circ & \text{if} \quad \gamma_s > 0 \\ -90^\circ & \text{if} \quad \gamma_s \leq 0 \end{cases} \tag{3}$$

$$\tan\alpha_T = \frac{a_c}{b_c} = \tan\alpha_s\sqrt{1 + \frac{1}{\tan^2\gamma_s}} = \frac{\tan\alpha_s}{\sin\gamma_s} \tag{4}$$

where $0° \leq \alpha_T \leq 180°$, and $\sin\gamma_s = 0$ when $\alpha_T = 90°$. Other angles of practical interest in concentration systems are the longitudinal angle of incidence θ_l and transversal angle of incidence θ_t. Both angles are shown schematically in Figure 3b, determined by Equations (5) and (6).

$$\theta_l = \arctan\left(\frac{|\sin\gamma_s|}{\tan\alpha_s}\right) \tag{5}$$

$$\theta_t = \arcsin(\cos\gamma_s \cos\alpha_s) \tag{6}$$

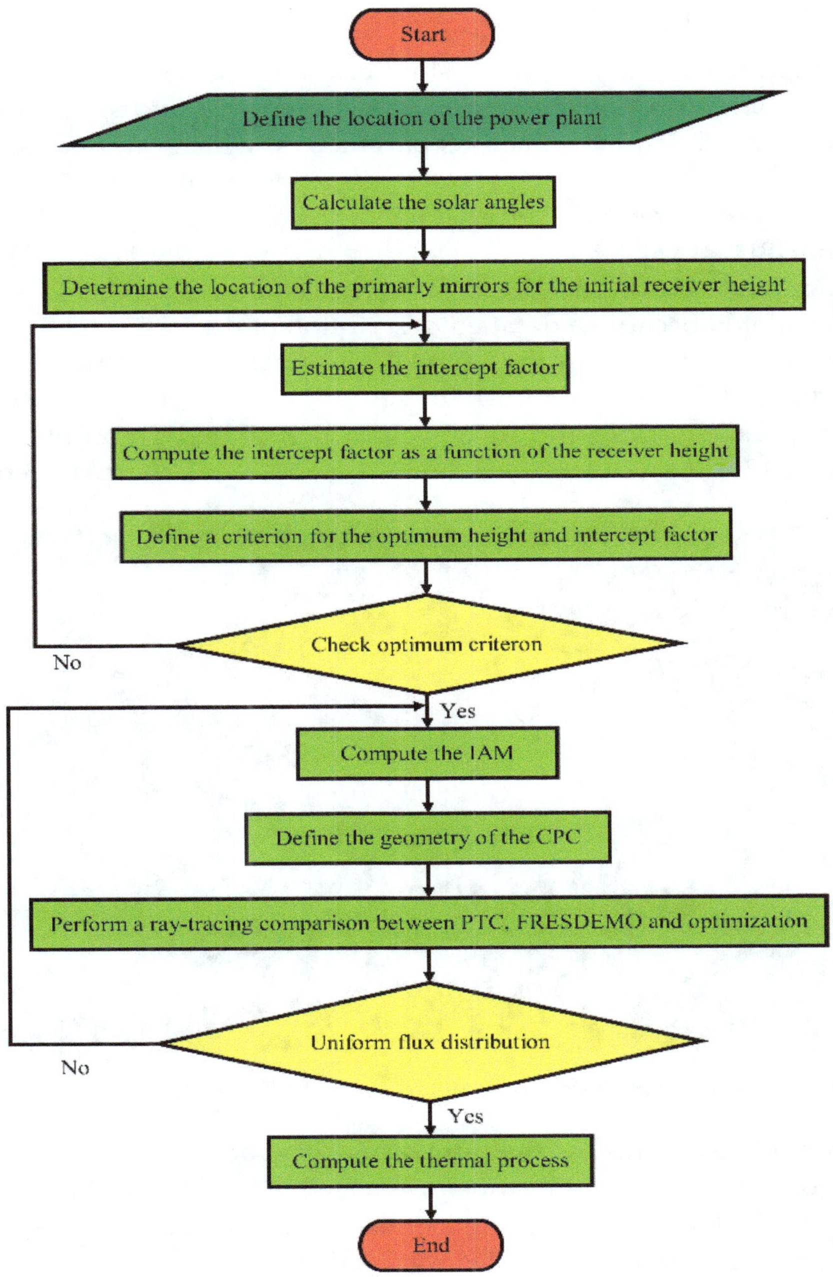

Figure 2. Methodology for the optimization of the linear Fresnel reflector (LFR) field.

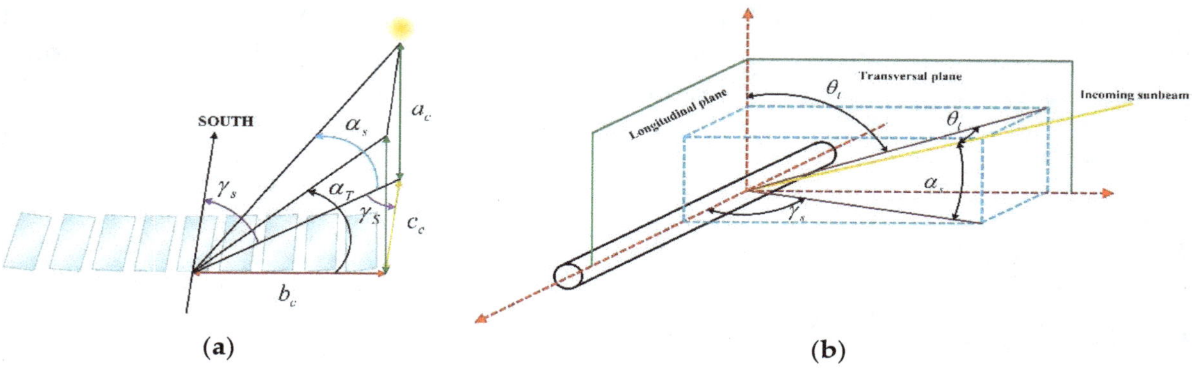

Figure 3. Angles in a LFR: (**a**) Determination of the transverse solar altitude angle; (**b**) Determination of the incidence angles.

2.2. Opto-Geometric Description of the LFR Field

Although the optical nature of LFRs is simple, there are few data available to describe a Fresnel reflector field [7]. However, even ignoring the dimensions and specifications of the receiver, the geometry of a LFR is defined by numerous parameters, which can be changed independently, at least in principle: the width, the position, and the focal length of each mirror are all independent variables. Therefore, the geometry of the solar field depends on 3 N_m degrees of freedom, where N_m is the number of mirrors [8].

Considering some descriptions for different Fresnel reflectors designs; a similar configuration (a similar field is considered because a detailed description of the FRESDEMO field is not available (mainly the location of the mirrors)) of the FRESDEMO's field; located in the Plataforma Solar de Almería, Spain, is proposed; this LFR is shown in Figure 4. The field is composed of several modules each with 25 mirrors of 0.60 m width and 100 m length, distributed in a 21 m wide-field oriented in N-S direction [6–8,19,23–28]. As a summary in Table 1, the parameters considered are shown.

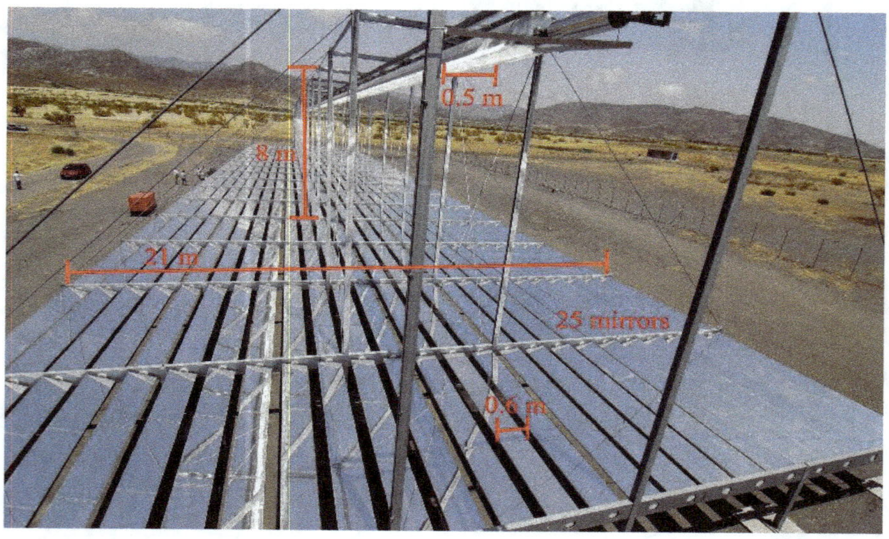

Figure 4. Main geometrical parameters of the FRESDEMO prototype [24].

The filling factor (FF) is defined as the ratio of useful space of concentration occupied by the primary mirrors to the whole area of the solar field, as stated in Equation (7). It is important to say that the main objective of the LFRs is to be able to maximize the filling factor ($FF \rightarrow 1$); this means better use of land while capturing as much as possible solar radiation.

$$FF = \frac{A_{pm}}{A} \tag{7}$$

where:
$A_{pm} = N_m W_m L_m = 1500\ m^2$: Effective area of the primary mirrors, $A = W_{sf} L_m = 2100\ m^2$: Area of the LFR module.

Thus, with the data described in Table 1, the filling factor for the proposed field is 0.7143; which is identical to the Fresdemo field [24,27].

Table 1. Geometric data of a FRESDEMO similar field.

Characteristic	Value
Number of primary mirrors (N_m)	25
Total width of the solar field (W_{sf})	21 m
Total length of primary mirrors (L_m)	100 m
Width of the primary mirrors (W_m)	0.6 m
Height of the receiver to the primary mirrors (h_{fr})	8 m
Receiver width (a')	0.5 m
Orientation of the mirror axis	N-S

Another factor that influences the performance of the LFRs is the height at which the receiver is placed (independent if only one absorber tube or several tubes are used). To determine the inclination of each mirror, in concordance with the Fermat's Principle, the light that reaches each mirror will be reflected towards a target zone (which in this case will necessarily be the receiver); however, the further away the mirrors are, the longer the path the light must travel, and in addition, not all the light will directly hit the receiver. The intercept factor (Γ) is a quantitative measure of this condition.

The intercept factor ideally should have a unit value; which would mean that all the radiation that enters the opening area (A) will be redirected entirely to the receiver; however, since the LFR is not an ideal (it should not be forgotten that a Fresnel reflector simulates the optical behavior of a parabolic channel, so in essence, it could not be a concentrator without optical losses) concentrator, there will be losses in this captured radiation. Although there are a great variety of methodologies to determine the intercept factor, such as directly applying the concept of the conservation of the étendue [29,30] or using an advanced ray-tracing technique [31], it is possible to use simpler methods such as that described by Pettit et al. [32], or, in its simplest form, only the quotient between the area illuminated by the redirected rays and the area of the receiver, as stated in Equation (8) and shown schematically in Figure 5.

$$FF = \frac{A'}{A_{il}} = \frac{a' L_m}{\left(a_l' + a' + a_r\right) L_m} \tag{8}$$

where

A': Receiver area
A_{il}: Total lighting area
a_l: Left receiver spillage
a_r: Right receiver spillage
a': Receiver width

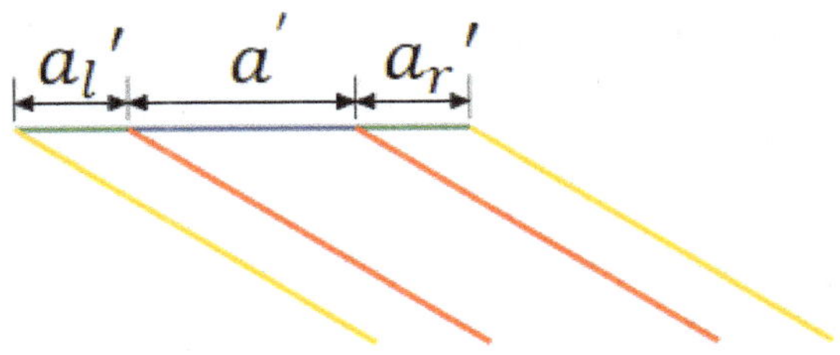

Figure 5. Schematic diagram of the cross-section of the receiver area.

The intercept factor is evaluated instantaneously; this means that as the Sun moves over the sky, the primary mirrors rotate to reflect the greatest amount of radiation to the receiver, so it can be concluded that the intercept factor is a function of the incidence angles. While the intercept factor can be maximized for any time of the day, a good alternative is maximizing it at solar noon (when the irradiation reaches its highest value), so the concentrated irradiance in the receiver is maximized. The position where the mirrors are aligned maximizing the intercept factor is called the reference position; under the hypothetical assumption that the field is placed at the equator.

To establish the geometric description of the Fresnel reflector, the determination of the inclination angle ψ_i, is required, which varies according to the position of the Sun, and is different for each mirror, as shown in Figure 6.

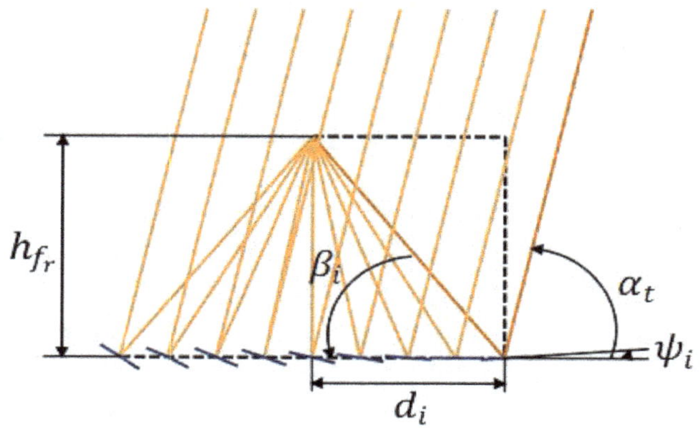

Figure 6. Angle of inclination of the *i*-th mirror and its determinants.

The inclination angle of each mirror ψ_i is determined by the angle β_i and the transverse solar altitude angle α_T; additionally, the β_i angle is determined by the distance of the respective mirror row from the central axis of the field, and by the distance of the real focus to the primary mirrors h_{f_r} through:

$$\psi_i = \frac{\alpha_T - \beta_i}{2} \tag{9}$$

$$\beta_i = \arctan\left(\frac{h_{f_r}}{d_i}\right) \tag{10}$$

From Equation (9), the inclination depends only on the temporal variation of the angle α_T, that is, in the apparent movement of the Sun, not in the position of the mirror. The different mirror rows have different angular positions but they move at the same angular velocity. Theoretically, this allows all mirrors to be connected using a simple four-bar mechanical coupling and driven with a single motor (rocker mechanism). However, the current designs of LFRs decide to move each mirror individually, because a more precise tracking is obtained, and it also allows a gradual defocusing if necessary [5]. Fresnel field configuration of FRESDEMO places the receiver of 0.6 m (a') at a distance of 8 m of the primary mirrors (h_{f_r}).

For the reference position, the intercept factor of the FRESDEMO field has a maximum value of 0.5753. Thus, the location of the primary mirrors can be defined from the last mirror. For the reference position (solar noon), the mirrors are located and tilted as described in Equation (9). Table 2 shows the location and tilt of the mirrors for the design point, and, due to the symmetry that exists in the system, only twelve mirrors are considered (the other ones have the same layout). In Figure 7 the first reference position is shown schematically, together with a simplified ray-tracing. In Figure 8 the results of the ray-tracing are shown, in Figure 8a it can be seen that for the area of the receiver for this reference position, not all the reflected rays will hit the receiver, so that the intercept factor will be less than one, while in Figure 8b the blockage that occurs between the last and second last mirror is appreciated.

Table 2. Position of the primary mirrors at solar noon ($\omega = 0$; $\Gamma = 0.5753$).

i-th Mirror	Longitudinal Location (d_i) [m]	Inclination on the Horizontal (ψ_i) [deg]
0	0	0
1	0.84	2.9
2	1.68	5.80
3	2.52	8.55
4	3.36	11.15
5	4.2	13.57
6	5.04	15.80
7	5.88	17.83
8	6.72	19.18
9	7.56	21.35
10	8.4	22.86

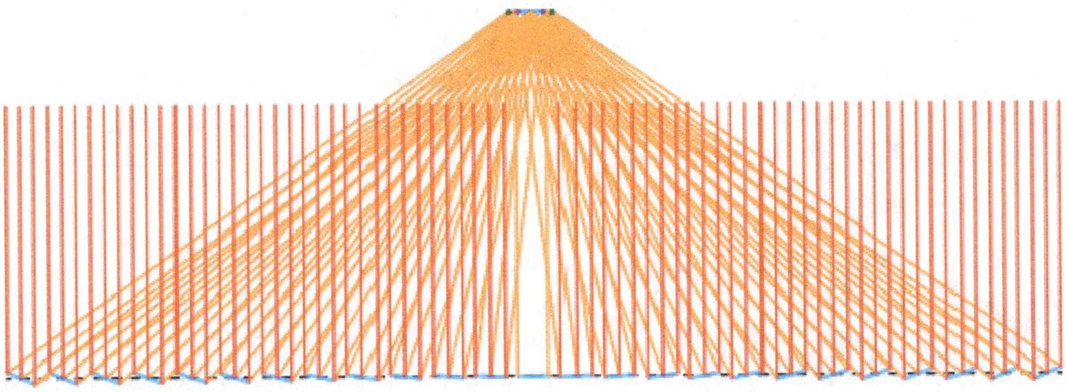

Figure 7. Arrangement of the Fresnel reflector mirrors.

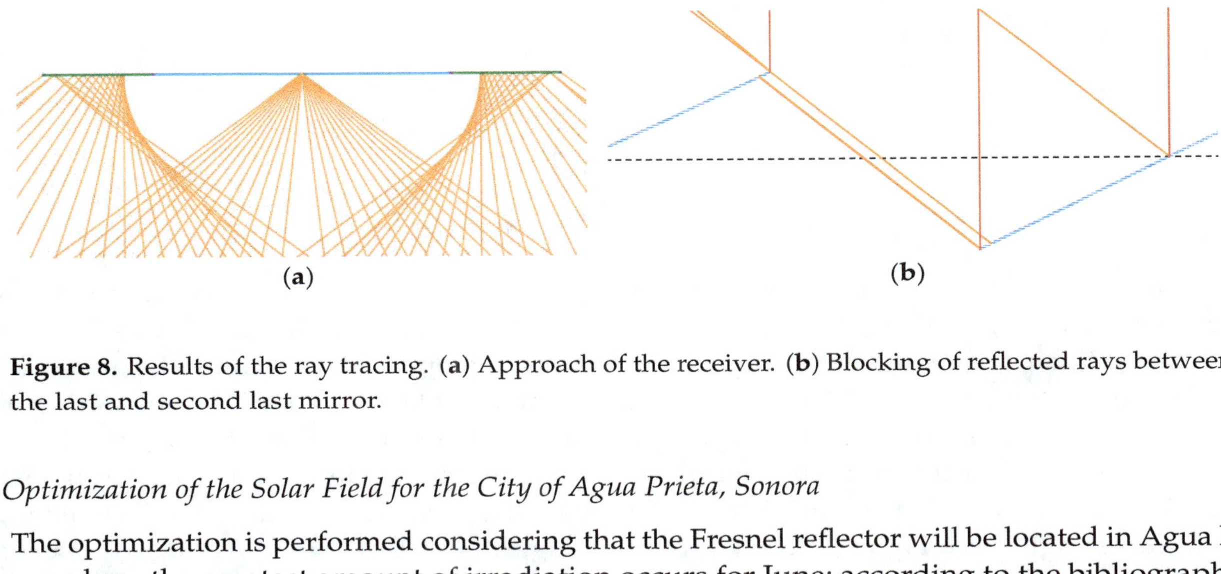

Figure 8. Results of the ray tracing. (**a**) Approach of the receiver. (**b**) Blocking of reflected rays between the last and second last mirror.

2.3. *Optimization of the Solar Field for the City of Agua Prieta, Sonora*

The optimization is performed considering that the Fresnel reflector will be located in Agua Prieta, Sonora, where the greatest amount of irradiation occurs for June; according to the bibliography, the 21st is the representative day of the month [14].

The city of Agua Prieta, Sonora, has a geographical location of 31.33 N;–109.55 E, so once the day has been defined (21 June, day 171 of the year), it is possible to determine the solar angles that will allow optimizing the dimensions of the receiver. Using the Equations (1)–(6), the solar angles reported in Table 3 are obtained.

Table 3. Solar angles for solar noon in Agua Prieta, Sonora.

Parameter	Value
Latitude	31.3°
Day	171°
Declination	23.4498°
Solar hour angle	0°
Zenith angle	7.8502°
Solar azimuthal angle	7.8502°
Incidence angle	7.85020°
Solar azimuthal angle	90°

3. Results

With the results of Table 3, the intercept factor for systems with variable height can be determined using the methodology described in the previous section, keeping fixed the layout and size of the primary mirrors. However, even the intercept factor is optimized as a function of height, the tilt of the mirrors must change in order to keep the absorber tube illuminated.

The plot in Figure 9 shows the variation of the intercept factor concerning the increase in receiver height. It is remarkable that after 20 m, the intercept factor does not increase greatly; however, working with receivers with such height would complicate the maintenance of the installation.

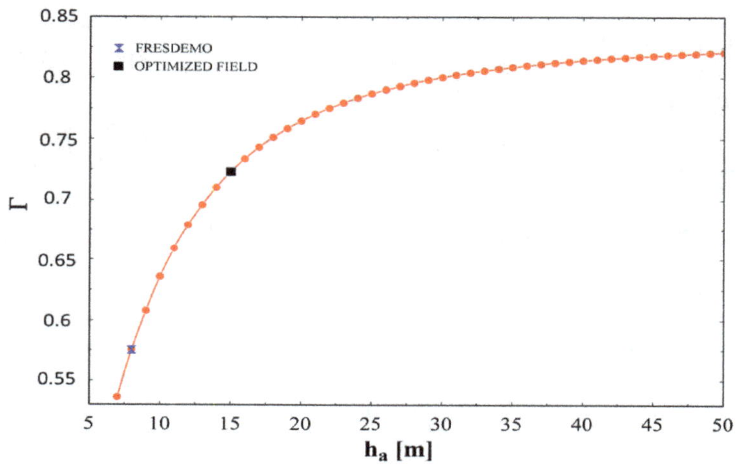

Figure 9. Intercept factor as a function of receiver height.

The criterion used to determine the height was simple. The point of the curve where the rate of change stops increasing to a large extent (which can be defined as an extreme point) was determined. Under this condition, the value is a little more than 15 m. As an optimization point, a height of 15 m is chosen, so that the previous graph determines that the intercept factor as 0.7231. The location of the mirrors remains by what is described in Table 2 since the width of the field remains constant.

Once the intercept factor has been determined, it is important to determine a geometric parameter called the incident angle modifier (IAM). The IAM for Fresnel reflectors is a function of both longitudinal and transverse angles (see Figure 3 to identify both angles), i.e., $\text{IAM} = \text{IAM}(\theta_l, \theta_t)$. It is usual to define the incidence angle modifier as the product of the longitudinal modifier K_l and the transverse modifier K_t so that it is possible to obtain both parameters independently [14,33].

Thus, the IAM represents the change in optical performance for angles of incidence other than perpendicular [14,34]:

$$\text{IAM} = \frac{\eta_o(\theta)}{\eta_o(\theta = 0^\circ)} \tag{11}$$

$$\text{IAM} = K_l K_t \tag{12}$$

From Equation (11), it is possible to plot the incidence angle modifiers for the longitudinal and transverse direction, as shown in Figure 10. The Equations (13) and (14) both model K_l and K_t as a function of the longitudinal (θ_l) and transverse (θ_t) incidence angle, plotted as a continuous curve for each case shown in Figure 10; where the coefficient of determination R^2 is 100% and 98.80%; and a statistical bias of 1.38×10^{-17} and 1.48×10^{-18} for the longitudinal and transversal directions respectively, ensuring a proper correlation between the data and the adjustment made.

$$K_l = 0.9990 - 0.0038\theta_l - 0,0002\theta_l^2 + 0.000001\theta_l^3 - 3.3597 \times 10^{-7}\theta_l^4 + 4.0205 \times 10^{-9}\theta_l^5 - 1.8542 \times 10^{-11}\theta_l^6 \quad (13)$$

$$K_t = 0.9871 + 0.0174\theta_t - 0.0025\theta_t^2 + 0.0002\theta_t^3 - 0.000003\theta_t^4 + 4.2733 \times 10^{-8}\theta_t^5 - 1.8163 \times 10^{-10}\theta_t^6 \quad (14)$$

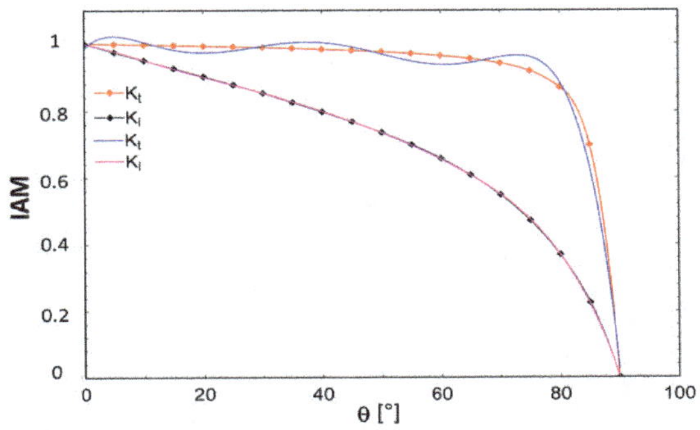

Figure 10. Incidence angle modifiers for longitudinal (K_l) and transverse (K_t) directions.

As described above, there are various alternatives in the design of the systems in terms of the number of absorber tubes and cavity systems for the receiver. The simplest design is based on the use of a single tube (with a diameter generally in the range of 7 cm to 18 cm). This tube is usually located inside a cavity provided with a secondary concentrator—usually a compound parabolic concentrator, CPC—and may have a glass cover on the bottom of the cavity [35–38].

Since it is considered a field similar to FRESDEMO's, a modified CPC will be located in the receiver to redirect all the radiation that does not hit the absorber directly, and also to increase the geometric concentration coefficient; so, it will be working with only one tube.

Regardless of the type of absorber that the CPC may have, the rays that enter the concentrator with a maximum semi-angle θ_{max} (extreme rays) must be reflected by the mirror so that they tangentially hit the absorber; while all rays entering at an angle less than the maximum semi-angle θ (i.e., within the angular acceptance window $2\theta_{max}$), are directed to the absorber after passing through the internal optics of the CPC (reflection in this case) [39].

The function of an unmodified CPC, raised by Winston [29,40], satisfy the parametric equations stated in Equation (15). The curve AB described in Equation (16) corresponds to an involute in polar coordinates, while the curve BC is an anticaustic, whose polar equation is described in Equation (17).

$$\begin{cases} x_1 = \rho_o \cos t - r \sin t \\ x_2 = \rho_o \sin t + r \cos t \end{cases} \quad (15)$$

$$AB : \rho_o = r\,t, \quad 0 \le t \le \frac{\pi}{2} + \theta_0 \quad (16)$$

$$BC : \rho_o = \frac{r\left[t + \theta_0 + \frac{\pi}{2} - \cos(t - \theta_0)\right]}{1 + \sin(t - \theta_0)}, \frac{\pi}{2} + \theta_0 \le t \le \frac{3\pi}{2} - \theta_0 \quad (17)$$

A modified CPC will not properly fulfill the condition of redirecting the rays tangentially towards the absorber, but the rays of light will hit it in a secant way. So, it will be necessary to propose a function that allows considering the separation of the absorber tot to the cusp of the involute.

To modify the functions proposed in Equations (16) and (17), consider Figure 11. To move the absorber a distance $g = r_2 - r_1$, the involute should be rotated by an angle β_i and start when $x_2 = 0$. Thus, the equation of an involute rotated by an angle β_i is defined in Equation (18).

$$\beta_i = \sqrt{\left(\frac{r_2}{r_1}\right)^2 - 1} - \arccos\left(\frac{r_1}{r_2}\right) \tag{18}$$

$$AB : \rho_o = r(t + \beta_i), \arccos\left(\frac{r_1}{r_2}\right) \leq t \leq \frac{\pi}{2} + \theta_0 \tag{19}$$

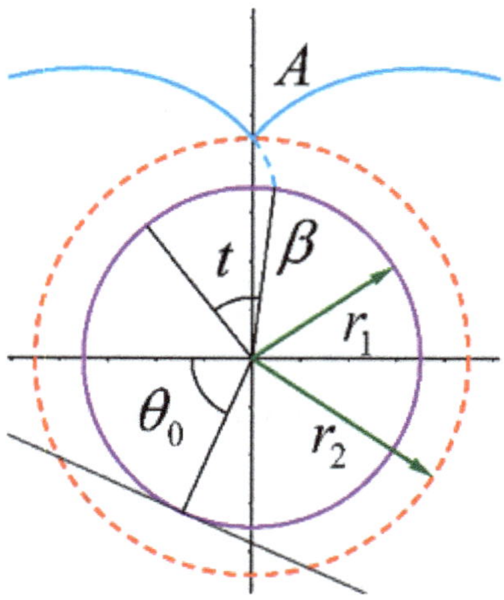

Figure 11. Displacement of the absorber of the involute cusp.

Similarly, the function of the anticaustic must be rotated $2\beta_i$, so that the modified function is described in Equation (20). Under this modification, the CPC profile is now modified with a gap between the absorber and the cusp of the involute, schematically shown in Figure 12.

$$BC : \rho_o = \frac{r_1\left[t + \theta_0 + \frac{\pi}{2} + 2\beta_i - \cos(t - \theta_0)\right]}{1 + \sin(t - \theta_0)}, \frac{\pi}{2} + \theta_0 \leq t \leq \frac{3\pi}{2} - 3\theta_0 \tag{20}$$

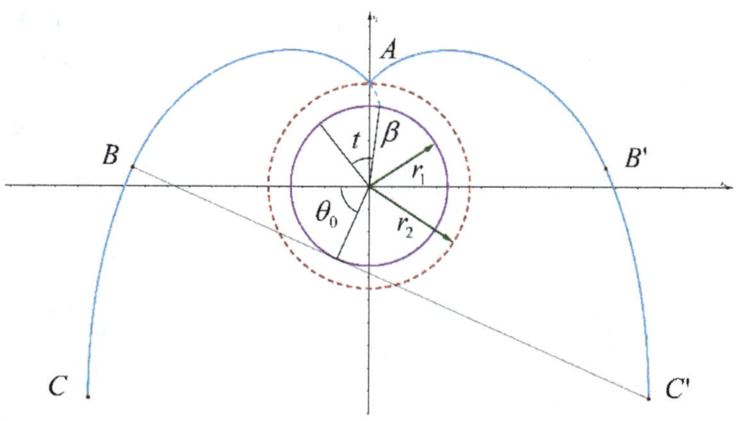

Figure 12. Modified geometry of the compound parabolic concentrator (CPC) with a separate absorber.

The FRESDEMO's field uses a 0.14 m external diameter and 0.125 m internal diameter tube; in which it is considered that the absorber will be displaced 0.02 m from the cusp. So, according to the definition of the geometric concentration, the relation of areas can be expressed as:

$$C_g = \frac{A}{A'} = \frac{2x_1\left(\frac{3\pi}{2} - \theta_0\right)}{2\pi r_1} \tag{21}$$

where

A = 0.5 m: Aperture length of the CPC, $A' = 2\pi(0.07\ \text{m})$: Perimeter of the tube absorber.

Note that the geometric concentration of the CPC is low ($C_g \cong 1.1368$), so that optimizing the height of the CPC is not convenient [41,42], and also because the acceptance angle for this concentration is higher than the minimum established by the Rincón's criterion (the maximum allowable acceptance half-angle for optimal truncation is $\pi/4$. For a more detailed description of the truncation criterion, as well as their implications, consult [39]) [43]; so the CPC will only redirect the rays that are not intercepted directly by the absorber, ensuring the etendue conservation and minimizing the optical losses [29,30,39].

The half acceptance angle θ_0 is limited according to the reflection given by the primary mirrors. Considering the design point, the angle θ_0 with which the rays enter the CPC must be equal to or less than the angle at which the extreme rays of the last mirror impinge on the receiver, according to the geometry of the field, it is determined that the half acceptance angle θ_0 should be equal to or greater than $33.90° \cong 0.59$ rad.

Solving Equation (21) for the acceptance half-angle, we have that $\theta_0 = 66.30° \cong 1.16$ rad; so, as discussed before, it is not advisable to truncate the CPC. Figure 13a shows the CPC that will be placed in the area of the receiver, and in Figure 13b the incident of the ray reflected by the last mirror is schematized. In Table 4, the description of the field is shown as a summary, following the methodology described.

As a result of the ray tracing (shown in Figure 13), all the extreme rays that enter the receiver will be redirected without optical leakage to the absorber, guaranteeing the conservation of the etendue [29,30], so that for the CPC, the intercept factor remains unitary.

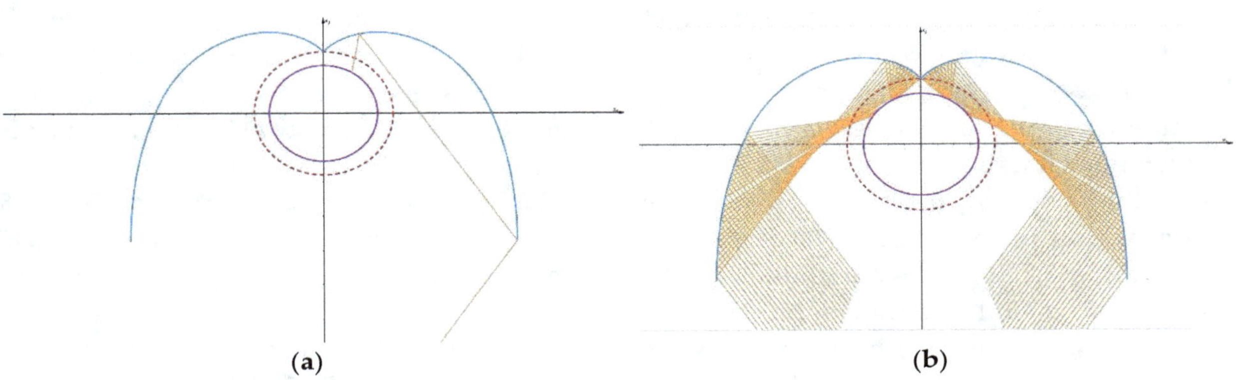

Figure 13. Ray-tracing in the CPC. (**a**) Incidence of the ray of light reflected at the end by the last mirror. (**b**) Tracing rays of the rays coming from the last mirror.

Table 4. Comparison of the description of the Fresnel reflectors and the optimized solar field for Agua Prieta, Sonora.

Parameter	FRESDEMO Original	FRESDEMO in Agua Prieta	FRESDEMO Optimized	Optimization for Agua Prieta
Number of primary mirrors	25	25	25	25
Solar field width [m]	21	21	21	21
Total length of the primary mirrors [m]	100	100	100	100
Width of the primary mirrors [m]	0.6	0.6	0.6	0.6
Filling factor	0.7143	0.7143	0.7143	0.7143
Receiver height [m]	8	8	15	15
Receiver width [m]	0.5	0.5	0.5	0.5
Outer diameter of the absorber tube [m]	0.14	0.14	0.14	0.14
Inner diameter of the absorber tube [m]	0.125	0.125	0.125	0.125
Semi-angle acceptance of the CPC [°]	56	56	66.30	66.30
Intercept factor	0.5753	0.4994	0.6285	0.7231
Receiver length [m]	100	100	100	100
Geometric concentration of the CPC	1.1368	1.1368	1.1368	1.1368
Geometric concentration of the entire field	34.1046	34.1046	34.1046	34.1046
Longitudinal incidence angle modifier	-	-	-	See Equation (13)
Transverse angle of incidence modifier	-	-	-	See Equation (14)

4. Comparison of the Optimized Field with FRESDEMO and A PTC

The comparison of the systems is carried out in two aspects. The first is carried out through a ray-tracing to know the distribution of radiation in the opening of the receiver between the FRESDEMO field, the optimized field and with a parabolic trough with the same geometric concentration and same receiver width. The second comparison between the FRESDEMO and the optimized field is done through a study of a thermal process of direct steam generation.

4.1. Flux Distribution

To determine the flux distribution in the aperture of the receiver, the SolTrace software, developed by the National Renewable Energy Laboratory, was used. This software works through solar geometry which is based on the ray-tracing methodology [44]. In SolTrace, each concentrator was simulated with a concentration ratio of 30, in other words, it just simulated the flux distribution in the aperture of the receiver, ignoring the CPC; and tracing 30, 000, 000 rays with a Gaussian Sun shape of 4.65 mrad.

Under this condition, the PTC has a half acceptance angle of 0.939 deg, an aperture of 15 m, a focus of 9.2674 m and a rim angle of 44.0609 deg is considered as the optimal concentrator. The FRESDEMO and the optimized LFR fields were simulated considering the distribution of the primary mirrors stated in Table 2, with 8 m and 15 m of height, respectively.

In Figure 14, the flux distribution is shown for each case, where the red colour stands for high flux zones and the blue for the low flux zone. In the PTC, the peak flux is of 76.2 kW/m^2 with an average flux of 28.1 kW/m^2; for the optimized LFR field, the peak flux is of 21.6 kW/m^2 with an average flux of 18.3 kW/m^2; and for the FRESDEMO the peak flux is of 19.7 kW/m^2 with an average flux of 18 kW/m^2.

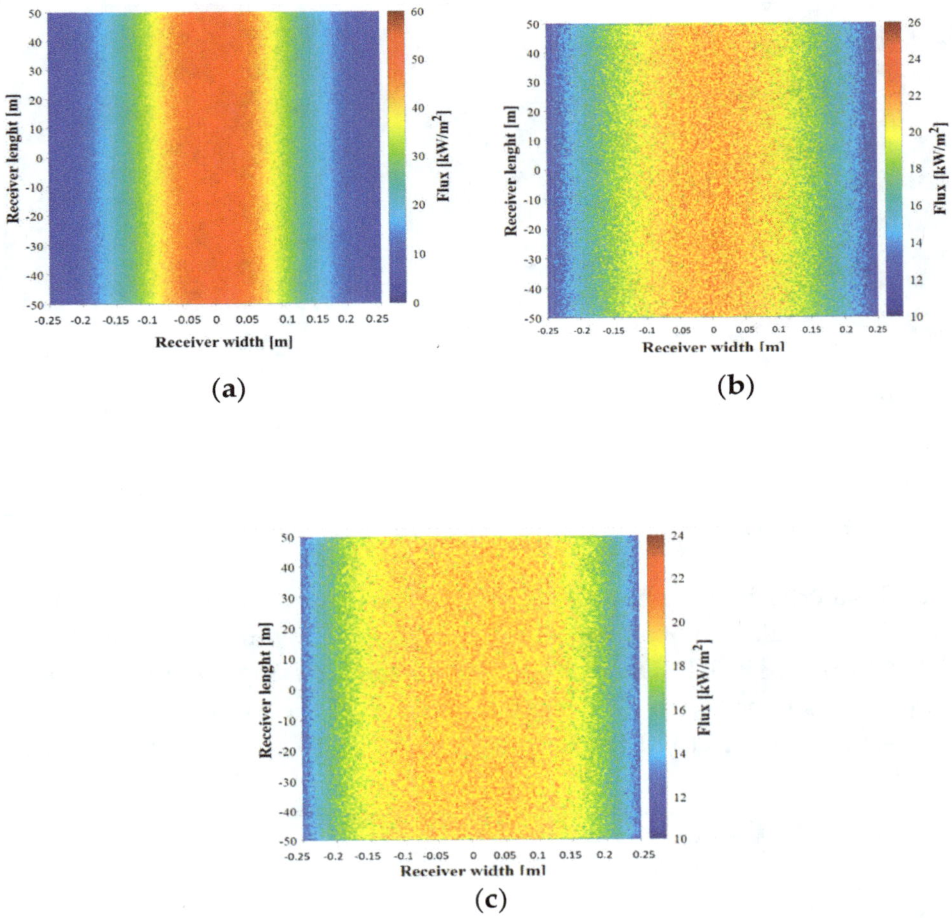

Figure 14. Flux distribution in the receiver aperture. (**a**) Parabolic trough collectors PTC. (**b**) Optimized Fresnel. (**c**) FRESDEMO.

As can be seen, the optimization leads to a flux distribution more similar to the one in the PTC, even though the FRESDEMO's distribution is more uniform. Although the FRESDEMO field homogenizes the radiative flux at the receiver's aperture, using a PTC or the optimized field implies that the concentrated radiation (flux with greater intensity) directly hits the receiver, and the rest will be distributed by the CPC.

The above is beneficial, considering that the power of the rays decreases according to the number of reflections in a CPC [29,40]. So, for a Fresnel reflector with a CPC as the second stage and one absorber tube, a more uniform distribution is not advisable.

4.2. Thermal Process

To carry out an adequate comparative analysis, a suitable framework and the nominal values of the parameters for the Fresnel loop have been previously defined, where the description of the primary mirrors is similar to those reported for FRESDEMO; however, it is of interest to determine the variation of the average optical concentration in the plane of the absorber (the geometric concentration is decreased by a factor of π against the average optical concentration) with respect to the incident angle modifiers. The values of the average optical concentration as a function of the IAM are shown in Figure 15.

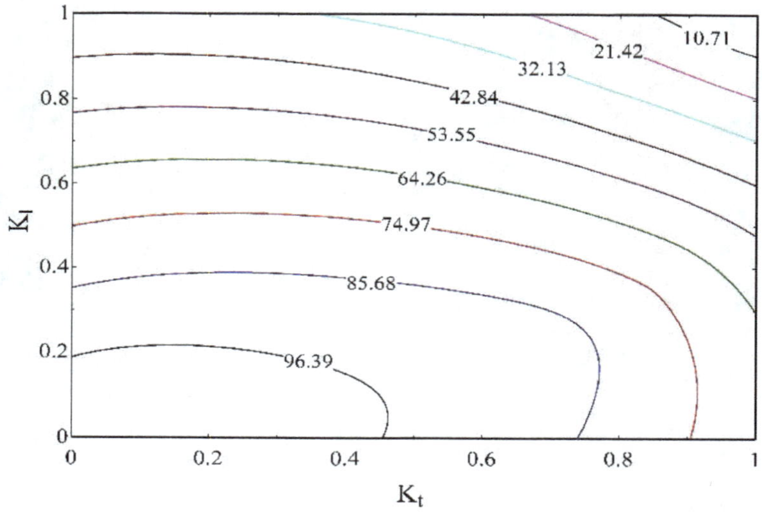

Figure 15. Average optical concentration in the receiver plane.

To ensure that the increase in the intercept factor of the optimized field will reduce the final length of the loop, consider the circulation of a mass flow of 4.25 kg/s of water entering 518 K, and that should leave at 673 K for both loops (FRESDEMO and optimized). Using the methodology described by Montes Pita et al. [12], it is possible to determine the length of the loop. In Figure 16, the temperature of the water is plotted for the length of the loop for FRESDEMO and the optimized field.

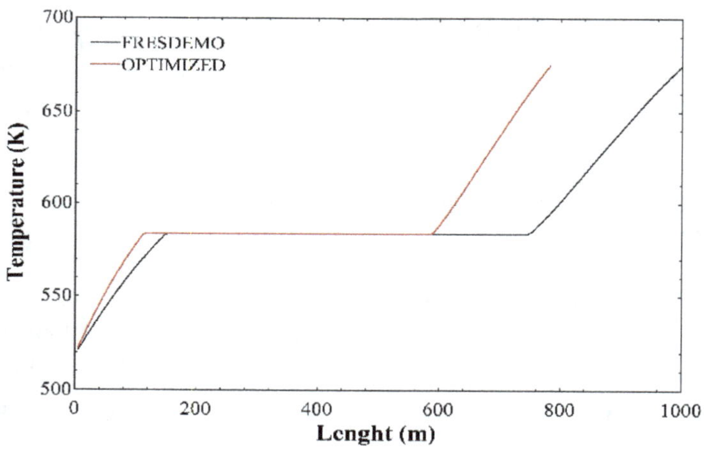

Figure 16. Water/steam temperature along the Fresnel loop in once-through operation mode.

After the simulation to determine the water temperature increase along the loop, it is observed that the FRESDEMO field requires 1000 m to perform the described thermal process, as described by Montes Pita et al. [12], while the optimized field requires only 782.4 m; which results in a decrease in length of 21.76%.

5. Conclusions

A methodology for the opto-geometric optimization of a LFR is described in the most generalized way, so that it can be used interchangeably to optimize any Fresnel field for the desired location, only affecting the solar angles that are involved in the system, which are described in Section 2.1.

As a result of having increased the height of the receiver, the dimensions of the CPC have been modified in such a way that the angular window is increased by 10°, which provides beneficial results because there is a larger window through which the rays can enter and be redirected to the CPC.

The main advantage of increasing the intercept factor will lead to a significant reduction in the total length of the loops of the field, since the spillage of the concentrated light is reduced, and more energy will hit the receiver achieving a greater heat flux per square meter.

Thus, using the methodology described applied to the location of Agua Prieta, Sonora, the Fresnel reflector has been optimized by increasing the intercept factor from 0.5753 to 0.7231, which implies an increase of just over 25%, only by increasing the height of the receiver, maintaining the width of the field as well as the dimensions of the mirrors and the receiver, as detailed in Table 4.

Finally, when comparing the performance of both loops of Fresnel fields for the same thermal process (an increase of 155 K in the water), it is verified that the total length of the field decreases from 1000 m to 782.4 m, which implies a reduction of 21.76% of the length of the loop, keeping the same filling factor, width of the primary mirrors, and absorber tube diameter.

Author Contributions: E.G.-M. conceived the paper and drafted it, developed the methodology and performed the ray-tracing; M.D.D.G. has done the project administration, supervision, and review. All authors have read and agreed to the published version of the manuscript

Nomenclature

Symbols

A	area [m^2]
a'	receiver width [m]
C_g	geometric concentration
d_i	longitudinal location of the *i*-th mirror [m]
g	gap between the absorber and the CPC cusp [m]
h_{fr}	height from the receiver to the primary mirrors [m]
K	angle modifier
L	length of primary mirrors [m]
FF	filling factor
N	number of mirrors
R	radius [m]
t	angle parameter [rad]
W	width [m]
x_1	abscissa [m]
x_2	ordinate [m]

Greek Symbols

α	solar altitude angle
β_s	angle of rotation of the CPC curves
β_s	slope
Γ	intercept factor
γ	azimuth angle
δ	declination
η_o	optical efficiency
θ	angle of incidence
θ_0	half acceptance angle
θ_{max}	half angular acceptance window
θ_z	zenith angle
ρ_o	ray from the pole
ψ	inclination angle of the primary mirrors
ω	hour angle

Subscripts

i	generic counter
il	illuminated
l	longitudinal / spillage left

m	primary mirrors/mirrors
r	spillage right
sf	solar field
T/t	transversal

Acronyms

CPC	Compound Parabolic Collector
DSG	Direct Steam Generation
HTF	Heat Transfer Fluid
IAM	Incident Angle Modifier
LFR	Linear Fresnel Reflector
PTC	Parabolic Trough Collector

References

1. Norton, B. *Harnessing Solar Heat*, 1st ed.; Springer: New York, NY, USA, 2014; Volume 18, ISBN 978-94-007-7274-8.
2. Breeze, P. *Solar Power Generation*, 1st ed.; Elsevier: Amsterdam, The Netherlands, 2014; ISBN 9780128040041.
3. Blanco, M.J.; Miller, S. 1—Introduction to concentrating solar thermal (CST) technologies. In *Advances in Concentrating Solar Thermal Research and Technology*; Woodhead Publishing: Amsterdam, The Netherlands, 2017; pp. 3–25. ISBN 9780081005163.
4. Zhu, G.; Wendelin, T.; Wagner, M.J.; Kutscher, C. History, current state, and future of linear Fresnel concentrating solar collectors. *Sol. Energy* **2014**, *103*, 639–652. [CrossRef]
5. Mills, D.R. 6—Linear Fresnel reflector (LFR) technology. In *Concentrating Solar Power Technology*; Elsevier: Amsterdam, The Netherlands, 2012; pp. 153–196. ISBN 9781845697693.
6. Abbas, R.; Martínez-Val, J.M. Analytic optical design of linear Fresnel collectors with variable widths and shifts of mirrors. *Renew. Energy* **2015**, *75*, 81–92. [CrossRef]
7. Boito, P.; Grena, R. Optimization of the geometry of Fresnel linear collectors. *Sol. Energy* **2016**, *135*, 479–486. [CrossRef]
8. Boito, P.; Grena, R. Optimal focal length of primary mirrors in Fresnel linear collectors. *Sol. Energy* **2017**, *155*, 1313–1318. [CrossRef]
9. Qiu, Y.; Li, M.-J.; Wang, K.; Liu, Z.-B.; Xue, X.-D. Aiming strategy optimization for uniform flux distribution in the receiver of a linear Fresnel solar reflector using a multi-objective genetic algorithm. *Appl. Energy* **2017**, *205*, 1394–1407. [CrossRef]
10. Cui, W.; Li, H.; Li, L.; Liao, Q. Thermal stress analysis of DSG solar absorber tube in stratified two-phase regime. *AIP Conf. Proc.* **2013**, *1558*, 2217–2220.
11. Zarza Moya, E. 5—Innovative working fluids for parabolic trough collectors. In *Advances in Concentrating Solar Thermal Research and Technology*; Blanco, M., Ramirez Santiagosa, L., Eds.; Woodhead Publishing: Amsterdam, The Netherlands, 2017; pp. 75–106. ISBN 9780081005163.
12. Montes, M.J.; Barbero, R.; Abbas, R.; Rovira, A. Performance model and thermal comparison of different alternatives for the Fresnel single-tube receiver. *Appl. Therm. Eng.* **2016**, *104*, 162–175. [CrossRef]
13. Montes Pita, M.J. *Análisis Y Propuestas De Sistemas Solares De Alta Exergía Que Emplean Agua Como Fluido Calorífero*; Universidad Politécnica de Madrid: Madrid, Spain, 2008.
14. Duffie, J.A.; Beckman, W.A. *Solar Engineering of Thermal Processes*, 4th ed.; John Wiley & Sons, Inc.: Hoboken, NJ, USA, 2013; ISBN 1118418123.
15. Wu, C. *Thermodynamics and Heat Powered Cycles. A Cognitive Engineering Approach*, 1st ed.; Nova Science Publishers, Inc.: Hauppauge, NY, USA, 2007; ISBN 9781606926260.
16. Goswami, Y.D. *Principles of Solar Engineering*, 3rd ed.; CRC Press: Boca Raton, FL, USA, 2015; ISBN 1466563796.
17. Johnson, D.C. Red Rock Energy. Available online: http://www.redrok.com (accessed on 22 March 2017).
18. Hakkarainen, E.; Kannari, L. Dynamic Modelling of Concentrated Solar Field for Thermal Energy Storage Integration. In Proceedings of the 9th International Renewable Energy Storage Conference (IRES 2015), Düsseldorf, Germany, 9–11 March 2015.
19. Abbas, R.; Montes, M.J.; Piera, M.; Martínez-Val, J.M. Solar radiation concentration features in Linear Fresnel

Reflector arrays. *Energy Convers. Manag.* **2012**, *54*, 133–144. [CrossRef]
20. Sun, J.; Liu, Q.; Hong, H. Numerical study of parabolic-trough direct steam generation loop in recirculation mode: Characteristics, performance and general operation strategy. *Energy Convers. Manag.* **2015**, *96*, 287–302. [CrossRef]
21. Häberle, A. Linear Fresnel Collectors. In *Encyclopedia of Sustainability Science and Technology*; Meyers, R.A., Ed.; Springer: New York, NY, USA, 2012; pp. 6031–6037. ISBN 978-1-4419-0851-3.
22. Nixon, J.D.; Davies, P.A. Cost-exergy optimisation of linear Fresnel reflectors. *Sol. Energy* **2012**, *86*, 147–156. [CrossRef]
23. Abbas, R.; Muñoz-Antón, J.; Valdés, M.; Martínez-Val, J.M. High concentration linear Fresnel reflectors. *Energy Convers. Manag.* **2013**, *72*, 60–68. [CrossRef]
24. Bernhard, R.; Laabs, H.-J.; de LaLaing, J.; Eickhoff, M. Linear Fresnel Collector Demonstration on the PSA, Part I—Design, Construction and Quality Control. In Proceedings of the 14th International SolarPACES Symposium, Las Vegas, NV, USA, 4–7 March 2014; pp. 1–10.
25. Klychev, S.I.; Abdurakhmanov, A.A.; Kuchkarov, A.A. Optical-geometric parameters of a linear Fresnel mirror with flat facets. *Appl. Sol. Energy* **2014**, *50*, 168–170. [CrossRef]
26. Singh, P.L.; Ganesan, S.; Yàdav, G.C. Technical note: Performance study of a linear Fresnel concentrating solar device. *Renew. Energy* **1999**, *18*, 409–416. [CrossRef]
27. Montes, M.J.; Rubbia, C.; Abbas, R.; Martínez-Val, J.M. A comparative analysis of configurations of linear fresnel collectors for concentrating solar power. *Energy* **2014**, *73*, 192–203. [CrossRef]
28. Barale, G.; Heimsath, A.; Nitz, P.; Toro, A. Optical design of a linear fresnel collector for sicily. In Proceedings of the 16th SolarPACES International Symposium Perpignan, Perpignan, France, 21–24 September 2010.
29. Winston, R.; Miñano, J.; Benítez, P. *Nonimaging Optics*, 1st ed.; Elsevier: Amsterdam, The Netherlands, 2005; ISBN 0127597514.
30. Chaves, J. *Introduction to Nonimaging Optics*, 2nd ed.; CRC Press: New York, NY, USA, 2017; ISBN 9781138747906.
31. Zhu, G. Development of an analytical optical method for linear Fresnel collectors. *Sol. Energy* **2013**, *94*, 240–252. [CrossRef]
32. Pettit, R.B.; Vittitoe, C.N.; Biggs, F. Simplified Calculational Procedure for Determining the Amount of Intercepted Sunlight in an Imaging Solar Concentrator. *J. Sol. Energy Eng.* **1983**, *105*, 101–107. [CrossRef]
33. Yang, F.; Itskhokine, D.; Benmarraze, S.; Benmarraze, M.; Hofer, A.; Lecat, F.; Ferrière, A. Acceptance Testing Procedure for Linear Fresnel Reflector Solar Systems in Utility-Scale Solar Thermal Power Plants. *Energy Procedia* **2015**, *69*, 1479–1487. [CrossRef]
34. Karathanasis, S. *Linear Fresnel Reflector Systems for Solar Radiation Concentration: Theoretical Analysis, Mathematical Formulation and Parameters' Computation Using MATLAB*; Springer Nature: Berlin/Heidelberg, Germany, 2019; ISBN 9783030052799.
35. Morin, G.; Platzer, W.; Eck, M.; Uhlig, R.; Haberle, A.; Berger, M. Road map towards the demonstration of a linear Fresnel collector using single tube receiver. In Proceedings of the International Symposium on Concentrated Solar Power and Chemical Energy Technologies, Seville, Spain, 20–23 June 2006.
36. Morin, G.; Platzer, W.; Strelow, M.; Leithner, R. Techno-economic system simulation and optimization of solar thermal power plants. In Proceedings of the 14th International SolarPACES Symposium on Solar Thermal Concentrating Technologies, Las Vegas, NV, USA, 4–7 March 2008.
37. Platzer, W.; Heimsath, A.; Hülsey, H. Parameter identification technique for the determination of optical efficiency of concentrating collectors. In Proceedings of the 15th International SolarPACES Symposium on Solar Thermal Concentrating Technologies, Berlin, Germany, 15–18 September 2009.
38. Abbas, R. *Towards Cost Reduction in Concentrating Solar Power: Innovative Design for an Efficient Fresnel Based Solar Field*; Universidad Politécnica de Madrid: Madrid, Spain, 2015.
39. González-Mora, E.; Rincón-Mejía, E.A. *Conceptos Básicos De La Concentración Solar. Notas De Clase*; FI-UAEM: Toluca, México, 2018; p. 130.
40. Rabl, A. *Active Solar Collectors and Their Applications*; Oxford University Press: Oxford, UK, 1985; ISBN 9780195365214.
41. Rabl, A.; Goodman, N.B.; Winston, R. Practical design considerations for CPC solar collectors. *Sol. Energy* **1979**, *22*, 373–381. [CrossRef]

42. Carvalho, M.J.; Collares-Pereira, M.; Gordon, J.M.; Rabl, A. Truncation of CPC solar collectors and its effect on energy collection. *Sol. Energy* **1985**, *35*, 393–399. [CrossRef]
43. Rincón Mejía, E.A.; Durán García, M.D.; Lentz Herrera, Á. New Solar Air Heater Based on Non-Imaging Optics for High-Temperature Applications. In Proceedings of the ASME 2009 3rd International Conference on Energy Sustainability, San Francisco, CA, USA, 19–23 July 2009; Volume 2, pp. 839–844.
44. Wendelin, T. SolTRACE: A New Optical Modeling Tool for Concentrating Solar Optics. In Proceedings of the International Solar Energy Conference, Kohala Coast, HI, USA, 15–18 March 2003.

10

Establishment, Validation and Application of a Comprehensive Thermal Hydraulic Model for a Parabolic Trough Solar Field

Linrui Ma [1,2,3,4], Zhifeng Wang [1,2,4,*], Dongqiang Lei [1,2,4] and Li Xu [1,2,4]

1 Key Laboratory of Solar Thermal Energy and Photovoltaic System, Chinese Academy of Sciences, Beijing 100190, China
2 Institute of Electrical Engineering, Chinese Academy of Sciences, Beijing 100190, China
3 Institute of Electrical Engineering, University of Chinese Academy of Sciences, Beijing 100049, China
4 Beijing Engineering Research Center of Solar Thermal Power, Beijing 100190, China
* Correspondence: zhifeng@vip.sina.com.

Abstract: To better understand the thermal hydraulic characteristics of the parabolic trough solar field (PTSF), a comprehensive thermal hydraulic model (CTHM) based on a pilot plant is developed in this paper. All of the main components and thermal and hydraulic transients are considered in the CTHM, and the input parameters of the model are no longer dependent on the total flow rate. In this paper, we solve the CTHM by a novel numerical approach based on graph theory and the Newton-Raphson method, and then examine it by two tests conducted based on a pilot plant. Comparing the flow rate, temperature, and pressure drop results show good agreement and further validate the availability and accuracy of the CTHM under hydraulic and thermal disturbance. Besides, two applications of the CTHM are implemented for presenting its potential function. In the first application, two cases are simulated to reveal how the thermal effects influence the PTSF behavior, and in the second application, the CHTF is used for the study of control strategies under uniform and nonuniform solar irradiance. The results verify the feasibility of controlling the PTSF outlet temperature through the header and loop valves.

Keywords: parabolic trough solar field; thermal and hydraulic model; validation; control

---

1. Introduction

Through long-time operation experience accumulation and continuous research, the parabolic trough solar thermal power plant (PTSTPP) is considered as the ripest in technique and most commercialized concentrating solar power technology [1]. The parabolic trough solar field (PTSF), where solar energy is collected and converted to the thermal energy, is the component most worth exploring in a PTSTPP [2]. In a representative PTSF, the heat transfer fluid (HTF) is distributed by the cold header and harvested by the hot header. Several parallel loops connect the cold and hot headers; each loop consists of a series of parabolic trough collectors (PTC), and the cold HTF is heated to a high temperature as it flows through the PTC loop. Finally, the hot HTF is pumped into the power block to produce the steam. As the solar irradiance and thermal requirement change, three parameters are used to adjust the flow rate of the HTF to meet the outlet temperature demand: the opening of the control valve in headers, the opening of the control valve in PTC loops, and the pump frequency [3].

The majority of the published studies that research the PTSF place emphasis on a thermal perspective. Forristall et al. [4] developed a detailed thermal steady-state model of the PTC, which is validated by the test results of a SEGS LS-2 solar collector [5]. Stuetzle et al. [6] built a thermal dynamic model of the PTC to study the control algorithm of a plant, and their measured data validated this model. Camacho et al. [7] developed a simplified thermal dynamic model of the PTC, in which both the heat loss and heat convective coefficient are fitted as polynomials in the temperature. Padilla et al. [8] and Hachicha et al. [9] improved the thermal model in two ways: implementing a more comprehensive radiative analysis and considering that the solar flux distribution around the receiver is nonuniform, respectively. Yılmaz et al. [10] developed a comprehensive thermo-mathematical model, whose detailed optical and thermal analysis in this model allow the model to simulate more accurate results for the optical loss, heat loss, and thermal efficiency. Behar et al. [11] improved on the accuracy of thermal performance prediction of the PTC in another way, i.e., developing a novel model that can consider the flow rate variation by adapting an optimization procedure of the HTF outlet temperature.

In effect, in a large-scaled PTSF, because the flow rate in the individual PTC loop is unknown and determined by the layout, pump head, and flow resistance, a detailed thermal model of the PTC is insufficient to understand the performance of the PTSF comprehensively. Therefore, in some studies, the PTSF is regarded as a complex piping network, whose thermal hydraulic characteristics are researched. The symmetrical and uniform flow distribution of the PTSF was assumed to simulate the PTSTPP in the System Advisor Model (SAM), which was developed by the National Renewable Energy Laboratory (NREL) and Sandia National Laboratory [12]. Abutayeh et al. [13] spelled out the presence of unbalanced flow distribution in the PTSF and proposed a method to amend it. Giostri [3] studied the behavior of the PTSF under a variety of thermal transients based on a thermal hydraulic model, which contained a thermal dynamic part and a hydraulic steady pipe network part. A virtual solar field (VSF), which was developed by the German Aerospace Centre (DLR) [14], applied a thermal hydraulic model similar to [3] to simulate the behavior of the whole PTSF, and the operating data from Andasol-3 validated the model. Ma et al. [15] revealed the relationship between the flow distribution and flow resistance of the PTSF, and then verified the method proposed in [13] based on a thermal hydraulic dynamic model (THDM) and tested data obtained from a pilot plant.

Two aspects need to be improved in the existing thermal hydraulic models of the PTSF, i.e., the lack of pump module and the unaccounted hydraulic transients. The former leads to the dependence on the inlet header flow rate in simulation and validation, and the latter obscures the PTSF specific behavior under the thermal and hydraulic transients. There are two reasons for the difficulty of solving these problems by directly employing the transient hydraulic model in the water distribution system (WDS), of which the numerical method is relatively ripe [16–18]. One is that in the WDS, the water is often assumed to be incompressible, while in the PTSF, the expansion and contraction of the HTF are non-negligible due to the significant temperature difference. The other is that the flow resistance is considered as a constant or merely a function of the flow velocity in the WDS, while it is a more complex function under the influence of temperature variation in the PTSF.

Besides, developing an effective control strategy of the PTSF is also dependent on the integrality of the thermal hydraulic model. Ref. [19] pointed out that the header control valves (HCVs) are used to control the flow rate and outlet temperature of the subfield in a real operated PTSTPP. [13] indicated that the automatic balancing valves can be used for PTSF control when the spatial variation of solar irradiance is an issue. Refs. [14] and [20] indicated that the loop control valves (LCVs) have the potential to improve the PTSF performance. However, the lack of a refined model increases the difficulty of calculating the valves opening and simulating the PTSF characteristics under varied solar irradiance.

To overcome the above problems, as well as improve the operational performance and controllability of the PTSTPP, a comprehensive thermal hydraulic model (CTHM) of the PTSF based on a pilot plant in Beijing is presented in this paper. The CTHM is improved based on the THDM developed in [15]. The optical and thermal analysis of the PTC, the flow characteristics of pipe, pipe fitting, pump, and the valves are all considered in the CTHM. A novel numerical method, which is based on graph theory and the Newton–Raphson method, is developed to solve the CTHM. Moreover, a cold test and a hot test are implemented to validate the CTHM under thermal and hydraulic disturbance.

Compared with the THDM and other existing models, the CTHM has three main advantages. Firstly, adding the pump module establishes the relationship among the pump capacity, flow resistance, and the temperature. This improvement further makes the required input parameters get rid of the dependence on the total flow rate. Secondly, the CTHM contains both the hydraulic and thermal inertia, and this leads to more accurate results in the transient simulation. Finally, a more comprehensive model can widen and deepen the research on the PTSF; the CTHM can be applied to simulate the behavior of the PTSF under various disturbances and study the control strategy of the PTSF under uniform or nonuniform solar irradiance.

2. Description of the PTSF of a 1 MW Pilot Plant

The Badaling parabolic trough solar power pilot plant, which is the first operational PTC solar thermal plant on the MW scale in China, is situated in Yanqing at a latitude of 40.5°N and a longitude of 115.94°E. A view of the pilot plant is shown in Figure 1.

Figure 1. View of the Badaling parabolic trough solar power pilot plant.

As illustrated in Figure 2, two east–west layout loops (Loop 1 and Loop 3) with 4150 m long solar collector assemblies (SCAs) and one south–north layout loop (Loop 2) with 6100 m long SCAs together form the principal part of the PTSF. Besides, an extra pipe is placed in Loop 2 for a study of flow balance [15]. An HCV and three LCVs are used to manipulate the total flow rate and flow distribution. A kind of synthetic oil named Therminol VP-1, of which the physical properties can be fitted into some polynomials in temperature based on the tested data [13], is used as the HTF in the pilot plant, while the more detailed information about the physical properties of the HTF is summarized in Appendix A.

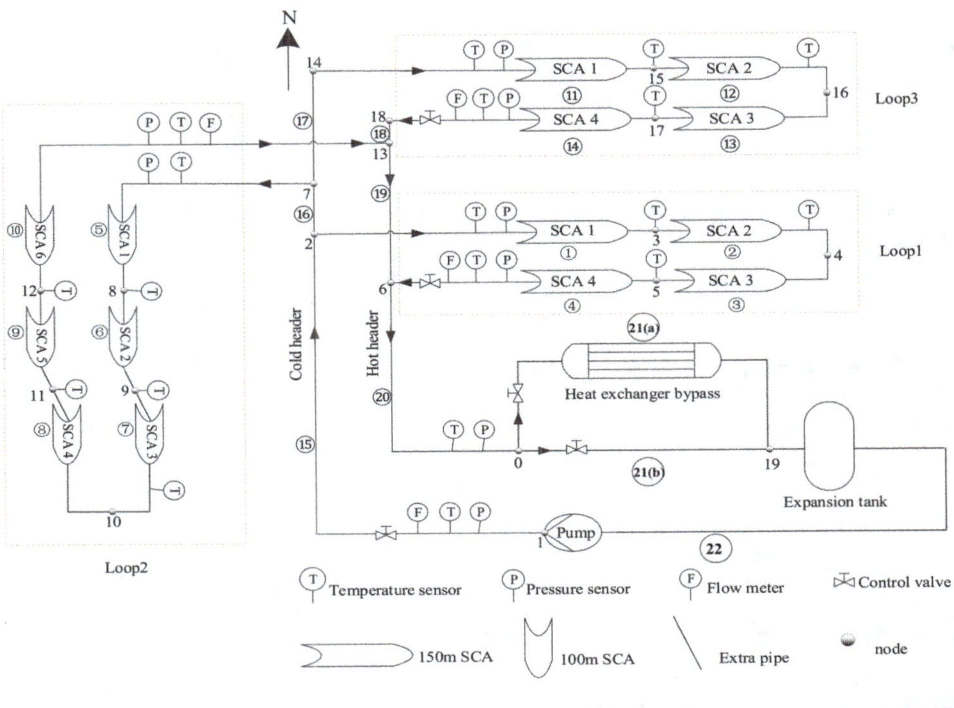

Figure 2. Flow diagram, nodes, and pipe section label of the parabolic trough solar field (PTSF) in the pilot plant.

The present study involves two kinds of pipes in the PTSF, i.e., the absorbed pipe covered with an evacuated glass envelope, and the insulated pipe (header). Hereafter, the symbols, which are concerned with the pipe, will represent the former with the subscript abs, and represent the latter with the subscript ins. Moreover, the symbols without any subscripts mean that they are universal for both types. Specific characteristics of the PTC, header, pump, and valve are described in more detail below.

2.1. The PTC and Header

In the relevant published studies [1–12], the thermal model of the PTC developed in [7] is applied in this paper. The effectiveness of this model has been validated by the tested data regardless of the simplified thermal loss of the PTC [7,15].

As shown in Figure 3a, the energy equation of the control volume with an axial length of Δx can be given by [7]:

$$\rho_{abs}A_{abs}c_{abs}\frac{\partial \overline{T}_{abs}}{\partial t} = P_{use} - H_{l,abs}(\overline{T}_{abs}) - D_{abs}\pi h(\overline{T}_{abs} - \overline{T}_f) \tag{1}$$

$$\Delta x A_f \overline{\rho}_f \overline{c}_f \frac{\partial \overline{T}_f}{\partial t} + c_{f,out}\rho_{out}Q_{f,out}T_{f,out} - c_{f,in}\rho_{f,in}Q_{f,in}T_{f,in} = \Delta x D_{abs}\pi h(\overline{T}_{abs} - \overline{T}_f) \tag{2}$$

where the subscript f represents the HTF, Q, ρ, T, and c represent the volume flow rate, density, temperature, and specific heat capacity, and D and A are the diameter and sectional area. $H_{l,abs}$ represents the heat loss of the absorber per meter, which is caused by the temperature difference between the absorber tube and the exterior glass envelope, and can be simplified as a polynomial in T_{abs}.

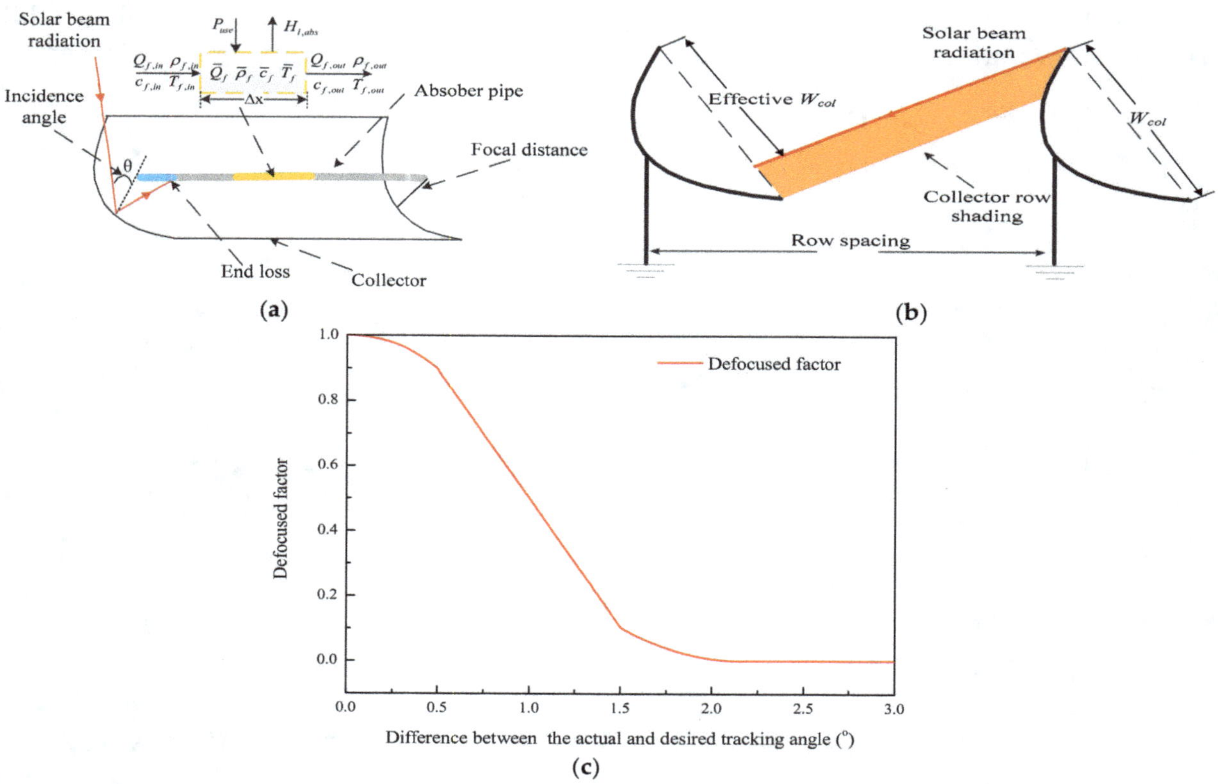

Figure 3. Thermal and optical analysis of the parabolic trough collectors (PTC) for (**a**) control volume and end loss; (**b**) row shading; and (**c**) trend of the defocused factor.

P_{use} in Equation (1) is the solar thermal power supplied to the absorbed pipe per meter, which is given by [21]:

$$\begin{aligned} P_{use} &= W_{col} I \kappa(\theta) \eta_o f_{shd} f_{end} \\ \eta_o &= \beta r_{cl} \tau \alpha \gamma \end{aligned} \tag{3}$$

where W_{col} represents the collector width, I represents the direct normal irradiance (DNI), and κ is the incidence angle (θ) modifier, which is used to account for all geometric and optical losses because of an incident angle of more than 0°. η_o is the optical efficiency of the PTC, which is the product of the clean mirror reflectivity (r_{cl}), the transmittance of the glass envelope (τ), the absorbance of the absorber pipe (α), and the intercept factor (γ). β is the defocused factor, which is used for considering the loss of efficiency caused by the difference between the actual and desired tracking angle. As shown in Figure 3c, the trend of β can be simulated by the Solartrace tool [3]. f_{end} and f_{shd} are two factors due to end loss and row shadowing; the schematic diagram of f_{end} and f_{shd} is shown in Figure 3a,b respectively. Expressions of these two factors can be given by [3]:

$$f_{end} = 1 - \frac{l_{fl} \tan(\theta)}{l_{SCA}} \tag{4}$$

$$f_{shd} = \min\left[\max\left[0; \frac{l_{rs}}{W_{col}} \frac{\cos(\omega)}{\cos(\theta)}\right]; 1\right] \tag{5}$$

where ω represents the zenith angle, and l_{SCA}, l_{rs}, and l_{fd} are the length of the SCA, row spacing, and focal distance, respectively.

h in equations (1) and (2) is the convection heat transfer coefficient between the absorber inner surface and the HTF, which can be given by [4]:

$$h_{conv} = Nu_f \frac{k_f}{D_{abs}} \tag{6}$$

$$Nu_{abs} = \frac{f_{abs}/8(Re_{abs} - 1000)Pr_f}{1 + 12.7\sqrt{f_{abs}}/8({Pr_f}^{2/3} - 1)} \left(\frac{Pr_f}{Pr_{abs}}\right) \tag{7}$$

$$f_{abs} = (1.82 \log_{10}(Re_{abs}) - 1.64)^2 \tag{8}$$

where Re, Nu, and Pr are the Reynolds number, Nusselt number, and Prandtl number, respectively. k is the thermal conductivity. In the normal operation of the pilot plant, the variation range of the Re and Pr are about $1 \times 10^4 < Re < 9 \times 10^5$ and $4.5 < Pr < 45$; these two parameters are within the range of application of Equation (8), which is valid for $2300 < Re < 5 \times 10^6$ and $0.5 < Pr < 2000$ [4].

More than the HCE absorber pipes, thermal losses exist in the insulated pipes (headers). In this work, the mineral wool is used as the thermal insulation material, and the thermal losses of the insulated pipes can be calculated under an assumption of neglecting the thermal resistances of the exterior air film and pipe walls [21], which is given by:

$$\begin{aligned} & A_{ins} l \bar{\rho}_f \bar{c}_f \frac{\partial \overline{T}_f}{\partial t} + c_{f,out} \rho_{f,out} Q_{f,out} T_{f,out} - c_{f,in} \rho_{f,in} Q_{f,in} T_{f,in} \\ &= 2\pi l \frac{\bar{k}_{ins}}{\ln((D_{ins}+2\delta)/D_{ins})} (T_a - \overline{T}_f) \end{aligned} \tag{9}$$

where $\bar{k}_{ins}$ is the thermal conductivity of the pipe insulation at the average temperature of $\overline{T_f}$ and T_{ins}, and T_a and δ are the insulation thickness and ambient temperature, respectively. The value of $\bar{k}_{ins}$ can be fitted into a polynomial at the average temperature [21].

The thermal and optical parameters, the geometry of the absorber and the header are listed in Table 1, where formulas of the incidence angle modifer is given by [21], and formulas of the heat loss r is supplied by the manufacturer. Other header parameters that are not included in the table, such as the length from pump to Loop 1, the length from Loop 3 to expansion, the expression of $\overline{k}_{ins}$, and so on, can be measured from the design drawing.

Table 1. Main parameters of the collector, absorber, and header.

Pipe	Item	Value/Expression
Collector	Length (m)	100/150
	Width (m)	5.776
	Focal length (m)	1.71
	Clean mirror reflectivity	0.96
	Ideal intercept factor	0.8268
	Incidence angle modifier	$\cos(\theta) - 5.25027 \times 10^{-4}\theta - 2.859621 \times 10^{-5}\theta^2$ [21]
	Row spacing (m)	20
Absorber	Inner diameter of the absorber tube (m)	0.064
	Outer diameter of the absorber tube (m)	0.07
	Absorber absorptivity	0.95
	Transmittance of the glass envelope	0.95
	Heat loss (W/m)	$5.2569 \times 10^{-6} {T_{abs}}^3 - 0.0012 {T_{abs}}^2 + 0.1343 T_{abs} + 13.7484$ [15]
Header	Inner diameter of the header (m)	0.125
	Total extra pipe length (m)	190
	Inner diameter of the extra pipe (m)	0.069
	Insulation thickness (m)	0.15
	Thermal conductivity of the mineral wool (W m^{-1} K^{-1})	$6.02 \times 10^{-7} {\overline{T}_f}^3 - 2.146 \times 10^{-5} {\overline{T}_f}^2 + 0.194 \overline{T}_f + 26.5$ [21]

Equations (1)–(9) together form the thermal part of the CTHM, this part is applied to simulate the outlet temperature and update the properties of the HTF.

2.2. Pump and Valve

As shown in Figure 4a, a canned centrifugal pump is chosen to circulate the HTF in the pilot plant, the capacity, head, and efficiency at design condition are 99 m^3/h, 200 m, and 45%, respectively. In normal hydraulic calculations, the pump head is usually fitted into a polynomial in the capacity according to the pump's performance curve, and this method has been applied to the PTSF by defining the polynomial with a series of universal coefficients [22,23]. In this study, as shown in Figure 4b, the polynomial of the pump head can be given in a more specific way based on the pump's performance curve supplied by the manufacturer.

(a) (b)

Figure 4. (**a**) View of pump; (**b**) pump's performance curve.

A typical control valve in the PTSF is shown in Figure 5a, the electric actuator manipulates the opening of the valve. In general, as shown in Figure 5b, the control valve has three inherent flow characteristics [24], i.e., linear, quick open, and equal percentage. In the pilot plant, both the LCVs and HCVs are chosen to have the flow characteristics with equal percentage.

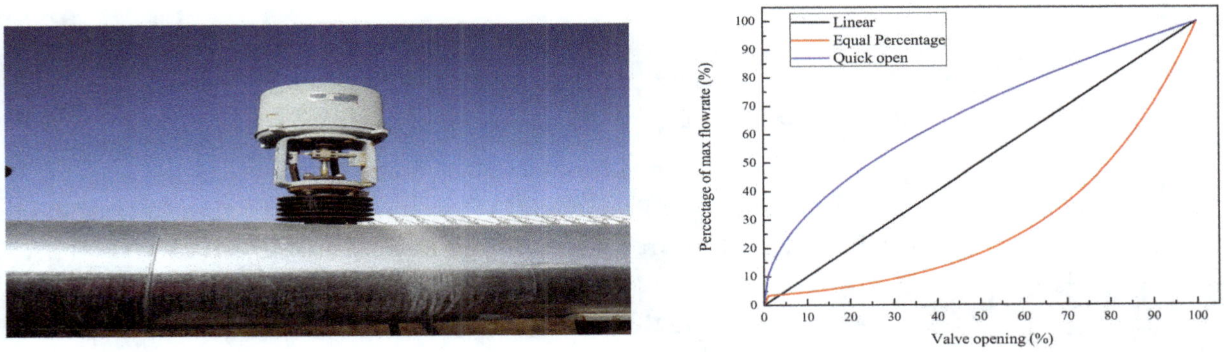

(a) (b)

Figure 5. (**a**) View of the control valve; (**b**) three inherent flow characteristics of the control valve.

2.3. *Measurement and Uncertainty*

In the PTSF of the pilot plant, the DNI, flow rate, temperature, and pressure of the HTF in the header and three loops, the opening of the valve, the actual tracking angle, and the ambient temperature are the main parameters that need to be measured; the information of measurements and their uncertainty are summarized in Table 2.

Table 2. Information about the measurements. DNI: direct normal irradiance.

Parameters	DNI	Temperature	Pressure	Flow Rate	Valve Opening	Tracking Angle
Measurements	Pyrheliometer model CH1 by Kipp and Zonen	Pt 100 class B	Pressure transmitters	Yewflo flowmeter	Electrical signals	Inclinometer
Uncertainty/Accuracy	0.5%	<2.3 °C	0.2%	1%	5%	<0.001 °C

3. Hydraulic Model Description

3.1. *Hydraulic Model in the Single Pipe*

The schematic of the control volume of the hydraulic model is shown in Figure 6, in which p and υ represent the pressure and kinematic viscosity of the HTF, respectively. The continuity equation of the control volume with a length of Δx can be given by [25]:

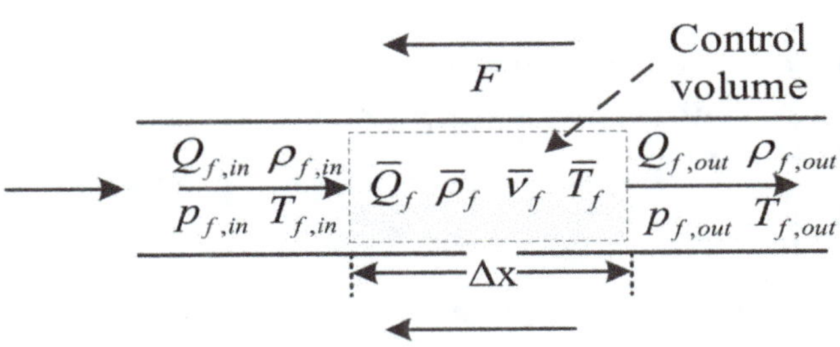

Figure 6. Control volume of the hydraulic model.

$$A\Delta x \frac{\partial \overline{\rho}_f}{\partial t} = \rho_{f,in} Q_{f,in} - \rho_{f,out} Q_{f,out} \tag{10}$$

where the left-hand side is caused by the thermal transients.

The momentum equation of the control volume in Figure 6 can be given by [25]:

$$\begin{aligned} &\left(\tfrac{\Delta x}{A}\right)\overline{Q}_f \frac{\partial \overline{\rho}_f}{\partial t} + \left(\tfrac{\Delta x}{A}\right)\overline{\rho}_f \frac{\partial \overline{Q}_f}{\partial t} + \frac{\rho_{f,out} Q_{f,out}^2}{A^2} - \frac{\rho_{f,in} Q_{f,in}^2}{A^2} \\ &= p_{f,in} - p_{f,out} - F + \overline{\rho}_f g H \end{aligned} \tag{11}$$

The first two terms on the left-hand side of Equation (11) are caused by the thermal transients and hydraulic transients, respectively. F, which represents the flow loss, can be given by:

$$\begin{aligned} F &= S\overline{Q}_f^2 \\ S &= S_p + \sum_{i=1}^{N_{fit}} S_{fit,i} + S_{va} \end{aligned} \tag{12}$$

where S represents the total flow resistance in the control volume, which can be caused by the pipe wall (S_p), pipe fitting (S_{fit}), and valve (S_{va}). S_p can be calculated using the Darcy–Weisbach equation [26]:

$$\begin{aligned} S_p &= \tfrac{1}{2}\overline{\rho}_f f\left(\tfrac{l}{DA^2}\right)\tfrac{1}{2} \\ f &= 8\left(\left(\tfrac{8}{Re}\right)^{12} + \tfrac{1}{B^{1.5}}\right)^{\frac{1}{12}} \\ B &= \left(2.457\ln\left(\frac{1}{(7/Re)^{0.9} + 0.27(\varepsilon/D)}\right)\right)^{16} + \left(\tfrac{37530}{Re}\right)^{16} \end{aligned} \tag{13}$$

where f is the inner friction factor of the pipe, which is an empirical formula for the Reynolds number (Re) and absolute roughness of the pipe (ε) [27].

Analogously, S_{fit} can be given by [26]:

$$S_{fit} = \frac{1}{2} K_{fit} \frac{\overline{\rho}_f}{A^2} \tag{14}$$

where K_{fit} is the pressure drop coefficient of the fitting. The elbow and ball joint are the two main fittings in the PTSF; their K_{fit} values are 0.75 and 0.6 according to the manufacturer.

S_{va} is a function of the opening of the control valve (op), for the LCVs and HCVs, of which the equal percentage of flow characteristics, S_{va}, can be calculated by [24]:

$$S_{va} = 100\overline{\rho}_f \left(\frac{3600}{R^{(op-1)} G_{\max}}\right)^2 \tag{15}$$

where G_{max} represents the max flow coefficient of the valve, and R represents the rangeability. The G_{max} values of LCVs and HCVs are 68 m^3/h and 158 m^3/h, respectively. The R of both the LCVs and HCVs are 30.

3.2. Matrix Representation of Pipe Network

Graph theory has been one of the most powerful tools for pipe network analysis due to its convenience and effectiveness [18]; therefore, this method is used to calculate the hydraulic part of the PTSF in the present study.

Before building the specific equation, a key concept in the graph theory, which is called the incidence matrix, must be introduced first. If a pipe network contains n nodes and b pipe sections, accordingly, the incidence matrix $\boldsymbol{M}$ is an $n \times b$ order matrix that can be expressed as [18]:

$$\boldsymbol{M}^{(ij)} = \begin{cases} +1 \text{ if pipe } j \text{ starts at node } i \\ 0 \text{ if pipe } j \text{ is not connected to node } i \text{ (18)} \\ -1 \text{ if pipe } j \text{ ends at node } i \end{cases} \tag{16}$$

In the PTSF, the nodes can be divided into two types; one is that of which the pressure is a reference pressure (type 0 with the subscript 0), the other is that the node pressure is unknown (type 1 with the subscript 1). Hence, there are four types of incidence matrix in the PTSF: $\boldsymbol{M}_{0,in}$, $\boldsymbol{M}_{0,out}$, $\boldsymbol{M}_{1,in}$, and $\boldsymbol{M}_{1,\text{out}}$. Moreover, all of the equations will be discretized with the implicit difference method in time. For the pilot plant, the detailed information about point and pipe labels is shown in Figure 2; choosing pipe 21 (a) or 21 (b) depends on the outlet temperature and thermal requirement.

According to the graph theory, the continuity equation of a type 1 node can be expressed by:

$$\boldsymbol{M}_{1,in}\boldsymbol{Q}^t_{f,in} + \boldsymbol{M}_{1,out}\boldsymbol{Q}^t_{f,out} = \boldsymbol{q}_1{}^t \tag{17}$$

where $\boldsymbol{Q}_{f,in}$ and $\boldsymbol{Q}_{f,out}$ are the inlet and outlet flowrate vectors, respectively. $\boldsymbol{q}_1$ is the type 1 nodal out flow rate vector.

According to Equation (10), the continuity equation of the pipe section can be expressed as:

$$\mathrm{D}\left(\boldsymbol{\rho}^t_{f,out}\right)\boldsymbol{Q}^t_{f,out} - \mathrm{D}\left(\boldsymbol{\rho}^t_{f,in}\right)\boldsymbol{Q}^t_{f,in} = \mathrm{D}(\boldsymbol{V})\frac{\left(\bar{\boldsymbol{\rho}}_f - \bar{\boldsymbol{\rho}}_f^{\,t}\right)}{\Delta t} \tag{18}$$

where D() in this paper means diagonalizing the vector in the bracket. $\boldsymbol{\rho}_{f,in}$, $\boldsymbol{\rho}_{f,out}$, and $\bar{\boldsymbol{\rho}}_f$ are the inlet, outlet, and average density vector of the HTF, respectively. $\boldsymbol{V}$ is the volume vector, of which the ith element is:

$$\boldsymbol{V}^{(i)} = \boldsymbol{l}^{(i)}\boldsymbol{A}^{(i)} \tag{19}$$

where $\boldsymbol{l}$ and $\boldsymbol{A}$ are the length vector and sectional area vector of the pipe.

Analogously, based on Equation (11), the momentum equation of the pipe network can be expressed as:

$$\begin{aligned} &\mathrm{D}(\boldsymbol{A}_c)\bar{\boldsymbol{Q}}_f^{\,t} - \mathrm{D}(\boldsymbol{B}_c)\bar{\boldsymbol{Q}}_f + \frac{\mathrm{D}\left(\boldsymbol{\rho}^t_{f,out}\right)\mathrm{D}\left(\left|\boldsymbol{Q}^t_{f,out}\right|\right)\boldsymbol{Q}^t_{f,out}}{\mathrm{D}^2(\boldsymbol{A})} - \frac{\mathrm{D}\left(\boldsymbol{\rho}^t_{f,in}\right)\mathrm{D}\left(\left|\boldsymbol{Q}^t_{f,in}\right|\right)\boldsymbol{Q}^t_{f,in}}{\mathrm{D}^2(\boldsymbol{A})} \\ &= (\boldsymbol{M}_{1,in} + \boldsymbol{M}_{1,out})^T\boldsymbol{p}^t_1 + (\boldsymbol{M}_{0,in} + \boldsymbol{M}_{0,out})^T\boldsymbol{p}^t_0 - \mathrm{D}(\boldsymbol{S})\mathrm{D}\left(\left|\bar{\boldsymbol{Q}}_f^{\,t}\right|\right)\bar{\boldsymbol{Q}}_f^{\,t} + g\mathrm{D}\left(\bar{\boldsymbol{\rho}}_f\right)\boldsymbol{H} \\ &\boldsymbol{A}_c^{(i)} = \frac{\boldsymbol{l}^{(i)}}{\boldsymbol{A}^{(i)}}\frac{2\bar{\rho}_f{}^{t,(i)} - \bar{\rho}_f{}^{(i)}}{\Delta t} \\ &\boldsymbol{B}_c^{(i)} = \frac{\boldsymbol{l}^{(i)}}{\boldsymbol{A}^{(i)}}\frac{\bar{\rho}_f{}^{t,(i)}}{\Delta t} \end{aligned} \tag{20}$$

where $\bar{\boldsymbol{Q}}_f$ is the average flow rate vector, $\boldsymbol{p}_1$ and $\boldsymbol{p}_0$ are the pressure vector of node type 1 and type 0, $\boldsymbol{S}$ is the flow resistance vector, and $\boldsymbol{H}$ is the pump head vector.

3.3. Numerical Method

Equations (10)–(20) form the hydraulic part of the CTHM, which can be used to simulate the flow distribution, pressure drop, and pump head of the PTSF. The hydraulic part together with the thermal part mentioned in Section 2 constitute an entire CTHM. Two difficulties must be solved first before simulating with the CTHM. One is how to decouple the thermal part and the hydraulic part, and the other is how to solve the complex matrix equations in the hydraulic part conveniently and quickly. An iterative approach referred to in [15] is applied to solve the first question. Firstly, we calculate the temperature used the thermal model at an initial flow rate distribution; then, we plug the updated temperature to the hydraulic part to update the physical properties of the HTF and calculate the hydraulic parameters. This process will continue until the calculated flow distribution converges to a given tolerance. For the second difficulty, a method based on matrix theory and the Newton–Raphson method is applied to solve the hydraulic matrix equations; this method is derived from [18] and improved in this work through adding the expansion and extraction of the HTF and the varying flow resistance due to the temperature change. The detailed procedure of this method is presented in Appendix B as an innovation of this study.

According to the above content, a detailed flow chart of the CTHM is shown in Figure 7. Besides containing both the hydraulic and thermal transients, another advantage of the CTHM is that the simulation no longer depends on the inlet flow rate of the header due to the newly added pump module.

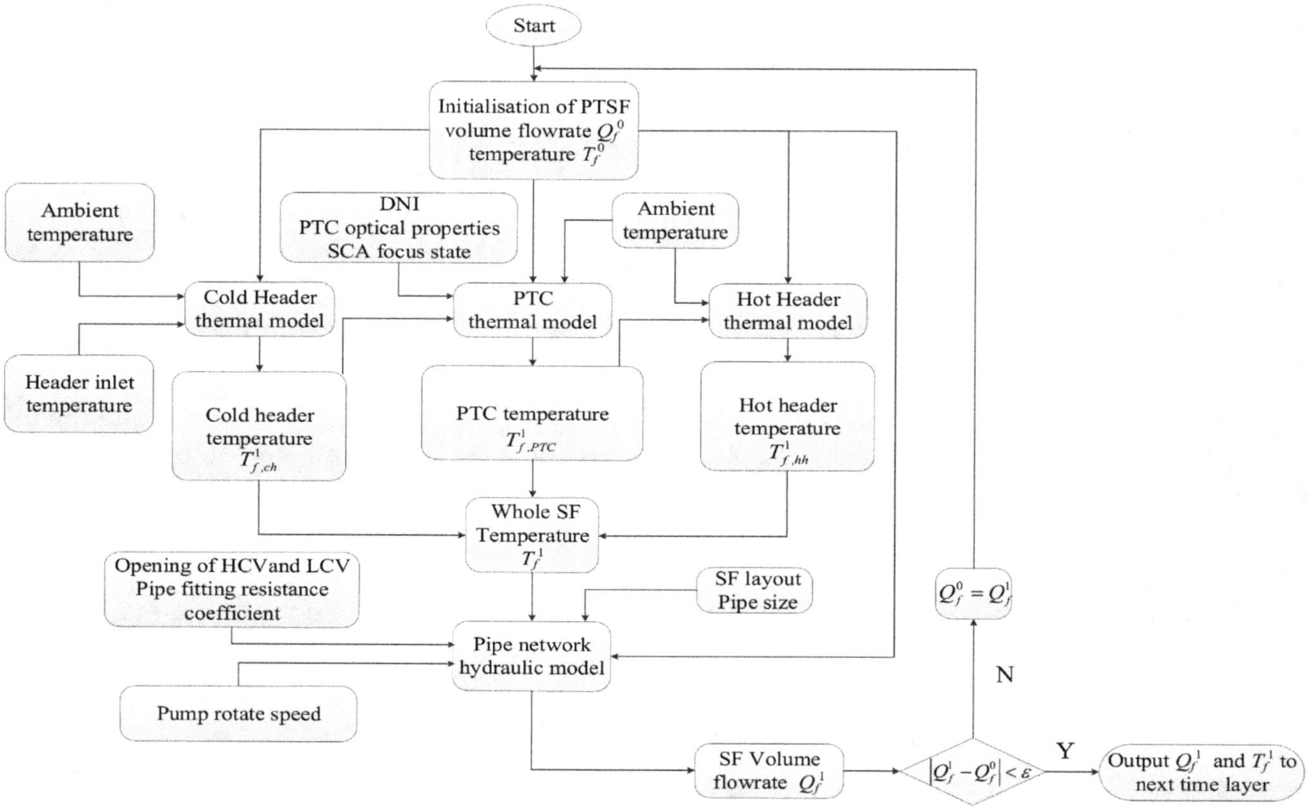

Figure 7. Flow chart of the comprehensive thermal hydraulic model (CTHM).

4. Validation Results and Discussion

In present work, a cold test and a hot test based on the pilot plant were conducted to validate the CTHM. The cold test was used to validate the hydraulic characteristics of the CTHM; the transient hydraulic effects are caused by periodically varying the opening of the LCVs. Both the hydraulic and thermal characteristics of the CTHM are validated through the hot test; hence, in addition to the periodically varying opening of the LCVs, the loop collectors are operating at concentrating status to cause the thermal transient effects. More detailed information about these two tests is summarized in Table 3. The simulation is implemented in MATLAB®. For the CTHM, Figure 7 shows the main input parameters: the header inlet temperature, DNI, ambient temperature, and the actual tracking angle of the SCA; the main output parameters that need to be validated are the total flow rate, flow rate distribution in each loop, pressure drop, and the outlet temperature of each loop. The uncertainty or accuracy of the measurements of all these parameters can be found in Table 2 for future discussion.

Table 3. The detailed summarized information of the two tests.

	Cold Test	Hot Test
Validation target	Transient hydraulic characteristics	Both the transient hydraulic and thermal characteristics
Collectors status	All loops defocused totally	Loop 1 and Loop 3 concentrated totally Loop 2 defocused totally
Loop control valves status	Loop 1 and Loop 3 varying the opening of the LCV periodically Loop 2 keeping the opening of the LCV fully open	
Pump operating condition	Full load	
Parameters to be validated	The inlet flow rate of header and all three loops Pressure drop between cold and hot header	The inlet flow rate of header and all three loops The outlet temperature of all three loopsPressure drop between cold and hot header

4.1. Cold Test

The cold test was conducted from 9:41:39 to 11:30:00 on 22 August 2018; as mentioned in Table 3 and shown in Figure 8, the opening of LCVs in Loop 1 and Loop 3 varied between 50–100% with a change rate of about 5% every 12 s, which is considered a moderate frequency (a too high frequency may cause a water hammer event, and a too-low frequency will obscure the transient hydraulic effect). Due to the lack of a valve automatic control device, the manual operations reduced a bit of the accuracy and simultaneity of the periodic variation.

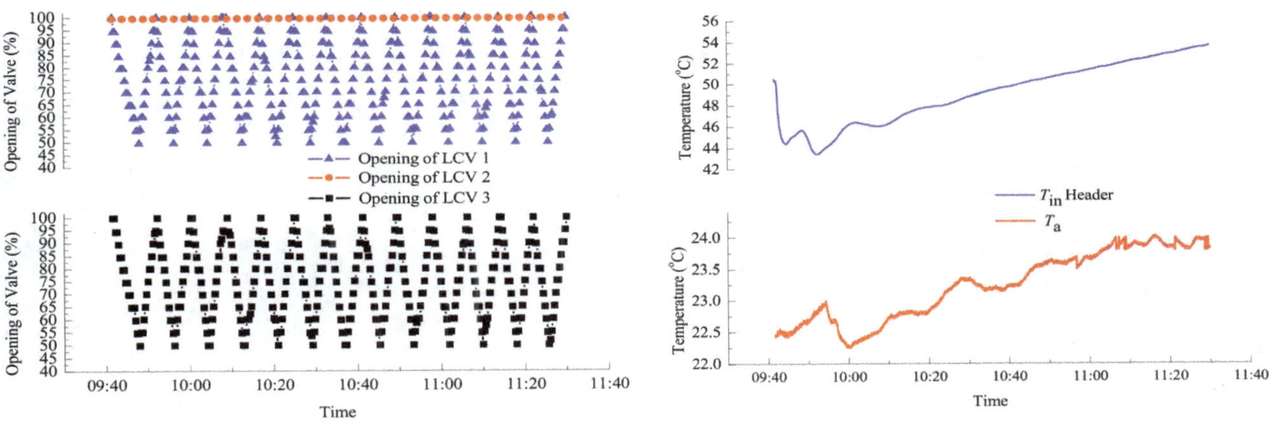

(a) (b)

Figure 8. Measured initial data in the cod test: (**a**) opening of three LCVs; (**b**) ambient and header inlet temperature.

Figure 9a shows that the calculated header inlet flow rate is slightly greater than the measured rate with a small root mean square error (RMSE). The error is likely caused by an underestimation of the pressure drop, because some local pressure loss such as the contraction or expansion of the pipe is not considered in the model. This view can be verified by Figure 9b, where the calculated pressure drop of the header is lower than the measured value. Figure 9 also presents the inlet flow rate, and the pressure drop varies at the same frequency, which is similar to the opening of the LCVs. This is because the periodical change of the valve opening varies the total flow resistance of the PTSF and further impacts the pump head and capacity.

Besides the total flow rate in the header, simulating the flow distribution in loops is another essential function for the CTHM that needs to be calculated and validated. Good agreement between the calculated results and measured data is shown in Figure 10a–c; the RMSEs for the outlet flowrate of loops 1 to 3 are 1.75 m^3/h, 0.86 m^3/h, and 0.97 m^3/h, respectively. The uncertainty of the electric control valve actuators with a value of 5% (shown in Table 2) is considered as the primary error sources. Besides, Figure 10 shows that the discrepancy is changing with the flow rate; this is due to Equation (16), which can cause different errors in the different flow rates and Re values [27]. The relative errors shown in Figure 10d can make this clear: when the valve opening varied, the relative errors present almost simultaneous fluctuations. Another noticeable feature presented in Figure 10b is that the outlet flow rate of Loop 2 has a lower value and opposite trend compared to the other two loops. The lower value is caused by an extra loop in Loop 2 that can increase the flow resistance of Loop 2 and diminish the flow rate. The opposite trend occurs because when the valve openings in Loop 1 and Loop 3 decreased, the relative value of the flow resistance of Loop 2 changed from the larger one to the smaller one; although the header inlet flow rate decreased during this process, Loop 2 could be distributed a greater flow rate due to its smaller flow resistance.

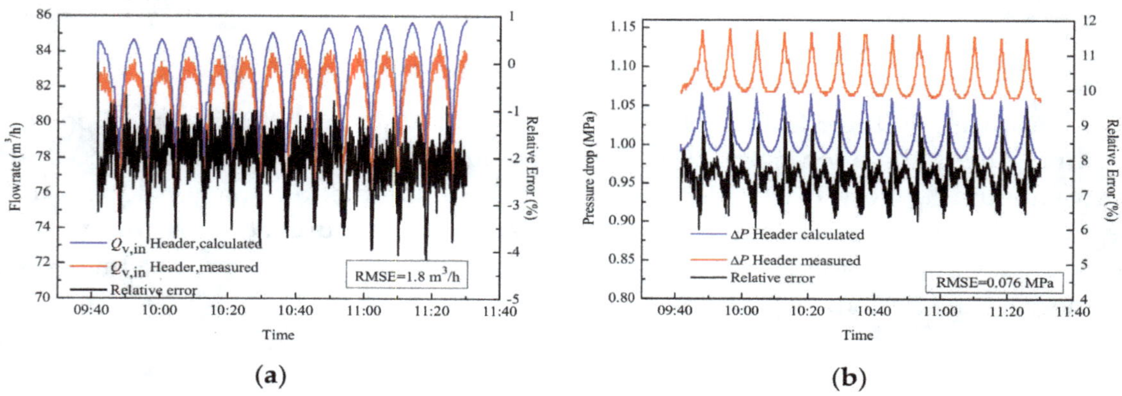

Figure 9. Compared results between calculated and measured data in a cold test for the (**a**) header inlet flow rate, and (**b**) header pressure drop.

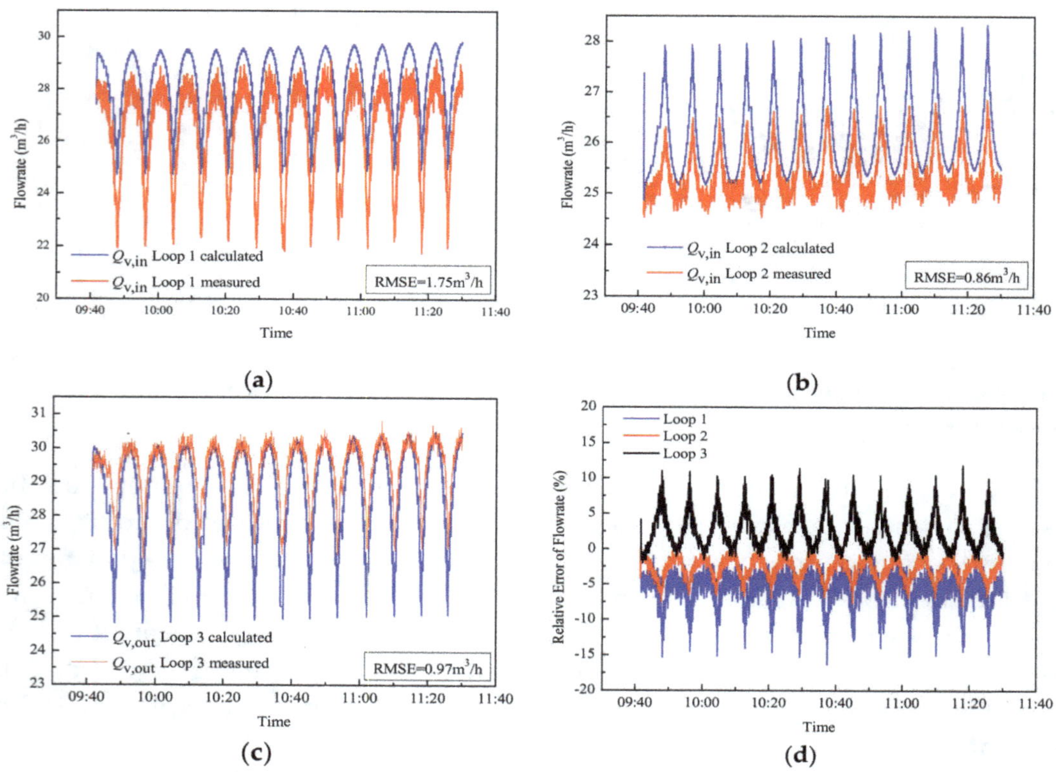

Figure 10. Compared results between calculated and measured data in cold test for (**a**–**c**): the outlet flow rate of loops 1 to 3; and (**d**) the relative errors of flow rate.

4.2. Hot Test

The hot test was conducted from 09:50 to 11:30 on 23 August 2018. As shown in Figure 11a, the opening of LCVs varies similarly to the cold test in addition to an intermission at about 10:50; this is because the operator should open the heat exchange bypass in case of overheating the HTF at this moment, and this operation also leads to the decreasing of the inlet temperature of the header, as shown in Figure 11b.

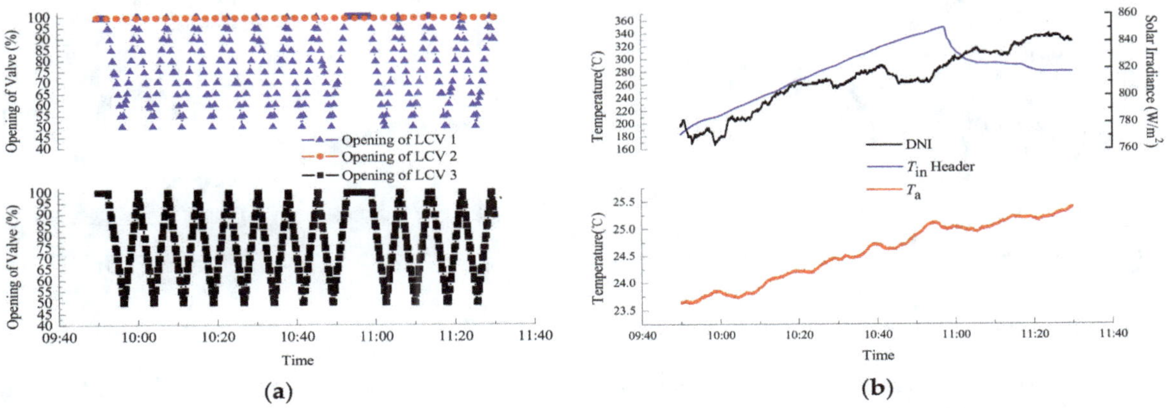

Figure 11. Measured data in the hot test: (**a**) opening of three LCVs and DNI; (**b**) ambient and header inlet temperature.

Compared with the cold test, the parameters refering to the solar irradiance absorption must be measured firstly in the hot test. The measured DNI is shown in Figure 11b, and the defocused factor of the SCAs in Loop 1 and Loop 3, which are measured by the inclinometer and calculated according to Figure 3c, are presented in Figure 12.

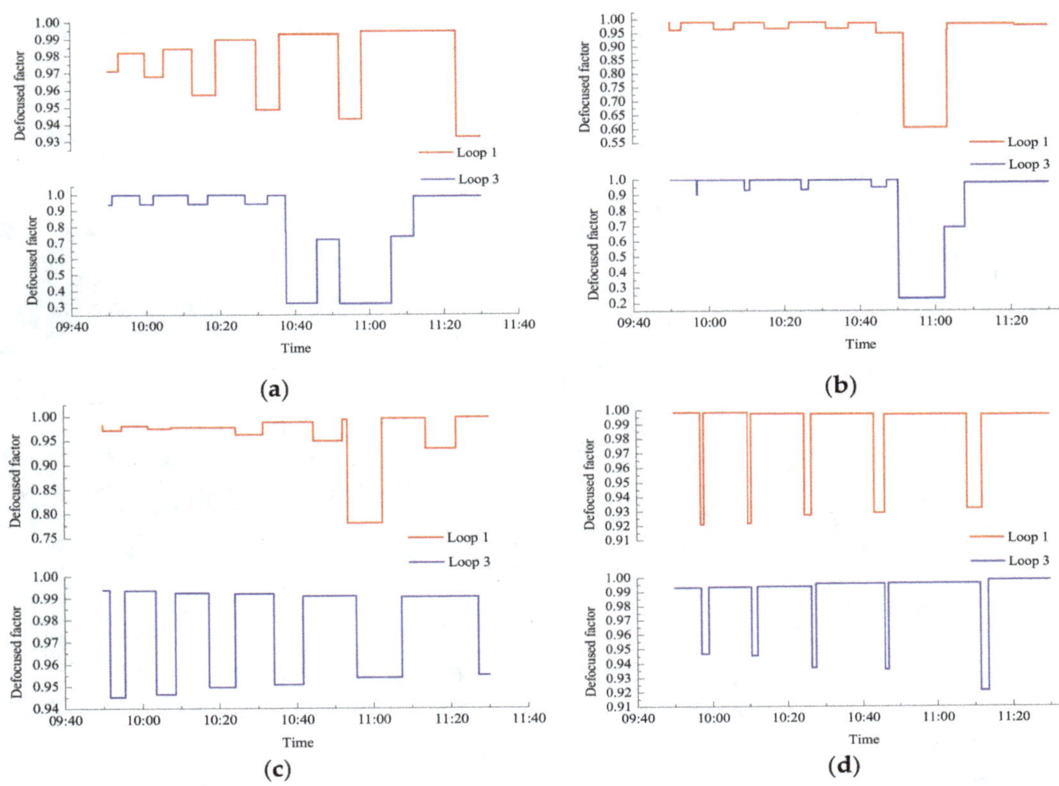

Figure 12. Defocused factor of the solar collector assemblies (SCAs) in Loop 1 and Loop 3: (**a**–**d**) SCA 1 to 4.

As shown in Figure 13, errors regarding the compared inlet flow rate and pressure drop present obvious differences before and after the opening of the exchange bypass; this is due to the misestimate of the flow resistance in the exchange bypass pipe. Compared with the cold test (Figure 9), the hot test has a greater inlet flow rate and lower pressure drop, which is mainly because the HTF has a lower viscosity at a higher temperature.

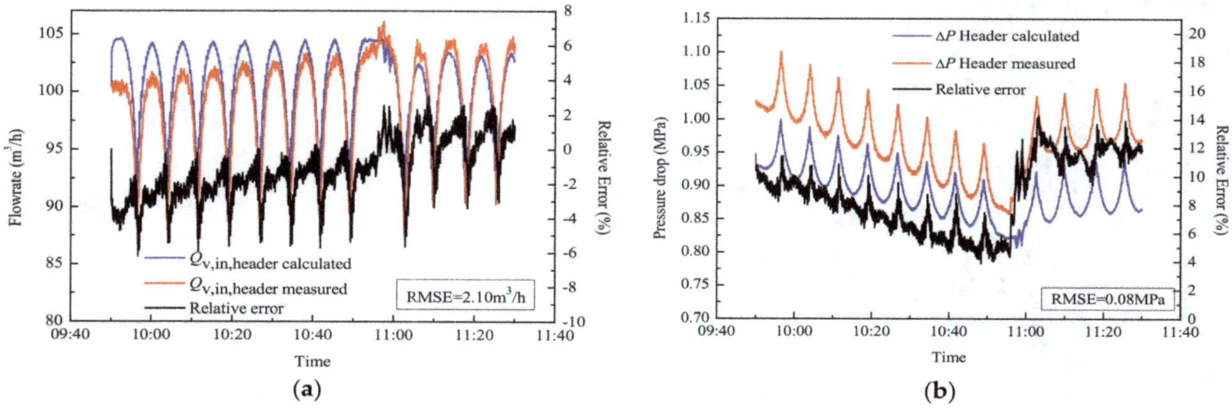

Figure 13. Compared results between calculated and measured data in a hot test for the (**a**) header inlet flow rate, and (**b**) header pressure drop.

As shown in Figure 14, the calculated flow rate distribution maintains good consistency with the measured data in the hot test, the RMSEs for the outlet flow rate of loops 1 to 3 are 1.82 m^3/h, 1.01 m^3/h, and 1.50 m^3/h, respectively. Compared with the cold test, there are two extra sources that cause the error. The one is that the larger outlet flow rate of the header will magnify the error caused by the uncertainty of electric control valve actuators; another is that the error of the calculated temperature will influence the flow resistance distribution of the PTSF, and then increase the error of the flow rate.

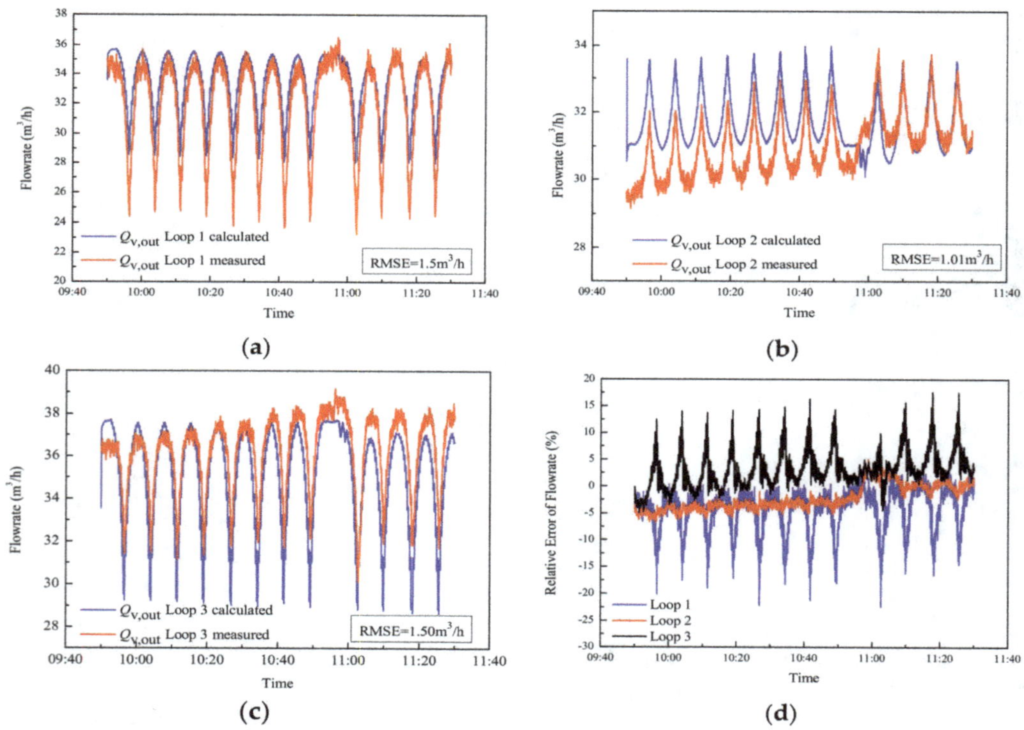

Figure 14. Compared results between calculated and measured data in a hot test for (**a–c**): the outlet flow rate of loops 1 to 3; and (**d**) the relative errors of the flow rate.

Figure 15 shows that the RMSEs for the calculated outlet temperatures of three loops were 8.54 °C, 6.30 °C, and 8.14 °C, respectively. This good agreement mainly benefited from the precisely calculated flow rate distribution and the accuracy of the thermal part in the CTHM. Despite the fluctuated flow rate, the great thermal inertia of the PTSF made the temperature trend present a relatively smooth curve. Besides the initialization temperature, which caused the discrepancies within the first 10 min, there are two reasons for the error in calculating the outlet temperature. One is that the flow rate error will accumulate along with the results of the outlet temperature. Another one is that the installation error of the SCAs will cause the discrepancies of outlet temperature within some short periods of time [28], such as Loop 1 from 10:50 to 11:10 and Loop 3 from 10:40 to 11:00.

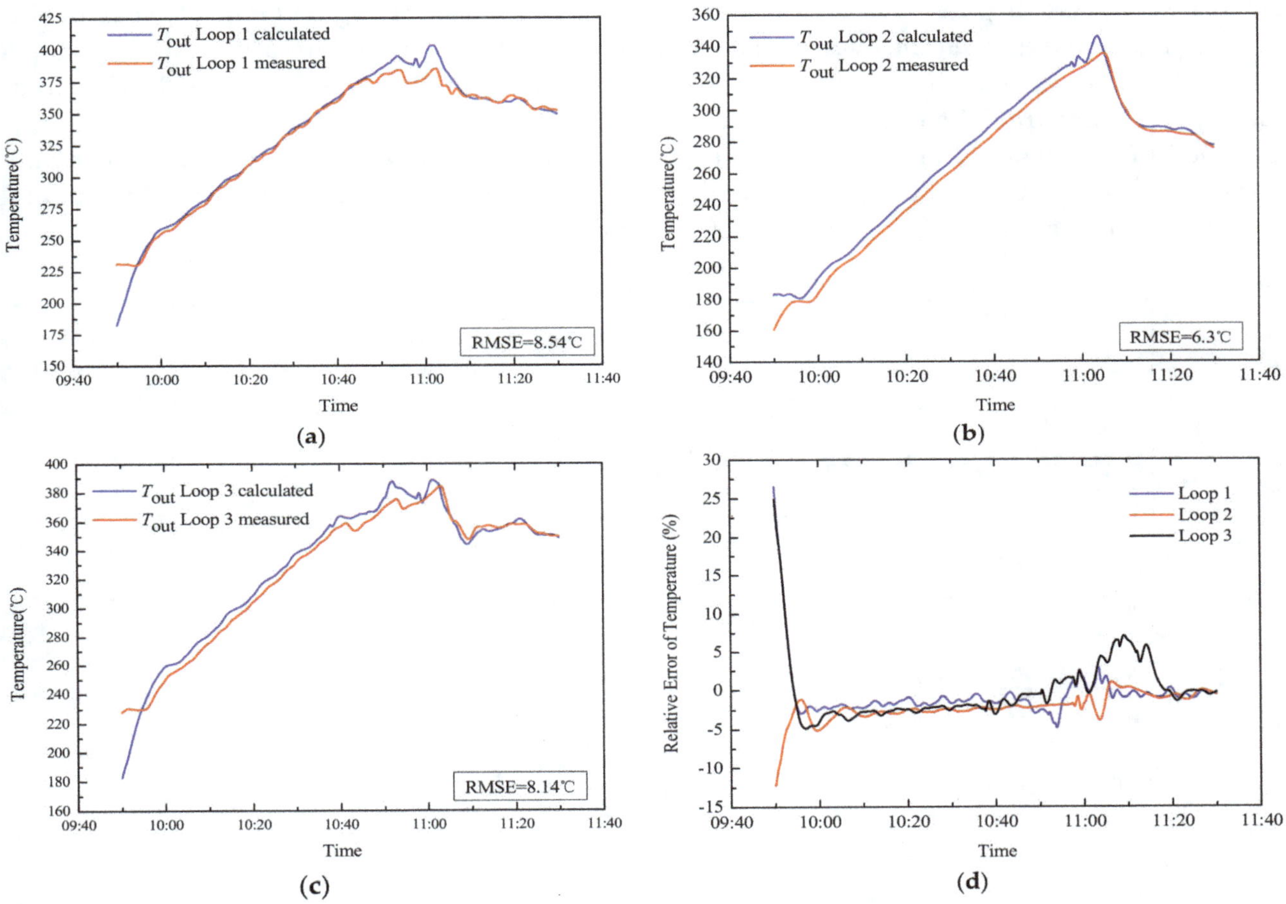

Figure 15. Compared results between calculated and measured data in hot test for (**a**–**c**): the outlet temperature of loops 1 to 3; and (**d**) the relative temperature errors.

5. Model Applications

After the accuracy and availability of the CTHM are validated, two potential abilities of the model are presented in this section. First, the relationship between the thermal and hydraulic transients are revealed by two simulating cases. Secondly, two simple feedforward control strategies are introduced and verified through the CTHM. In all of the simulations, the ambient temperature and inlet temperature of the header remain 25 °C and 290 °C, respectively.

5.1. *Transients Simulating*

As a complex piping network, the thermal effect is the biggest difference between the WDS and the PTSF. Two cases are conducted to simulate how the thermal transients influence the hydraulic state. Ma et.al [15] pointed out that the PTSF of the pilot plant can achieve balanced flow distribution when the openings of the three LCVs are 58%, 100%, and 60%, respectively; these values will be maintained in these two cases.

Case 1, which lasts for 1.3 h, is implemented for simulating the impacts of DNI saltation on the total flow rate and pump pressurizing of the PTSF. As shown in Figure 16a, the DNI changes from 0 to 800 W/m^2 at 0.4 h and turns back into 0 at 0.9 h, and these two disturbances cause a drastic fluctuation in the header flow rate, as shown in Figure 16b. At 0.4 h, as shown in Figure 16b, the inlet and outlet header flow rate present two different trends. The peaking of heat gain causes an expansion of the HTF and further leads to the increasing velocity; this will result in an increase of the pump pressurizing (as shown in Figure 16c) and a reduction of pump capacity (inlet header flow rate). Meanwhile, the viscosity of the HTF decreases with the temperature rise (as shown in Figure 16d). Together with the density and viscosity, the total flow resistance and pump pressurizing will reach a local maximum and begin to decrease; when the effect of expansion surpasses the effect of viscosity reduction, the total pressure drop increases again, and the pump capacity decreases until it reaches a steady state. Similar reasons can demonstrate the fluctuation of the header inlet flow rate, and pump pressurizing happens when the DNI vanishes. For the header outlet flow rate, according to Equation (12), the dramatic expansion and contraction of the HTF caused by the DNI saltation will be embodied in the outlet flow rate, which causes an opposite trend between the inlet and outlet flow rate.

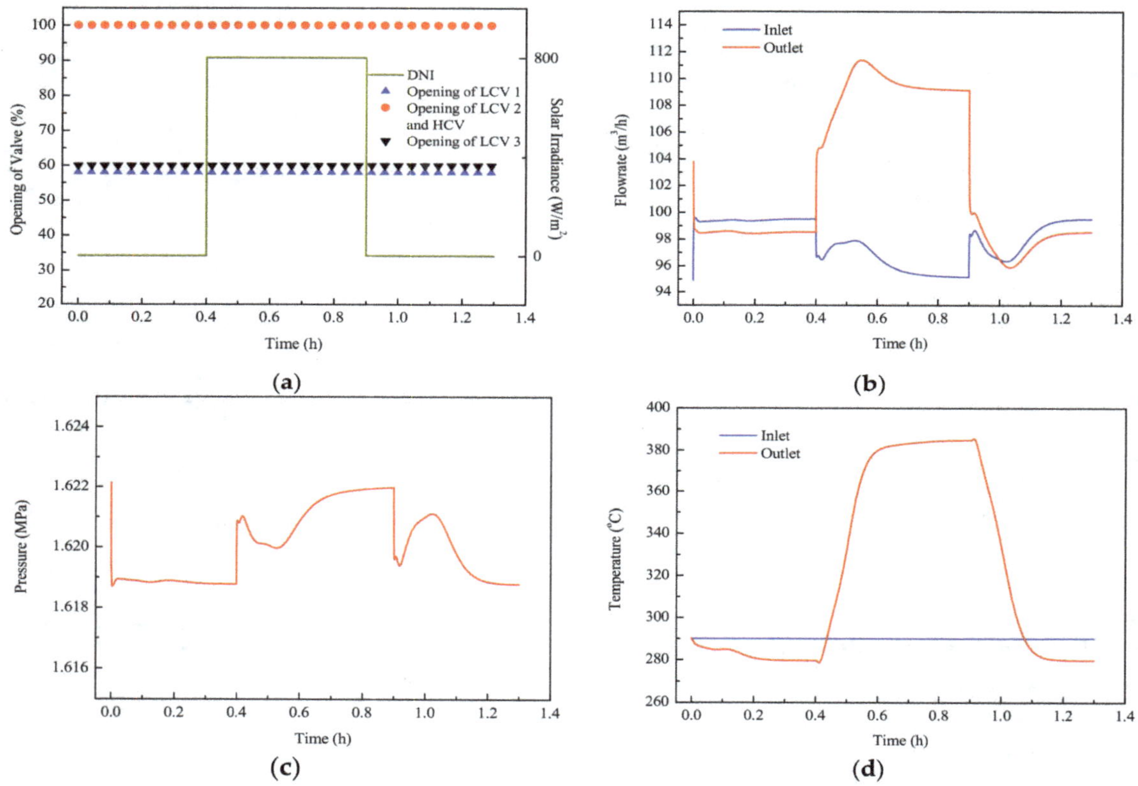

Figure 16. Simulating Case 1 for (**a**) initial parameters, (**b**) inlet and outlet header flow rate results; (**c**) pump pressurizing results, and (**d**) inlet and outlet header temperature results.

Case 2, which lasted for 1.7 h, was conducted for simulating how the thermal effect influences the flow distribution of the PTSF; in this case, the DNI is maintained at 800 W/m^2 throughout the whole process. As shown in Figure 17a, the SCAs in Loop 1 will be defocused in a positive (from one to four) and a negative (from four to one) sequence. The reasons for the fluctuation of flow rate and pump pressurizing at the time of defocus are demonstrated in Case 1. As shown in Figure 17b,c, both defocus sequences cause an increase of the header flow rate and a reduction of pump pressurizing. This is because the HTF will stop expanding and has a lower velocity in the defocused SCAs compared with a focused one; this further leads to a lower pump pressure and higher pump capacity. This effect is presented in a more obvious way in the comparison of flow rate among the three loops shown in Figure 17b; the flow rate in the defocused Loop 1 increases step-by-step compared with the other two focused loops due to its lower pressure drop. Besides, compared with the negative defocused sequence, the cold HTF will flow over longer distances when the SCAs are defocused in a positive sequence. As a result, the higher flow rate and lower pump pressure exist in the positive defocused sequence due to the lower flow resistance. Finally, the HTF in the SCAs, which is closer to the inlet, has greater thermal inertia than the SCAs near the outlet; this causes the difference of outlet temperature between the two opposite sequences of defocus shown in Figure 17d.

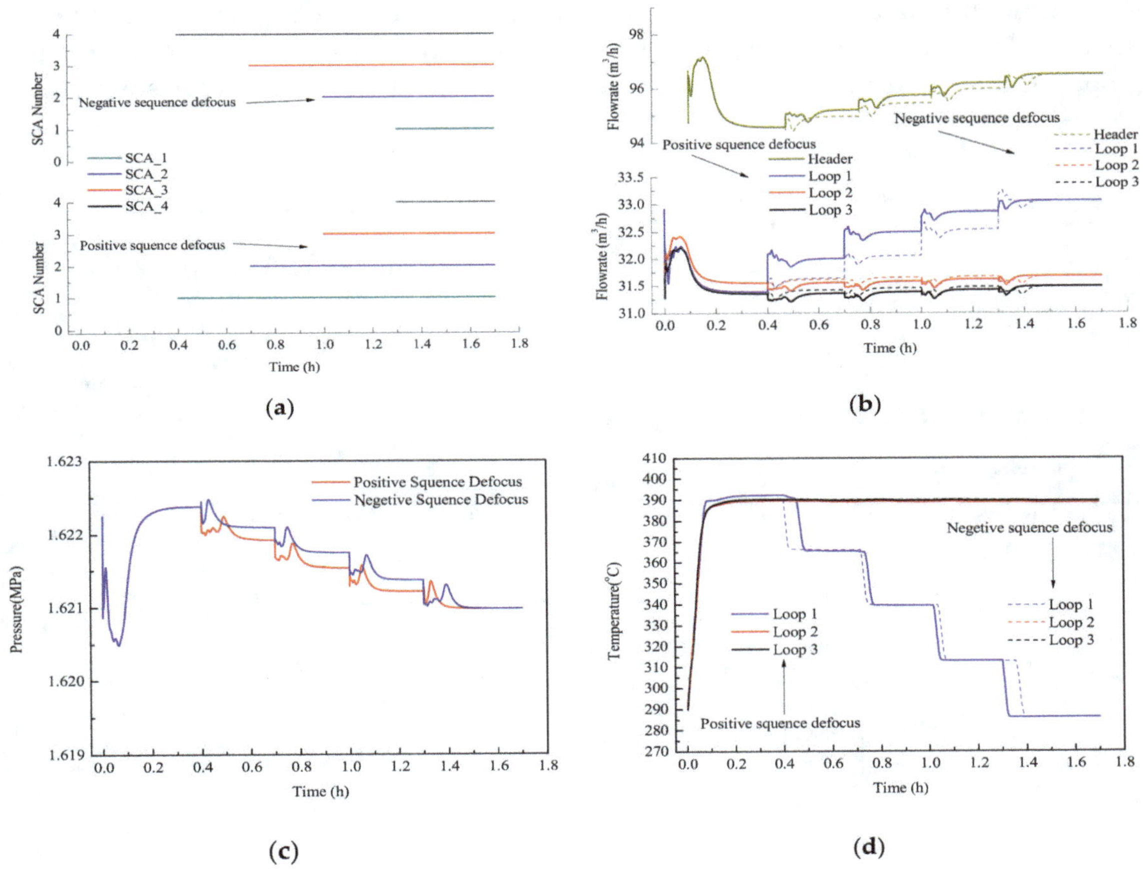

Figure 17. Simulating Case 2 for (**a**) initial parameters; (**b**) inlet flowrate results of the header and three loops; (**c**) pump pressurizing results; and (**d**) outlet temperature results of three loops.

The density, specific heat capacity, and the viscosity are the most relevant properties to the thermal hydraulic characteristics of the PTSF. Based on the simulation results of the above two cases, it can be found that when the inlet temperature of the header stays the same, the effect of density is the most influential property, followed by the specific heat capacity and viscosity. From a thermal perspective, although the specific heat capacity increases with the temperature, the thermal inertia decreases with the temperature because of the reduction of density. From a hydraulic perspective, although the viscosity decreases with the temperature, the pressure loss increases with the temperature because the expansion of the HTF causes a greater velocity.

5.2. *Feedforward Control Strategy*

An important function of the CTHM is being used for study of control strategy; two feedforward control strategies based on the CTHM are introduced in this section. In both strategies, the header inlet temperature is set at 290 °C, and opening of the HCV and the LCVs are the control variables that control the outlet temperature of the header and three loops at 390 °C.

The first strategy is studied under a uniform solar irradiance; as shown in Figure 18a, the DNI of the whole PTSF varies with a cosine disturbance from 800 W/m^2 to 400 W/m^2, of which the period is 0.2 h. Due to the uniformity of the DNI, a balanced flow distribution must be maintained for the same outlet temperature of the three loops, so the openings of the three LCVs are kept constant. The ideal flow rate under different DNI values can be calculated according to the thermal part of the CTHM, while the opening of the HCV and pump pressurizing can be solved according to the hydraulic part of the CTHM and the calculated ideal flow rate; these results are shown in Figure 18b. The control results of the outlet temperature are shown in Figure 18c, the outlet temperatures of the three loops and header are all very close to 390 °C, and the small-range fluctuations of the outlet temperature in the three loops and headers are caused by the heat loss in the header, which varies with the flow rate.

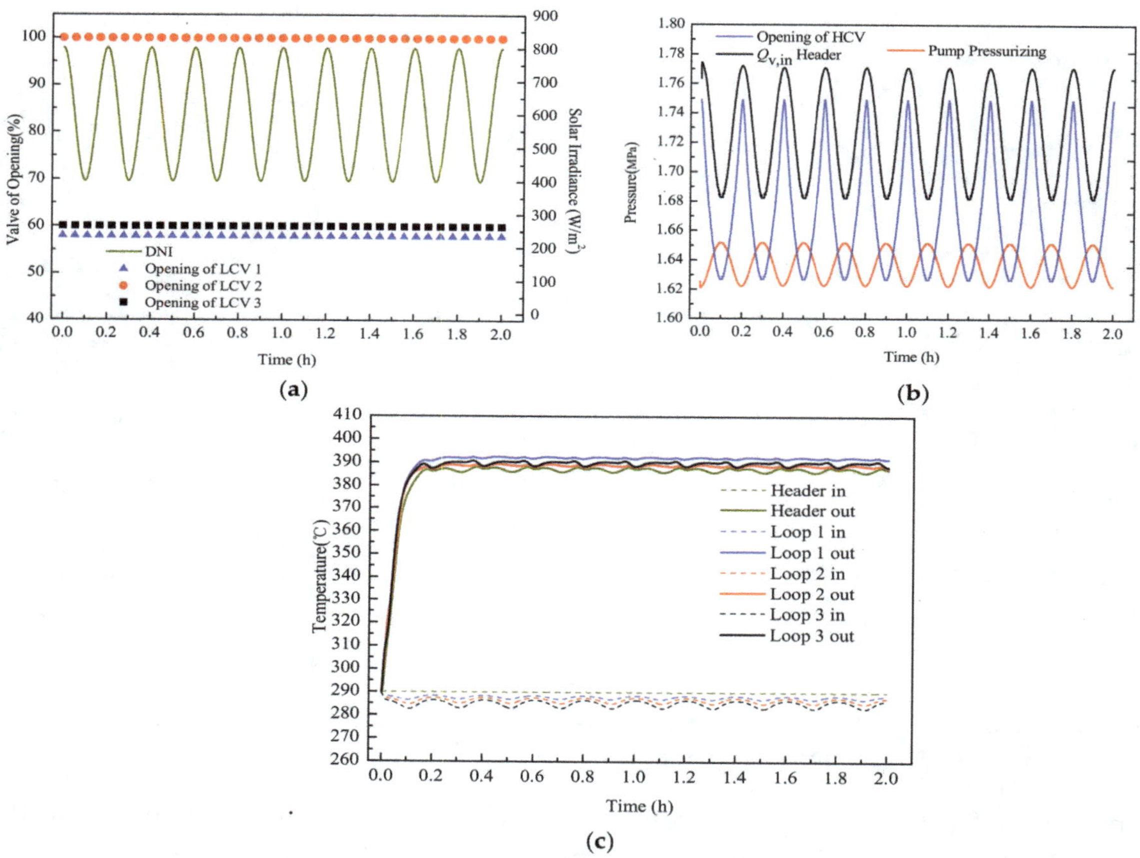

Figure 18. Header control valve (HCV) control for: (**a**) Initial parameters and DNI disturbance; (**b**) Results of HCV opening, header inlet flow rate, and pump pressurizing; (**c**) Results of inlet and outlet temperature of header and three loops.

The second strategy is researched under a nonuniform solar irradiance. As shown in Figure 19a, the DNI values in Loop 1 and Loop 3 are kept at 800 W/m^2, while for Loop 2, the DNI varied in a similar way as shown in Figure 18a. The balance of flow distribution must be thrown off due to the nonuniformity of DNI, so the opening of three LCVs should be recalculated to meet the varied DNI. The ideal flowrate distribution can also be solved by the thermal part of the CTHM, while the openings of the HCV and LCVs can be calculated in two steps. First, we set the opening of the LCV in Loop 3 at

60% and calculated the opening of the LCVs in Loop 2 and Loop 3 with the method referenced in [13] and [15]. Second, we work out the opening of the HCV and pump pressurizing by the hydraulic part of the CTHM. The above results are shown in Figure 19b: the opening of the LCVs in Loop 1 varied within a tiny scale, and the varied opening of the HCV and LCV in Loop 2 are the major variables for controlling the flow rate.

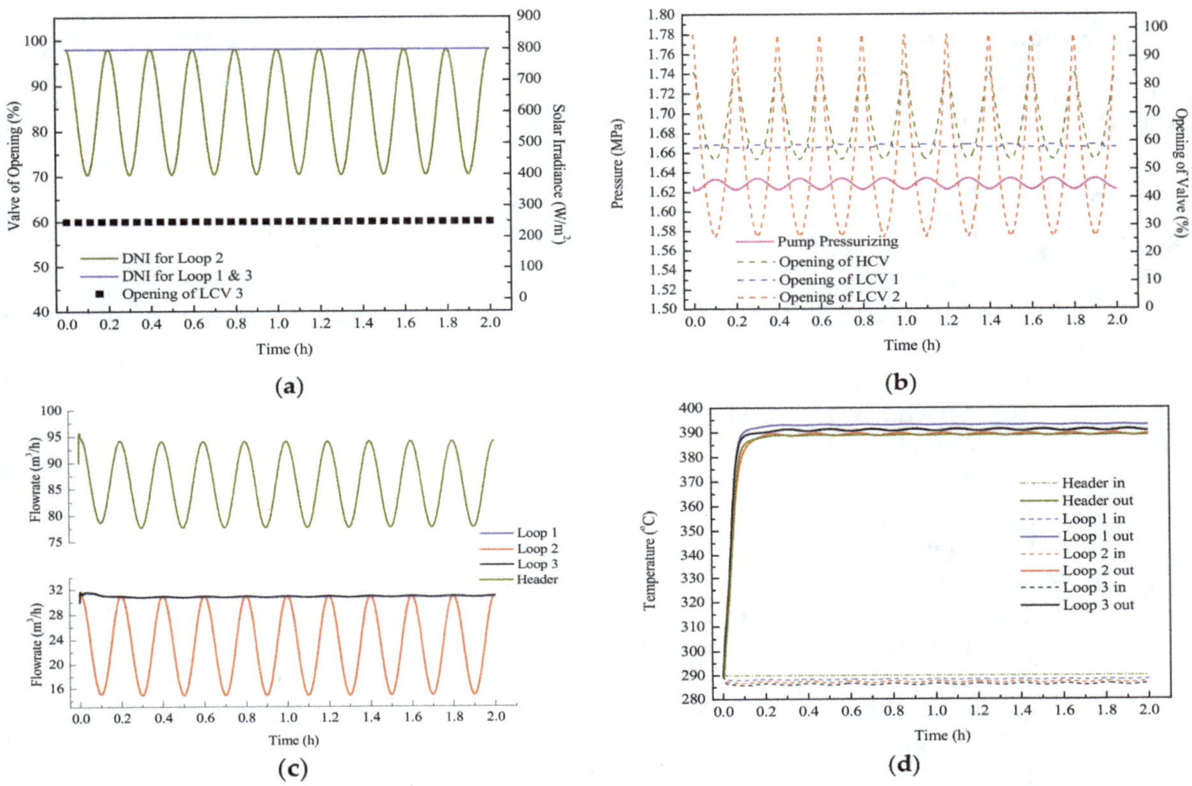

Figure 19. LCV control for (**a**) Initial parameters and DNI disturbance; (**b**) Results of HCV opening, LCV opening, and pump pressurizing; (**c**) Inlet flow rate results of the header and three loops; (**d**) Inlet and outlet temperature results of the header and three loops.

The good control results for the flow rate and outlet temperature are shown in Figure 19c,d; the key presented in the results is that the fluctuation of the header flow rate can be centered on Loop 2 if the openings of the LCV and HCV vary in a similar way. These results verify the feasibility of PTSF control by the cooperation of LCVs and HCVs under a complex distribution of DNI and SCA performance in a large-scaled plant.

6. Conclusions

This paper mainly outlines a comprehensive thermal hydraulic model (CTHM) for the parabolic trough solar field (PTSF) based on a pilot plant. The CTHM is established and solved by a novel numerical method, validated by the experimental data based on the pilot plant, and applied to simulate the dynamic behavior and develop the control strategy of the PTSF under many types of disturbances. The main contribution of this paper is its development of a powerful model for the deeper study on PTSF. On the one hand, the total flow rate, flow distribution, pump head, and outlet temperature can be detailed and calculated by the CTHM under normal or disturbed conditions; on the other hand, when the value and distribution of the solar irradiance deviates from that of the design situation, the CTHM can also be used to calculate the opening of valves for the desired outlet temperature.

In this paper, all of work surrounds a 1 MW pilot plant, which is much smaller than a large-scaled commercial plant. However, all of the key factors (such as the unbalanced flow distribution) are included in the pilot plant, so the validity of the CTHM will always remain when it is applied to a large-scaled PTSF. In the future, the CTHM can play an important role in the further study of the commercial parabolic trough power plant, such as providing more detailed outlet parameters when combining the PTSF with a thermal storage system or power block, or using it to develop the optimal operation strategy.

Author Contributions: Conception, L.M.; Methodology, L.M. and L.X.; Software, L.M.; Validation, L.M.; Formal analysis, L.M.; Investigation, L.M.; Resources, L.M.; data curation, L.M.; Handwriting—Original draft preparation, L.M.; Handwriting—review and editing, L.M. and D.L.; Visualization, L.M.; Supervision, Z.W.; project administration, Z.W. and L.X.; funding acquisition, Z.W. and L.X.

Nomenclature

Latin symbols:	
$A/\mathbf{A}$	Area (m^2)/Area vector
b	Pipe number
B	Constant
$\mathbf{B}, \mathbf{C}, \mathbf{D}, \mathbf{E}, \mathbf{U}, \mathbf{W}, \mathbf{Y}, \mathbf{Z}$	Coefficient matrix
c	Specific heat capacity (J kg^{-1} K^{-1})
D	Diameter (m)
f	Friction factor or optical efficiency factor
F	Flow loss (MPa)
G	Flow coefficient
h	Convection heat transfer coefficient (W m^{-2} K^{-1})
$H/\mathbf{H}$	Heat (W m^{-1}) or pump head (m)/pump header vector
I	Direct normal irradiance (W m^{-2})
k	Thermal conductivity (W m^{-1} K^{-1})
K	Pressure drop coefficient
$l/\mathbf{l}$	Length (m)/Length vector
$\mathbf{M}$	Incidence matrix
n	Node number
Nu	Nusselt number
op	Opening of valve
P	Power flux (W m^{-1})
$p/\mathbf{p}$	Pressure (bar)/Pressure vector
P	Partial vector
Pr	Prandtl number
$Q/\mathbf{Q}$	Volume flowrate (m^3/h)/Volume flowrate vector
$\mathbf{q}$	Node volume flowrate vector (m^3/h)
R	Valve adjustable ratio
$Re/\mathbf{Re}$	Reynolds number/Reynolds number vector
r	Mirror reflectivity
$S/\mathbf{S}$	Flow resistance/Flow resistance vector
t	Time (s)
T	Temperature (°C)
$\mathbf{V}$	Volume vector
W	Width (m)
Greek:	
α	Absorbance
β	Defocused factor
γ	Intercept factor
δ	Thickness (m)

ε	Absolute roughness
η	Efficiency
θ	Incidence angle (°)
κ	Incidence angle modifier
$\rho/\boldsymbol{\rho}$	Density (kg m^{-3})/Density vector
τ	Transmittance
ω	Zenith angle (°)
Subscripts:	
abs	Absorber tube
cl	Clean mirror
col	Collecto
end	End loss
f	Fluid
fit	Pipe fitting
fl	Focal distance
ins	Insulated pipe
l	Heat loss
max	Max
o	Optical
P	Pipe
rs	Row spacing
sha	Shadowing
va	Valve
use	Useful
Superscript:	
i,j	Element number of the matrix
m	iterations
t	Time layer number
Abbreviations:	
CTHM	Comprehensive thermal hydraulic model
DNI	Direct Normal Irradiance
HCV	Header control valve
HTF	Heat Transfer Fluid
LCV	Loop control valve
PTC	Parabolic trough collector
PTSTPP	Parabolic trough solar thermal power plant
PTSF	Parabolic trough solar field
SCA	Solar collector assembly
THDM	Thermal hydraulic dynamic model

Appendix A

Table A1. Thermophysical properties of Therminol VP-1.

Property	Expression	Coefficients				
		c_1	c_2	c_3	c_4	c_5
Density (kg/m^3)	$c_1T + c_2T^2 + c_3T^3 + c_4$	−0.90797	7.8116×10^{-4}	2.367×10^{-6}	1083.25	-
Kinematic viscosity (mm^2 s^{-1})	$exp\left(\frac{c_1}{T+c_2} + c_3\right)$	544.149	114.43	−2.59578	-	-
Specific heat (J kg^{-1} K^{-1})	$c_1T + c_2T^2 + c_3T^3 + c_4T^4 + c_5$	0.002414	5.9591×10^{-6}	-2.9879×10^{-8}	4.4172×10^{-11}	1.498
Thermal conductivity (W m^{-1} K^{-1})	$c_1T + c_2T^2 + c_3T^3 + c_4T^4 + c_5$	-8.19477×10^{-5}	-1.92257×10^{-7}	2.5034×10^{-11}	-7.2974×10^{-15}	0.137743

Appendix B

In one of the iterative process, the hydraulic part can be combined to be a matrix equation of $Q_{f,in}$, $Q_{f,out}$, and p_1; this matrix equation can be expressed as:

$$X\left(Q^t_{f,in}, Q^t_{f,out}, p^t_1\right) = \begin{bmatrix} U_{11} & U_{12} & U_{13} \\ U_{21} & U_{22} & U_{23} \\ U_{31} & U_{32} & U_{33} \end{bmatrix} \begin{bmatrix} Q^t_{f,in} \\ Q^t_{f,out} \\ p^t_1 \end{bmatrix} - \begin{bmatrix} W_1 \\ W_2 \\ W_3 \end{bmatrix} = 0 \tag{A1}$$

where the elements of U and W are given by:

$$\begin{aligned} U_{11} &= M_{1,in}, U_{12} = M_{1,out}, U_{13} = 0 \\ U_{21} &= D\left(\rho^t_{f,in}\right), U_{22} = -D\left(\rho^t_{f,out}\right), U_{23} = 0 \\ U_{31} &= \frac{D(A_c)+D(S)D\left(\left|\bar{Q}^t_f\right|\right)}{2} - \frac{D\left(\rho^t_{f,in}\right)D\left(\left|Q^t_{f,in}\right|\right)}{D^2(A)} \\ U_{32} &= \frac{D(A_c)+D(S)D\left(\left|\bar{Q}^t_f\right|\right)}{2} + \frac{D\left(\rho^t_{f,in}\right)D\left(\left|Q^t_{f,in}\right|\right)}{D^2(A)} \\ U_{33} &= \left(M_{1,in} + M_{1,out}\right)^T \end{aligned} \tag{A2}$$

$$\begin{aligned} W_1 &= {q_1}^t, W_2 = D(V)\frac{\left(\bar{\rho}^t_f - \bar{\rho}_f\right)}{\Delta t} \\ W_3 &= \left(M_{0,in} + M_{0,out}\right)^T p^t_0 + D(B_c)\bar{Q}_f \end{aligned} \tag{A3}$$

X is a system of nonlinear equations; these kinds of equations are frequently solved by the Newton–Raphson iterative method, which can be shown as [17]:

$$J\left(x^{(m)}\right)\left(x^{(m+1)} - x^{(m)}\right) = -F\left(x^{(m)}\right) \tag{A4}$$

where J is the jacobian matrix of $F(x)$ and m is the iteration number.

Before calculating the jacobian matrix, the partial of the pressure loss and pump head with respect to $Q_{f,in}$, $Q_{f,out}$ must be introduced. The partials can be given by [29]

$$\begin{aligned} P_{in} &= \frac{\partial\left(D(S)D\left(\left|\bar{Q}_f\right|\right)\bar{Q}_f - D(\rho_f)gH\right)}{\partial Q_{in}} \\ &= \frac{\partial S_p}{\partial Q_{in}} D\left(\left|\bar{Q}_f\right|\right)D\left(\bar{Q}_f\right) + D(S)D\left(\left|\bar{Q}_f\right|\right) - D\left(\rho_f\right)g\frac{\partial H}{\partial Q_{in}} \\ \frac{\partial S_p}{\partial Q_{in}} &= \frac{\partial S_p}{\partial f}\frac{\partial f}{\partial Re}\frac{\partial Re}{\partial Q_{in}} \\ P_{out} &= \frac{\partial S_p}{\partial Q_{out}} D\left(\left|\bar{Q}_f\right|\right)D\left(\bar{Q}_f\right) + D(S)D\left(\left|\bar{Q}_f\right|\right) - D\left(\rho_f\right)g\frac{\partial H}{\partial Q_{in}} \end{aligned} \tag{A5}$$

where S_p, f, and Re are the vectors of pipe resistance, inner friction factor, and Reynolds number, respectively.

According to Equation (A4), the $(m + 1)$th iterative calculation of function X can be expressed as:

$$\begin{bmatrix} Y_{11} & Y_{12} & Y_{13} \\ Y_{21} & Y_{22} & Y_{23} \\ Y_{31} & Y_{32} & Y_{33} \end{bmatrix} \begin{bmatrix} Q^{[m+1]}_{f,in} \\ Q^{[m+1]}_{f,out} \\ p^{[m+1]}_1 \end{bmatrix} - \begin{bmatrix} Z_1 \\ Z_2 \\ Z_3 \end{bmatrix} = 0 \tag{A6}$$

where the elements Y and Z are given by:

$$\begin{aligned} Y_{11} &= M_{1,in}, Y_{12} = M_{1,out}, Y_{13} = 0, Y_{21} = D\left(\rho^t_{f,in}\right), Y_{22} = -D\left(\rho^t_{f,out}\right), Y_{23} = 0 \\ Y_{31} &= \frac{D(A_c)}{2} - \frac{2D\left(\rho^t_{f,in}\right)D\left(\left|Q^{|m|}_{f,in}\right|\right)}{D^2(A)} + P^{(m)}_{in} \\ Y_{32} &= \frac{D(A_c)}{2} + \frac{2D\left(\rho^t_{f,in}\right)D\left(\left|Q^{|m|}_{f,out}\right|\right)}{D^2(A)} + P^{(m)}_{out} \\ Y_{33} &= \left(M_{1,in} + M_{1,out}\right)^T \end{aligned} \tag{A7}$$

$$
\begin{aligned}
&\boldsymbol{Z}_1 = \boldsymbol{q}_1{}^t, \boldsymbol{Z}_2 = \mathrm{D}(\boldsymbol{V})\frac{\left(\bar{\boldsymbol{\rho}}_f^t - \bar{\boldsymbol{\rho}}_f\right)}{\Delta t} \\
&\boldsymbol{Z}_3 = -\mathrm{X}\left(\boldsymbol{Q}_{f,in}^{(m)}, \boldsymbol{Q}_{f,out}^{(m)}, \boldsymbol{p}_1^m\right) + \frac{\mathrm{D}(\boldsymbol{A}_c)}{2}\left(\boldsymbol{Q}_{f,in}^{(m)} + \boldsymbol{Q}_{f,out}^{(m)}\right) \\
&+\left(-\frac{2\mathrm{D}\left(\boldsymbol{\rho}_{f,in}^t\right)\mathrm{D}\left(\left|\boldsymbol{Q}_{f,in}^{|m|}\right|\right)}{\mathrm{D}^2(\boldsymbol{A})} + \boldsymbol{P}_{in}^m\right)\boldsymbol{Q}_{f,in}^{(m)} + \left(\frac{2\mathrm{D}\left(\boldsymbol{\rho}_{f,out}^t\right)\mathrm{D}\left(\left|\boldsymbol{Q}_{f,out}^{|m|}\right|\right)}{\mathrm{D}^2(\boldsymbol{A})} + \boldsymbol{P}_{out}^m\right)\boldsymbol{Q}_{f,out}^{(m)}
\end{aligned}
\quad \text{(A8)}
$$

Finally, the solution of the hydraulic model can be shown as:

$$
\begin{aligned}
\boldsymbol{p}_1^{(m+1)} &= \boldsymbol{E}_c{}^{-1}\left(\boldsymbol{M}_{1,out}\left(\boldsymbol{\rho}_{f,out}^t\right)^{-1}\boldsymbol{Z}_2 + \boldsymbol{q}^t - \left(\boldsymbol{M}_{1,in} + \boldsymbol{M}_{1,out}\mathrm{D}^{-1}\left(\boldsymbol{\rho}_{f,out}^t\right)\mathrm{D}\left(\boldsymbol{\rho}_{f,in}^t\right)\right)\boldsymbol{C}_c{}^{-1}\boldsymbol{D}_c\right) \\
\boldsymbol{Q}_{f,in}^{(m+1)} &= \boldsymbol{C}_c{}^{-1}\left(\boldsymbol{D}_c + \left(\boldsymbol{M}_{1,in} + \boldsymbol{M}_{1,out}\right)^T\boldsymbol{p}_1^{(m+1)}\right) \\
\boldsymbol{Q}_{f,out}^{(m+1)} &= \mathrm{D}^{-1}\left(\boldsymbol{\rho}_{f,out}^t\right)\left(D\left(\boldsymbol{\rho}_{f,in}^t\right)\boldsymbol{Q}_{f,in}^{(m+1)} - \boldsymbol{Z}_2\right)
\end{aligned}
\quad \text{(A9)}
$$

The whole expression of $\boldsymbol{C}_c$, $\boldsymbol{D}_c$, and $\boldsymbol{E}_c$ is given by:

$$
\begin{aligned}
\boldsymbol{C}_c &= \boldsymbol{Y}_{31} + \boldsymbol{Y}_{32}\mathrm{D}^{-1}\left(\boldsymbol{\rho}_{f,out}^t\right)\mathrm{D}\left(\boldsymbol{\rho}_{f,in}^t\right) \\
\mathrm{D}_c &= \boldsymbol{Z}_3 + \boldsymbol{Y}_{32}\mathrm{D}^{-1}\left(\boldsymbol{\rho}_{f,out}^t\right)\boldsymbol{Z}_2 \\
\boldsymbol{E}_c &= \left(\boldsymbol{M}_{1,in} + \boldsymbol{M}_{1,out}\mathrm{D}^{-1}\left(\boldsymbol{\rho}_{f,out}^t\right)\mathrm{D}\left(\boldsymbol{\rho}_{f,in}^t\right)\right)\boldsymbol{C}_c{}^{-1}\left(\boldsymbol{M}_{1,in} + \boldsymbol{M}_{1,out}\right)^T
\end{aligned}
\quad \text{(A10)}
$$

It is noteworthy that $\boldsymbol{E}_c$ will degrade into a definite matrix when $\rho_{f,in} = \rho_{f,out}$; i.e., the hydraulic model is not affected by the temperature, and this conclusion is consistent with [18].

References

1. Yılmaz, I.H.; Mwesigye, A. Modeling, simulation and performance analysis of parabolic trough solar collectors: A comprehensive review. *Appl. Energy* **2018**, *225*, 135–174. [CrossRef]
2. Cheng, Z.-D.; He, Y.-L.; Wang, K.; Du, B.-C.; Cui, F. A detailed parameter study on the comprehensive characteristics and performance of a parabolic trough solar collector system. *Appl. Therm. Eng.* **2014**, *63*, 278–289. [CrossRef]
3. Giostri, A. Transient Effects in Linear Concentrating Solar Thermal Power Plant. Ph.D. Thesis, Polytechnic University of Milan, Milan, Italy, 2014.
4. Forristall, R. *Heat Transfer Analysis and Modeling of a Parabolic Trough Solar Receiver Implemented in Engineering Equation Solver*; Technical Report NREL/TP-550-34169; National Renewable Energy Laboratory: Golden, CO, USA, 2003.
5. Dudley, V.E.; Kolb, G.J.; Mahoney, A.R.; Mancini, T.R.; Matthews, C.W.; Sloan, M.; Kearney, D.W. *Test Results: SEGS LS-2 Solar Collector*; Nasa Sti/recon Technical Report N; Sandia National Labs.: Albuquerque, NM, USA, 1994; p. 96.
6. Stuetzle, T.; Blair, N.; Mitchell, J.W.; Beckman, W.A. Automatic control of a 30 MWe SEGS VI parabolic trough plant. *Sol. Energy* **2004**, *76*, 187–193. [CrossRef]
7. Camacho, E.F.; Berenguel, M. Control of solar energy systems. *IFAC Proc. Vol.* **2012**, *45*, 848–855. [CrossRef]
8. Padilla, R.V.; Demirkaya, G.; Goswami, D.Y.; Stefanakos, E.; Rahman, M.M. Heat transfer analysis of parabolic trough solar receiver. *Appl. Energy* **2011**, *88*, 5097–5110. [CrossRef]
9. Hachicha, A.; Rodríguez, I.; Capdevila, R.; Oliva, A. Heat transfer analysis and numerical simulation of a parabolic trough solar collector. *Appl. Energy* **2013**, *111*, 581–592. [CrossRef]
10. Yılmaz, I.H.; Söylemez, M.S. Thermo-mathematical modeling of parabolic trough collector. *Energy Convers. Manag.* **2014**, *88*, 768–784. [CrossRef]
11. Behar, O.; Khellaf, A.; Mohammedi, K. A novel parabolic trough solar collector model—Validation with experimental data and comparison to Engineering Equation Solver (EES). *Energy Convers. Manag.* **2015**, *106*, 268–281. [CrossRef]
12. Wagner, M.J.; Gilman, P. *Technical Manual for the SAM Physical trough Model*; National Renewable Energy Laboratory: Golden, CO, USA, 2011.
13. Abutayeh, M.; Alazzam, A.; El-Khasawneh, B. Balancing heat transfer fluid flow in solar fields. *Sol. Energy* **2014**, *105*, 381–389. [CrossRef]

14. Noureldin, K.; Hirsch, T.; Pitz-Paal, R. Virtual Solar Field—Validation of a detailed transient simulation tool for line focus STE fields with single phase heat transfer fluid. *Sol. Energy* **2017**, *146*, 131–140. [CrossRef]
15. Ma, L.; Xu, E.; Li, J.; Xu, L.; Li, X. Analysis and validation of a thermal hydraulic dynamic model for the parabolic trough solar field. *Energy* **2018**, *156*, 430–443. [CrossRef]
16. Todini, E.; Pilati, S. A gradient method for the solution of looped pipe networks. In *Computer Applications in Water Supply*; Research Studies Press: Taunton, UK, 1988; Volume 1, pp. 1–20.
17. Ellis, D.J. The Behaviour of Pipe Network Analysis Solution Techniques. Ph.D. Thesis, University of Adelaide, Adelaide, Australia, 2001.
18. Nault, J.D.; Karney, B.W. Improved Rigid Water Column Formulation for Simulating Slow Transients and Controlled Operations. *J. Hydraul. Eng.* **2016**, *142*, 04016025. [CrossRef]
19. Colmenar-Santos, A.; Munuera-Pérez, F.-J.; Tawfik, M.; Castro-Gil, M. A simple method for studying the effect of scattering of the performance parameters of Parabolic Trough Collectors on the control of a solar field. *Sol. Energy* **2014**, *99*, 215–230. [CrossRef]
20. Almasabi, A.; Alobaidli, A.; Zhang, T.; Zhang, T. Transient Characterization of Multiple Parabolic Trough Collector Loops in a 100 MW CSP Plant for Solar Energy Harvesting. *Energy Procedia* **2015**, *69*, 24–33. [CrossRef]
21. Padilla, R.V. Simplified Methodology for Designing Parabolic trough Solar Power Plants. Ph.D. Thesis, University of South Florida, Tampa, FL, USA, 2011.
22. Karassik, I.J.; Messina, J.P.; Cooper, P.; Heald, C.C. *Pump Handbook*, 3rd ed.; McGraw-Hill: New York, NY, USA, 2001; p. 2.70.
23. Arias, D.A.; Gavilán, A.; Russel, M. Pumping Power Parasitics in Parabolic Trough Solar Fields. In Proceedings of the 15th International SolarPACES Symposium, Berlin, Germany, 15–18 September 2009.
24. Skousen, P.L. *Valve Handbook*, 1st ed.; McGraw-Hill: New York, NY, USA, 1997; pp. 24–33.
25. Chaudhry, M.H. *Applied Hydraulic Transients*, 3rd ed.; Van Nostrand Reinhold: New York, NY, USA, 2014; pp. 35–43.
26. Munson, B.R.; Young, D.F.; Okiishi, T.H. *Fundamentals of Fluid Mechanics*, 6th ed.; John Wiley & Sons. Inc.: New York, NY, USA, 2009; pp. 411–412.
27. Churchill, S.W. Friction-factor equation spans all fluid-flow regimes. *Chem. Eng.* **1977**, *84*, 91–92.
28. Zhao, D.; Xu, E.; Wang, Z.; Yu, Q.; Xu, L.; Zhu, L. Influences of installation and tracking errors on the optical performance of a solar parabolic trough collector. *Renew. Energy* **2016**, *94*, 197–212. [CrossRef]
29. Simpson, A.; Elhay, S. Jacobian Matrix for Solving Water Distribution System Equations with the Darcy-Weisbach Head-Loss Model. *J. Hydraul. Eng.* **2011**, *137*, 696–700. [CrossRef]

11

Characterization of Core-Shell Spherical Lens for Microtracking Concentrator Photovoltaic System

Masakazu Nakatani [1,2,*] and Noboru Yamada [1]

1 Department of Energy and Environment Science, Nagaoka University of Technology, 1603-1 Kamitomioka-machi, Nagaoka 940-2188, Japan; noboru@nagaokaut.ac.jp
2 Sun Marion Co., Ltd., 1603-1 Kamitomioka-machi, Nagaoka 940-2188, Japan
* Correspondence: nakatani@sun-marion.com

Abstract: The optical characteristics of a radially symmetrical core-shell spherical (CSSP) lens is analyzed for its suitability to application in microtracking concentrator photovoltaic systems (MTCPVs). The CSSP lens is compared to a conventional homogenous spherical lens through both ray-tracing simulations and outdoor experiments. Simulation results show that the CSSP lens is superior to the conventional homogenous spherical lens in terms of its optical efficiency for long focal lengths, for which the CSSP lens exhibits less spherical and chromatic aberrations. Outdoor experiments are conducted using test concentrator photovoltaic (CPV) modules with prototype CSSP and homogenous spherical lenses; the trend of the measured short circuit current agrees with the that of the simulated optical efficiency for both lenses. Furthermore, compared to the homogenous lens, the CSSP lens significantly increases module efficiency because of its better illumination uniformity at the solar cell surface. The optical characteristics of the CSSP lens are preferable for MTCPVs with a spherical lens array to achieve a higher module efficiency for a wider incidence angle although further studies on more practical system configurations are needed.

Keywords: photovoltaics; solar concentrators; solar cells; geometrical optics; concentrator photovoltaics; spherical lens

---

1. Introduction

Multijunction solar cells can achieve a high solar-cell efficiency of 46% [1]. Currently, the maximum efficiency of a concentrator photovoltaic (CPV) system that includes a multijunction solar cell and a concentrator is 36.7% [2]. Although a high-concentration CPV system has high solar-cell efficiency, it requires a solar tracking system. Pedestal solar trackers are commonly used in CPV systems; however, pedestal trackers are massive and unsuitable for building integrated photovoltaics (BIPV).

A microtracking concentrator photovoltaic system (MTCPV) that integrates a CPV and a solar tracker into a module is being developed [3]. Examples of MTCPV structures include (a) a two-lens structure on the top surface of a multijunction solar cell [4], (b) an upper waveguide and lens structure on a multijunction solar cell [5–11], (c) a structure combining an upside lens and a downside mirror on a multijunction solar cell [12,13], (d) a resin filled mirror under a solar cell stage [14], (e) a structure comprised of a wide angle aplanatic lens and a solar cell stage with three-dimensional control [15], and (f) a structure comprised of a gradient-index(GRIN) lens [16,17] and a solar cell stage with three-dimensional control [18]. Of these, structure (f) has yet to be experimentally demonstrated as it requires a gradient low-refractive index (RI) structure. The RI of Mg_2F, a conventional transparent low-RI material, is 1.39 [19]. Optical porous materials can achieve a lower RI [20]; however, they are still under development for practical use [21,22]. These reported MTCPV systems have demonstrated the technological potential of the MTCPV; however, further investigation is still needed to develop a practical optical system with better performance and a simpler structure.

Previously, we had proposed a spherical-lens microtracking CPV (SMTCPV) system [23]. This proposed SMTCPV system requires a spherical lens with a long focal length and a low spherical aberration. A radially symmetrical core-shell spherical (CSSP) lens has been studied for use in an antenna as well as in a CPV [24,25]. The CSSP lens is a kind of a simplified GRIN lens; because of the low-RI material used as the core material, it has a low spherical aberration and a long focal length. The CSSP lens may be suitable for application in a SMTCPV system. However, experimental studies of the SMTCPV with the CSSP lens have not been reported to date. Accordingly, here we analyze the optical characteristics of a CSSP lens made of commercially available materials, fabricate a prototype, evaluate the real-world performance, and compare its performance with a conventional homogenous spherical lens.

2. Configuration of the SMTCPV System

Figure 1 depicts the tracking motion of a SMTCPV system consisting of a spherical lens array and a solar-cell stage. Sunlight is concentrated onto the solar cell via mechanically controlling the relative position of the lens array and the solar-cell stage. As an example, the solar-cell stage can move in the lateral direction (XY movement), while the spherical-lens array moves in the vertical direction (Z movement), thus enabling a wide acceptance angle. Unlike a hemisphere or other lenses, the optical performance of the spherical lens does not depend on the angle of incidence unless interferences by adjacent lenses occurs in an array configuration. Thus, changing the lens angle, i.e., rotating the lens, is unnecessary, and the focal position is easily known, contributing to a simpler mechanical design and tracking control. A longer focal length is desirable for an SMTCPV system because the longer the focal length, the wider the mechanical angle limit. Figure 2 shows the schematic of the mechanical angle limit θ_{limit}, which is defined as the angle at which the top end surface of the solar-cell stage contacts the bottom end surface of the spherical lens. Thus, a longer focal length f enables a wider θ_{limit}, and contributes to widening the acceptance angle. Additionally, a shorter focal length would increase the reflection loss at the solar cell surface as the solar cell must receive the concentrated rays at a high incidence angle. If a micro spherical lens array is integrated with a compact internal tracking mechanism, a compact SMTCPV system can be designed. When a spherical lens with a diameter of 3–7 cm is used, the expected module thickness is 10–20 cm. Here, the optical performance of a single CSSP lens is examined as a first-step in considering its application to the SMTCPV system.

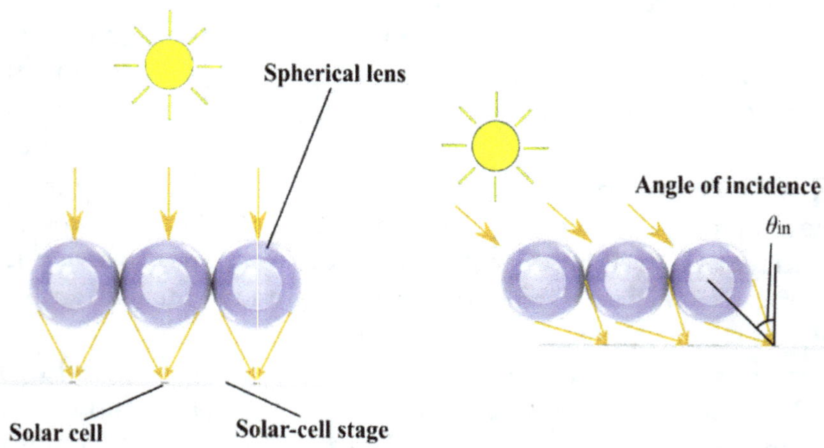

Figure 1. Tracking motion of the spherical-lens microtracking concentrator photovoltaic (CPV) (SMTCPV).

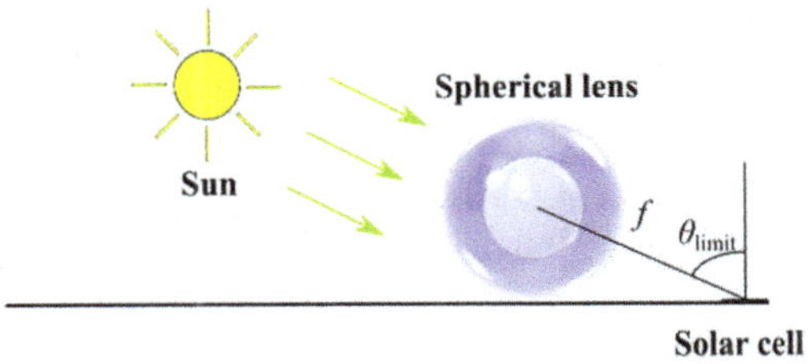

Figure 2. Mechanical angle limit of the microtracking concentrator photovoltaic systems (MTCPV) with a spherical lens.

3. Simulation

To comprehend the optical characteristics of the CSSP lens, ray-tracing analyses are conducted in the following steps: (1) Analysis of monochromatic light (sodium D line) without considering volumetric absorption by lens materials; (2) analysis of chromatic light (nearly full solar spectrum) without considering volumetric absorption by lens materials; (3) analysis of chromatic light (nearly full solar spectrum) with considerations of volumetric absorption by lens materials.

3.1. CSSP Lenses Characteristics for Monochromatic Light without Lens Material Absorption

Figure 3 shows the simulation model of the CSSP lens. Table 1 lists the simulation conditions A, which includes the specifications of the CSSP lens and the homogenous spherical lens, for comparison. This CSSP lens model includes a shell comprised of solid plastic (PMMA) and a core composed of liquid silicone oil. Three-dimensional ray-tracing was performed using commercial software (LightTools 8.7.0), in which the Fresnel reflection loss was considered. In this subsection, volumetric absorption was ignored.

The core RI, n_{core} = 1.403, is a silicone oil RI (Shin-Etsu Silicone KF-96-500cs [26]) at 589.5 nm. The core material requires the conditions listed in Table 2, which are satisfied by KF-96-500cs. The CSSP lens requires two different materials with different RIs. In the solid state, most materials have different thermal expansion coefficients and tend to cause a separation at the core-shell interface. Thus, one solution is to use a material in the liquid state with a high transparency, low RI, and less evaporation. The shell RI, n_{shell} = 1.491, is a PMMA RI at 587.6 nm [27]. The core radius, r_{core}, at which spherical aberration was effectively reduced was determined to be 10.2 mm [23]. For comparison, two homogenous spherical lenses composed of only PMMA and only KF-96-500cs are also simulated. The KF-96-500cs homogeneous spherical lens is a hypothetical lens since it is liquid. As explained in the Section 2, a higher optical efficiency and longer focal length are preferable for the SMTCPV system; thus, these two factors were analyzed.

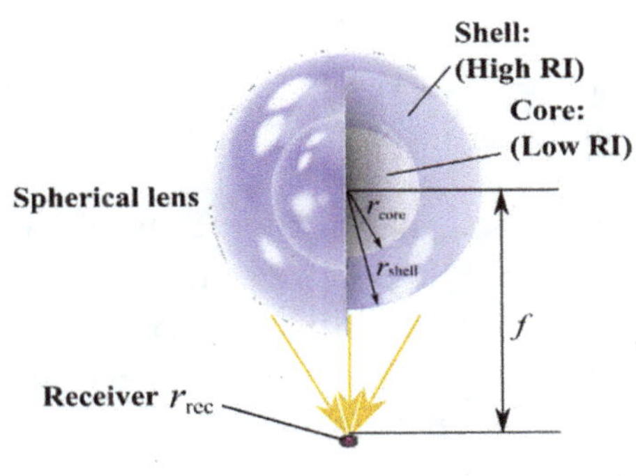

Figure 3. Simulation model of the core-shell spherical (CSSP) lens.

Table 1. Simulation conditions A.

Parameters	PMMA Homogenous Spherical Lens	KF-96-500cs Homogeneous Spherical Lens	CSSP Lens
Shell refractive index (RI), n_{shell}	1.491 @ 589 nm	1.403 @ 589 nm	1.491 @ 589 nm
Core RI, n_{core}	N/A	N/A	1.403 @ 589 nm
Shell radius, r_{shell}	15 mm	15 mm	15 mm
Core radius, r_{core}	N/A	N/A	10.2 mm (0.68 × r_{shell})
Geometrical concentration ratio, C		6.25× for Figure 8a 100× for Figures 8b and 9	
Ray spectrum		Sodium D line (589 nm)	
Absorption by lens materials		Ignored	

Table 2. Core material requirements and KF-96-500 cs specifications.

Required	Reason	KF-96-500cs [26]
Low refractive index	For long focal length	1.403 @ 589 nm
Liquid in ambient	Solid material tends to cause a separation at core-shell interface	Liquid
Low volatile matter content	No bubble formation	max 0.5% at 150 °C/24 h

Figure 4 shows the ray behavior of both the homogenous and the CSSP lenses, wherein Figure 4a,b show the homogeneous lenses and Figure 4c shows the CSSP lens. The rays of the homogenous spherical lenses are focused near the lens surface and spread out from this position, indicating that ray divergence due to spherical aberration is significant at a longer focal plane in the homogenous spherical lenses. On the other hand, the rays of the CSSP lens are focused slightly further away from the lens surface, and the focused spot size is larger than that of the homogenous lens. However, better ray convergence is observed at a longer focal plane, indicating reduced spherical aberration compared to that of the homogenous lens.

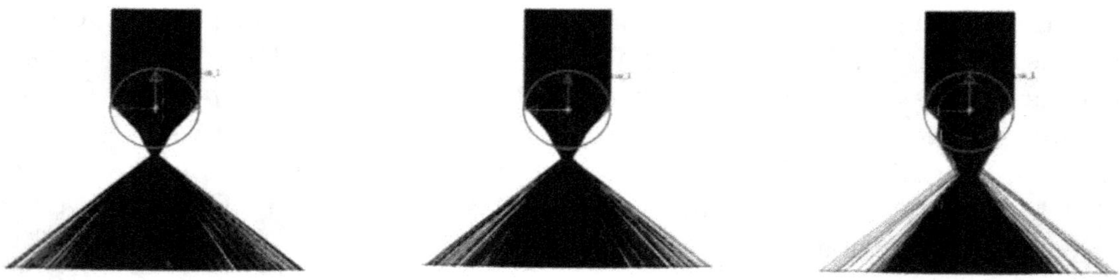

(a) (b) (c)

Figure 4. Ray behavior comparison of the (**a**) solid plastic (PMMA) homogenous spherical lens, (**b**) KF-97-500cs homogeneous spherical lens, and (**c**) CSSP lens.

Figures 5–7 depict the ray distribution, when the plane of the focal length, f, is varied for the PMMA homogeneous spherical lens, the KF-96-500cs homogeneous spherical lens, and the CSSP lens, respectively. The f = 15 mm plane is in contact with the surface of the spherical lens. For both of the homogenous lenses, the smallest spot size appears at the f = 15 mm plane and increases as f increases. The ray distribution is like that of a Gaussian distribution. For $f \geq 18$ mm, the ray convergence degrades. On the other hand, for the CSSP lens, even at f = 24mm, better concentration is achieved. The ray distribution is similar to a double circle. The outer circle shrinks at f = 24 mm, and both circles merge.

For $f > 24$ mm, the ray circle spreads, but the spot is still clearer than that of the homogenous lens at the same f.

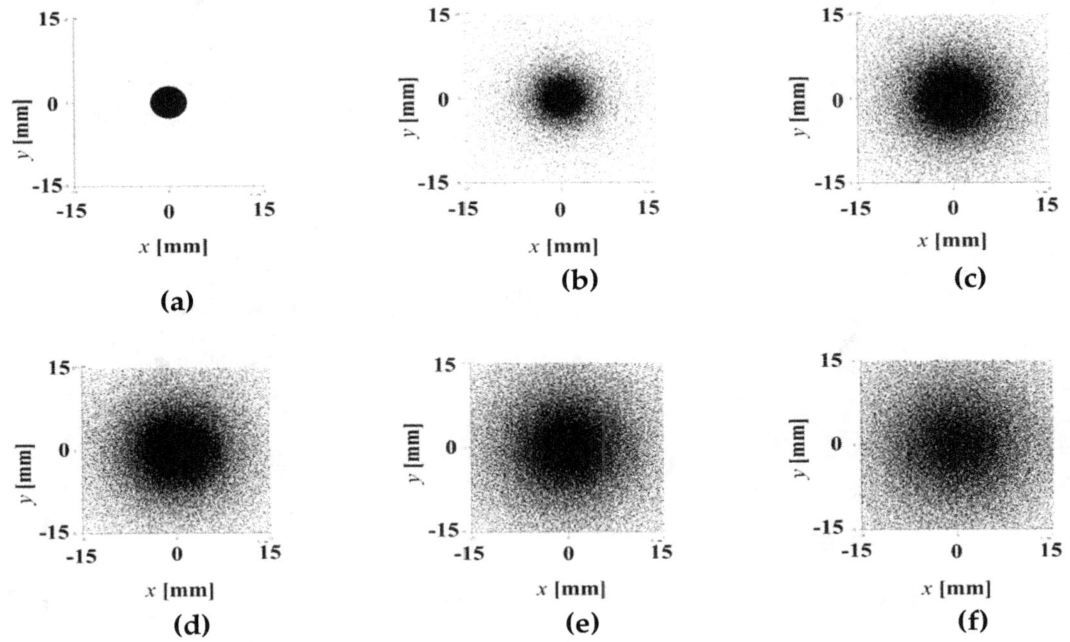

Figure 5. Simulated ray distribution of the PMMA homogenous spherical lens at the focal plane length f: f = (**a**) 15 mm, (**b**) 18 mm, (**c**) 24 mm, (**d**) 30 mm, (**e**) 36 mm, and (**f**) 42 mm.

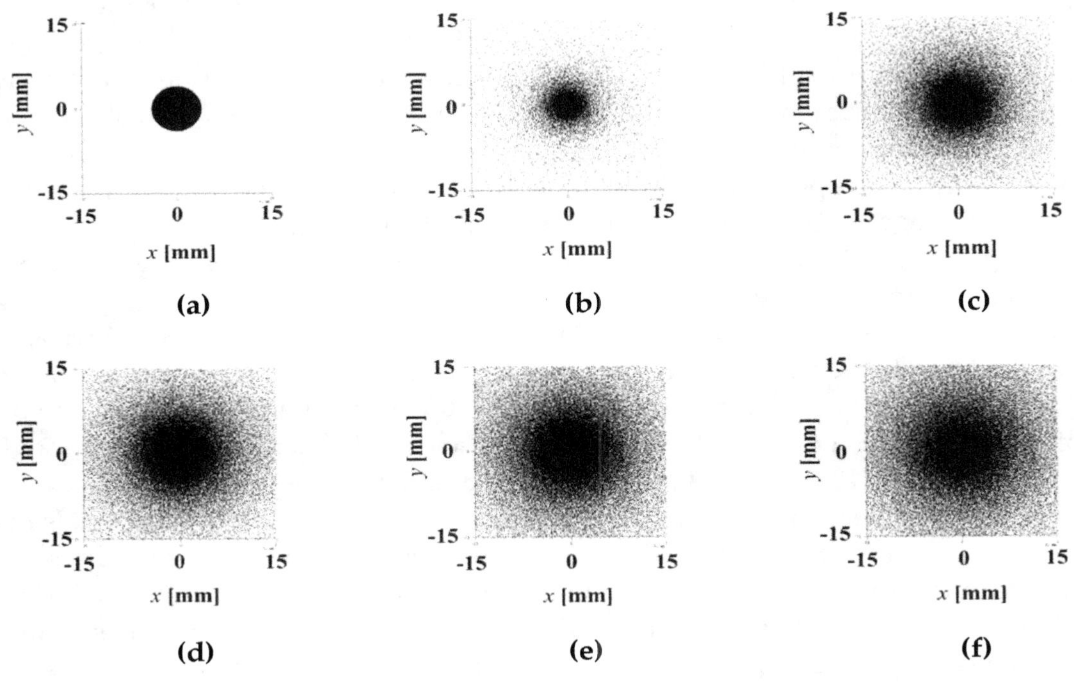

Figure 6. Simulated ray distribution of the KF-96-500cs homogenous spherical lens at the focal plane length f: f = (**a**) 15 mm, (**b**) 18 mm, (**c**) 24 mm, (**d**) 30 mm, (**e**) 36 mm, and (**f**) 42 mm.

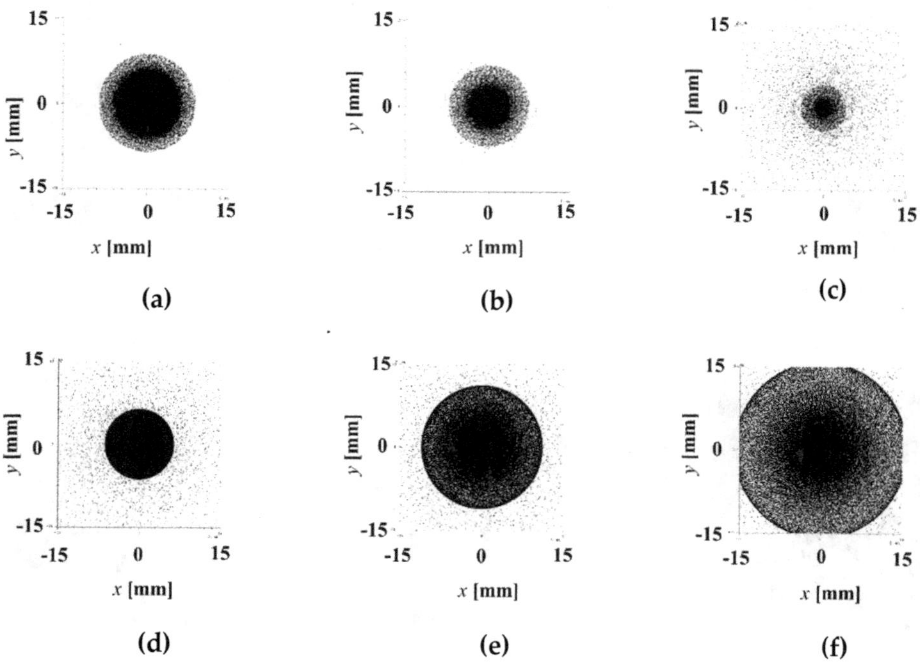

Figure 7. Simulated ray distribution of the CSSP lens at the focal plane length f: f = (**a**) 15 mm, (**b**) 18 mm, (**c**) 24 mm, (**d**) 30 mm, (**e**) 36 mm, and (**f**) 42 mm.

Figure 8 depicts variation in the optical efficiency, η_{opt}, with the change in the focal length, for the CSSP and the homogenous spherical lenses. The simulation conditions are the same as those listed in Table 1. Figure 8a shows the results of a low-concentration condition; the receiver radius, r_{rec} = 6 mm and the geometric concentration ratio, C = 6.25×, which is defined as the area ratio of the receiver to the lens aperture. Figure 8b shows the results of a high-concentration condition; r_{rec} = 1.5 mm and C = 100×.

In Figure 8a, the focal length range of the PMMA homogenous lens for η_{opt} > 75% is f = 15–18.3 mm (Δf = 3.3 mm), and that of the KF-96-500cs homogeneous lens is f = 15–23.1 mm (Δf = 8.1 mm), whereas that of the CSSP lens is f = 18.6–27.6 mm (Δf = 9 mm). Thus, in the CSSP lens, a wider focal length range with high optical efficiency exists at a longer focal length position. This characteristic is attributed to the low spherical aberration (as observed in Figure 7), which contributes to widening the mechanical angle limit of the SMTCPV. The homogeneous lenses exhibit a higher η_{opt} at shorter values of f where the beam spot is smaller than the CSSP lens. In Figure 8b, for η_{opt} > 75%, there is no focal length range in the homogenous lenses, while in the CSSP lens, there is a focal length range of f = 21.9–24.6 mm (Δf = 2.7 mm). The peak η_{opt} of the CSSP lens is even higher than that of the homogenous lenses for a higher value of f; the PMMA homogeneous lens shows a peak η_{opt} of 73.9% at f = 17.4 mm; the KF-96-500cs homogeneous lens, 73.8%, at 19.5 mm; the CSSP, 78.1%, at 24.6 mm. These results demonstrate the advantage of the CSSP lens, especially for high concentration conditions. However, even though the volumetric absorption loss by the lens material is ignored, the peak optical efficiency of the CSSP does not reach 80%. The low efficiency is attributed to the losses due to the Fresnel reflection at the Air-PMMA and PMMA-Silicone interfaces, where a part of the rays are directed to the outside of the receiver. The combination of refractive indexes of the lens materials in this study is not always the best; there may be a combination of refractive indexes that can further reduce this loss and improve efficiency. Thus, the best lens material should be further investigated in future work.

Figure 9 shows the simulated η_{opt} of the CSSP lens for varied r_{core} and f values. The simulation conditions are the same as those listed in Table 1. Here, r_{rec} = 1.5 mm and C = 100×. This result shows that there are two high-optical-efficiency regions around f = 16.5 mm at r_{core} = 0–9 mm and around f = 21 mm at r_{core} = 9–15 mm. The latter region shows the highest optical efficiency. This result suggests that the CSSP lens performs better than the homogenous spherical lens at a specific r_{core} and a longer focal length.

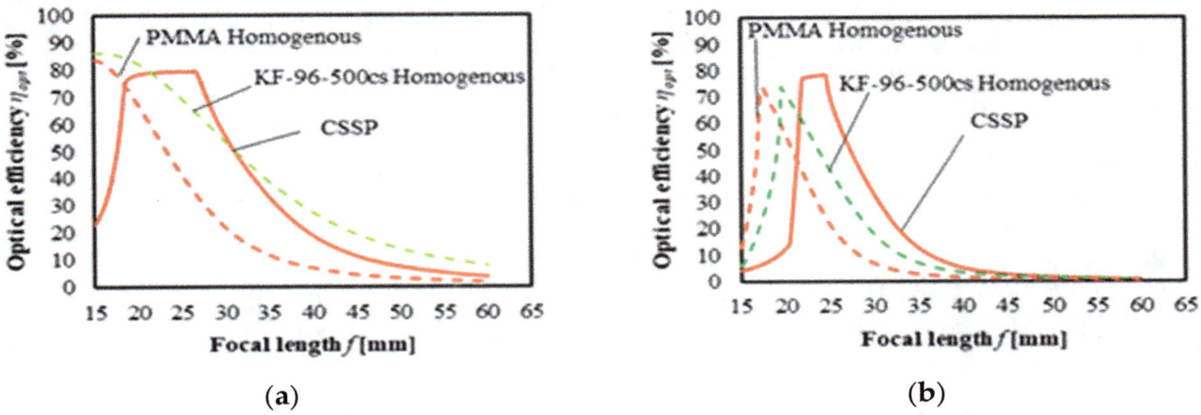

Figure 8. Simulated optical efficiency across varied focal lengths for the homogenous and core-shell spherical lenses with a the geometric concentration ratio C of (**a**) 6.25× and (**b**) 100×.

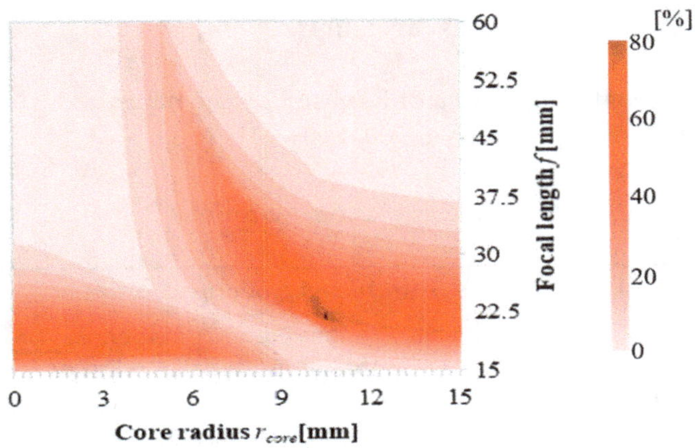

Figure 9. Simulated optical efficiency for varied core radii and focal lengths (C = 100×).

3.2. CSSP Lenses Characteristics for Full Solar Spectrum Without Lens Material Absorption

Prior to the simulation, the wavelength dispersion of RI and the transmittance of KF-96-500cs was measured, as that for PMMA was known that for KF-96-500cs was unknown. RI measurement was performed using an Abbe refractometer DR-M2 (Atago Co., Ltd.). Optical filters for 480, 546, 589, and 656 nm were used. The measuring temperature was 25.5 °C ± 0.1°. To extrapolate the RI profile for the full solar spectrum range, the measured RIs at the four wavelengths were adopted to the following Cauchy dispersion equation:

$$n = \alpha_0 + \frac{\alpha_2}{\lambda^2} + \frac{\alpha_4}{\lambda^4} + \ldots + \frac{\alpha_{2x}}{\lambda^{2x}} \quad (1)$$

Here, n is RI at wavelength λ, and α_{2x} are the Cauchy coefficients. Table 3 lists the estimated Cauchy coefficients. The absorption measurement was performed using a UV-VIS-IR spectrometer V-670 (JASCO Co., Ltd.). The optical path length was 10 mm. Figure 10 shows the extrapolated profile of RI, n_{KF} (solid red line), and transmittance, $\%T$ (solid blue line). The measured RIs are illustrated as square points. This result reveals that RI decreases with increasing wavelengths, which is consistent with the reported trend of the conventional silicone elastomer [28]. The transmittance for $\lambda = 300-1100$ nm is nearly 100%, and low around $\lambda = 1180, 1400, 1700$ nm.

From the wavelength dispersion data of RI, the ray-tracing analysis for chromatic light, i.e., the nearly full solar spectrum, is conducted under the simulation condition B, listed in Table 4. The power generation performance of a multijunction solar cell decreases when the generated current of each subcell of the multijunction solar cell is different (so-called current mismatch); thus, the effect of chromatic aberration on the concentrated light for spectral bandwidths corresponding to each subcell of the multijunction solar cell is analyzed.

Table 3. Estimated Cauchy coefficients for KF-96-500cs.

Cauchy Coefficients	Values
α_0	1.389
α_2	0.005296
α_4	−0.00018

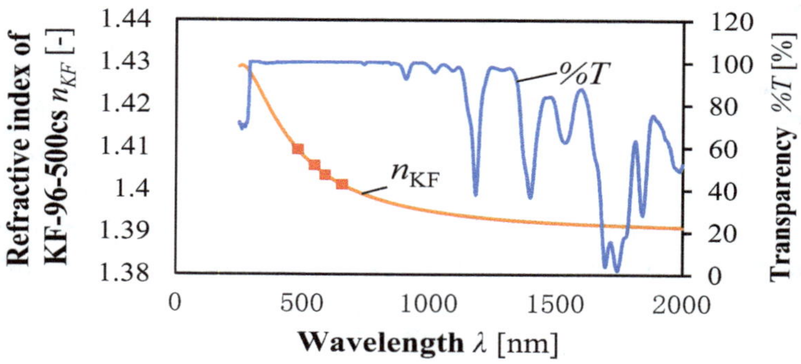

Figure 10. Estimated wavelength dispersion of RI based on the measured RI at four wavelengths and the measured spectral transmittance of KF-96-500cs.

Table 4. Simulation conditions B.

CSSP Lens Simulation Specifications	
Shell RI, n_{shell}	Wavelength dispersion of PMMA is considered
Core RI, n_{core}	Wavelength dispersion of KF-96-500cs is considered
Shell radius, r_{shell}	15 mm
Core radius, r_{core}	10.2 mm (0.68 × r_{shell})
Geometrical concentration ratio, C	100×
Ray spectrum	AM1.5D standard solar spectrum Top subcell range: 280–749 nm Middle subcell range: 750–949 nm Bottom subcell range: 950–2000 nm Full range: 280–2000 nm
Absorption by lens materials	Ignored

Figure 11 shows the simulated η_{opt} of the homogenous and CSSP lenses for varied values of f in comparison with different wavelength bandwidths corresponding to the top, middle, and bottom subcells of the triple-junction solar cell used in the experiment described in the Section 4, as well as the full solar spectrum. Table 5 summarizes the maximum optical efficiency for each spherical lens and wavelength range. As shown in Figure 11a,b, the homogeneous lenses shows sharp peaks at short f and the peak values of the three subcell-ranges are different. This difference is more emphasized in the KF-96-500cs homogeneous lens, and the η_{opt} for the full range is reduced. While in contrast, as shown in Figure 11c, it is confirmed that the η_{opt} of the CSSP lens for the top, middle, and bottom subcells is largely similar in the range of f = 21.9–23.8 mm, where the power generation loss due to the chromatic aberration is minimized and η_{opt} is nearly 80%. For f < 21.9 mm, η_{opt} of the CSSP lens for the top subcell is greater than that for the middle and bottom subcells; whereas, for f > 23.8 mm, η_{opt} of the CSSP lens for the bottom subcell is greater than that for the middle and top subcells. These characteristics are attributed to the change in the spot size due to chromatic aberration. Compared to Figure 8b, although the chromatic aberration reduces the high-η_{opt} range of f, the CSSP lens retains a high η_{opt} at a higher value of f than the homogenous lenses. In usual achromatic lenses, layered structures, e.g., achromatic doublet, reduce the chromatic aberration, however, unlike the spherical lenses, the achromatic lenses must be rotated to keep facing the sun for the solar concentration.

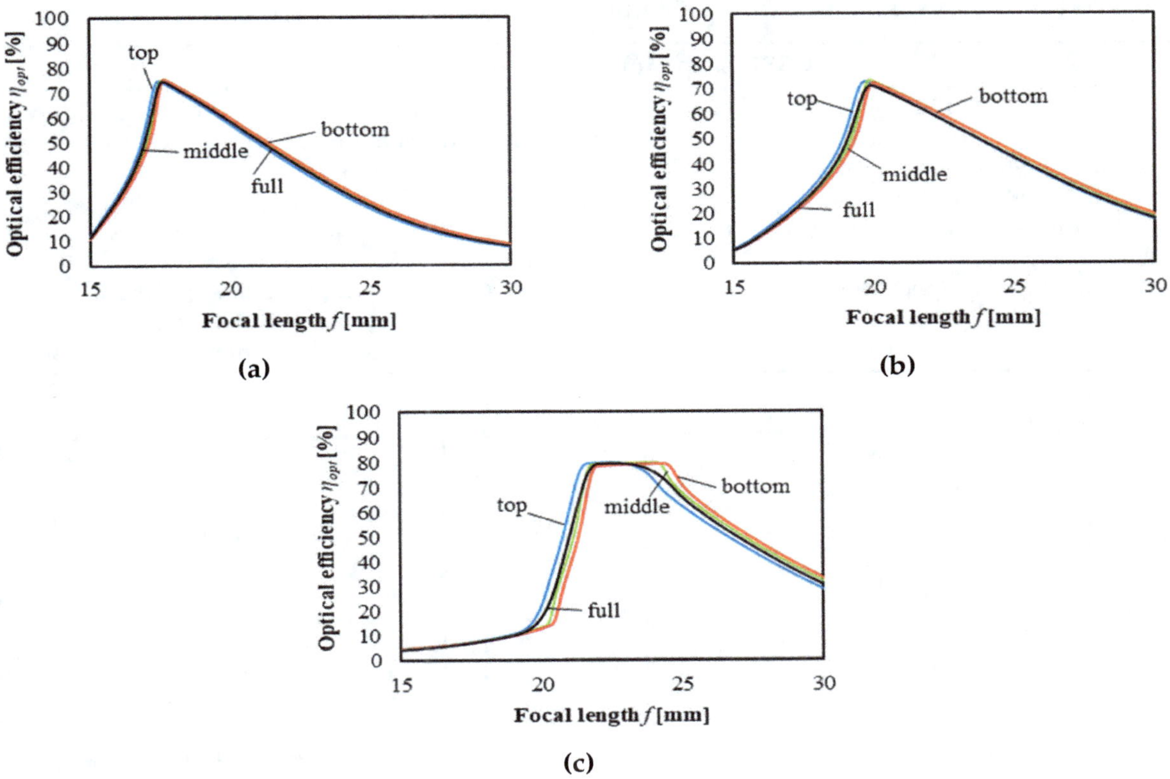

Figure 11. Simulated optical efficiency for varied focal lengths in comparison with different wavelength bandwidths corresponding to the top (blue), middle (green), bottom (red) subcell of the triple-junction solar cell, and the full solar spectrum at C = 100× (black). (**a**) PMMA homogenous spherical lens, (**b**) KF-97-500cs homogeneous spherical lens, and (**c**) CSSP lens.

Table 5. The maximum optical efficiency of the lenses corresponding to Figure 11.

Parameters	PMMA Homogeneous Lens	KF-96-500cs Homogeneous Lens	Cssp Lens
Top subcell range	74.7%	72.9%	79.8%
Middle subcell range	75.2%	73.5%	79.4%
Bottom subcell range	75.0%	72.4%	79.2%
Full range	74.4%	71.5%	79.3%

3.3. CSSP Lenses Characteristics for Full Solar Spectrum with Lens Material Absorption

Another ray-tracing simulation is performed under the simulation condition C listed in Table 6. Herein, not only the wavelength dispersion of RI but also the spectral transmittance of the lens materials are taken into account; additionally, the effect of the CSSP lens size is analyzed by varying the shell radius from 15–1 mm. To keep the similarity of the lens geometry, the ratio of the core radius to the shell radius is kept at 0.68.

Figure 12 shows the simulated η_{opt} of the CSSP lens for varied values of f in comparison with different lens sizes when considering the volumetric absorption by the lens materials. On the horizontal axis, f is normalized by the r_{shell}. This result shows that the optical efficiency is improved by reducing the shell radius. For C = 6.25×, the peak η_{opt} percentages are: 71.1% at r_{shell} = 15 mm, 73.4% at 10 mm, 76.7% at 5 mm, and 80.0% at 1 mm; for C = 100×, 68.6% at 15 mm, 70.9% at 10 mm, 74.5% at 5 mm, and 78.1% at 1 mm. The peak η_{opt} at r_{shell} = 15 mm is 8.9 percentage points (at C = 6.25×) and 9.5 percentage point (at C = 100×) less than that at r_{shell} = 1 mm. Thus, the CSSP lenses with smaller radii are more suitable for the SMTCPV. However, because it was difficult to fabricate small lenses, homogenous spherical and CSSP lenses with r_{shell} = 15 mm were tested in the experiment presented in the next section.

Table 6. Simulation conditions C.

CSSP Lens Simulation Specifications	
Shell RI, n_{shell}	Wavelength dispersion of PMMA is considered
Core RI, n_{core}	Wavelength dispersion of KF-96-500cs is considered
Shell radius, r_{shell}	1, 5, 10, 15 mm
Core radius, r_{core}	$0.68 \times r_{shell}$
Geometrical concentration ratio, C	6.25× for Figure 12a 100× for Figure 12b
Ray spectrum	AM1.5D standard solar spectrum Full range: 280–2000 nm
Absorption by lens materials	Spectral transmittance of lens materials is considered

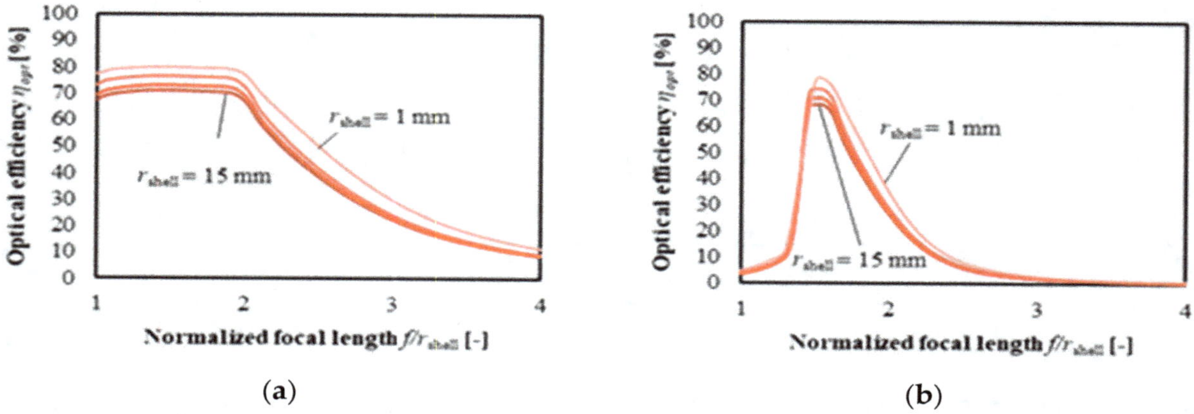

Figure 12. Simulated optical efficiency of the CSSP lens for varied normalized focal lengths across different shell radii lengths at (**a**) $C = 6.25\times$ and (**b**) 100×.

4. Experiment

4.1. Visual Comparison

Figure 13a,b display comparative photographs of the prototype homogenous and CSSP lenses, respectively (with the structure shown in Figure 3). The r_{shell} of both lenses was 15 mm. The r_{core} of the CSSP lens was 9.8 mm. The material of the homogenous lens was PMMA; whereas the materials of the CSSP lens were silicone oil KF-96-500cs (Shin-Etsu chemical) for the core and PMMA for the shell. These photographs were captured when three lines of letters were read the most clearly by the lenses. Comparing Figure 13a with Figure 13b, the letters at the center of both lenses are clear, whereas those at the edge of the homogenous lens are more distorted than those at the edge of the CSSP lens. As spherical aberration occurs at the edge of the spherical lens, the letters are strongly affected, as shown in Figure 13a. On the other hand, Figure 13b shows letters up to the edge of the lens. This result demonstrates that the CSSP lens reduces spherical aberration in the field of view.

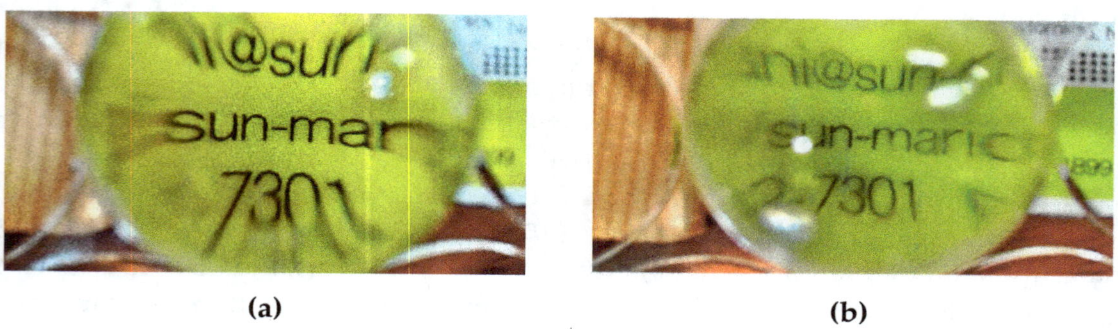

Figure 13. Comparative photographs of the prototype (**a**) homogenous spherical lens and (**b**) CSSP lens.

4.2. Outdoor Experiment

Using the prototype lenses shown in Figure 13, the module efficiency of the lens-cell CPV module was measured under outdoor conditions in Isezaki-city, Japan (latitude 36.1°, longitude 138.6°). The direct solar irradiation was measured by a pyrheliometer near the experimental setup (not shown in the figure). Figure 14a shows a solar cell receiver with a lens holder. The spherical lens was attached to this lens holder. By changing the height of the lens holder, the focal length was set as f = 18, 24, and 30 mm, corresponding to the analysis of Figures 5 and 7. Figure 14b shows the lens-cell module with the spherical lens attached to the solar cell receiver; Figure 14c shows a photograph of the measurement setup. The solar cell receiver used was the TCSC100A (Taicrystal Co., Ltd.), with a solar cell area of 100 mm^2. The type of solar cell was a GaInP/GaInAs/Ge triple-junction solar cell, which has been designed to achieve approximately 40% cell efficiency at 500 suns. The lens aperture area was 706.5 mm^2, which is the area of a circle with r_{shell} = 15 mm; hence, C = 7×. It is noted that this concentration ratio is much lower that of the designed concentration ratio of the solar cell used in the experiment. It is also noted that in this experiment, the solar cell can receive the light impinging directly to the solar cell (without going through the optics), which is mostly a part of the diffuse component of sunlight, and is very small compared to the concentrated direct component; thus, it does not have a significant impact on changing the trend of the results.

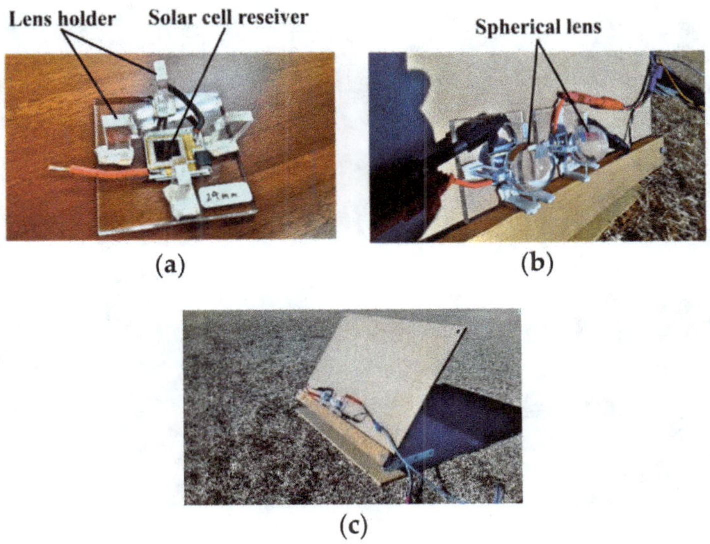

Figure 14. Test module photographs: (**a**) Solar cell receiver, (**b**) lens-cell module, and (**c**) measurement setup.

Figure 15 shows photographs of concentrated sunlight at the surface of the solar cell for both prototype lenses with different focal lengths. For the homogenous spherical lens, the light spot is concentrated sufficiently for f = 18 mm (left of Figure 15a); the light spot was distorted because of spherical aberration for f = 24 mm (left of Figure 15b); the light spot cannot be seen for f = 30 mm (left of Figure 15c). For the CSSP lens, a large light spot can be seen for f = 18 mm (right of Figure 15a); the light spot becomes smaller for f = 24 mm (right of Figure 15b). The light spot is still maintained because of low aberration for f = 30 mm (right of Figure 15c). The behavior of the light spot is consistent with that of the simulated result in Figures 5 and 7. Under the condition of C = 100× in Figure 8b, the homogenous lens with f = 18 mm and the CSSP lens with f = 24 mm achieve high optical efficiency. These high optical efficiency conditions correspond to the conditions of the figure, where the light spot is smaller and shrunk (left of Figure 15a and right of Figure 15b).

Figure 16 shows the short circuit current, I_{sc}, and the module efficiency, η_{module}, for the homogenous and CSSP lenses. The dashed line denotes the homogenous lens, whereas the solid line denotes the CSSP lens. Current-voltage curves were measured every 20 s under constant direct solar irradiation and I_{sc}, open circuit voltage, fill factor, and maximum power were obtained by averaging the measured data over 10 minutes. The η_{module} was then calculated as the ratio of maximum power to the direct solar irradiation incident on the lens aperture area. The results for f = 18, 24, and 30 mm were obtained under no-cloud condition at 11:03–11:14, 13:27–13:37, and 11:21–11:31, respectively. Though the solar spectrum was not directly measured, around noon on a clear sky day, the spectrum is generally neither too red-shifted nor too blue-shifted, according to long-term observation data of solar spectrums in Japan [29].

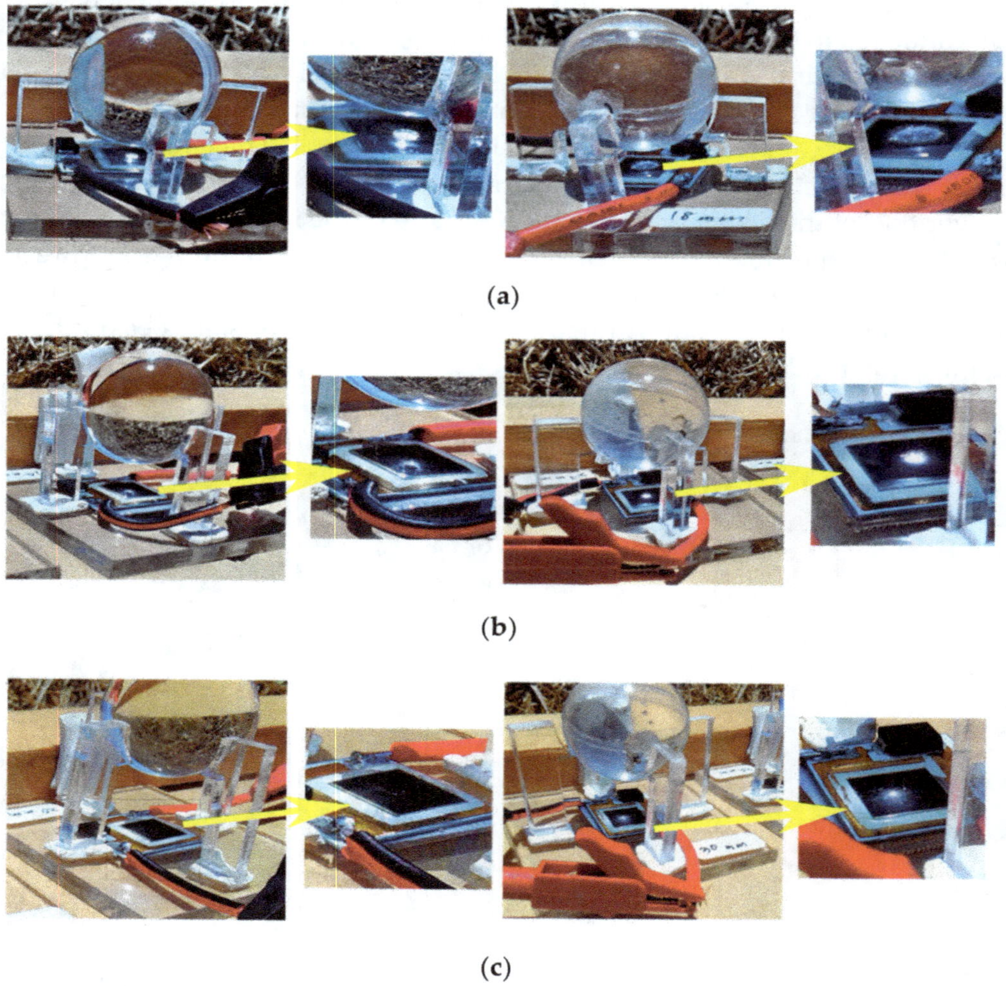

Figure 15. Photographs of concentrated sunlight by the homogenous (left) and the CSSP lenses (right) with different focal lengths f: f = (**a**) 18 mm, (**b**) 24 mm, and (**c**) 30 mm.

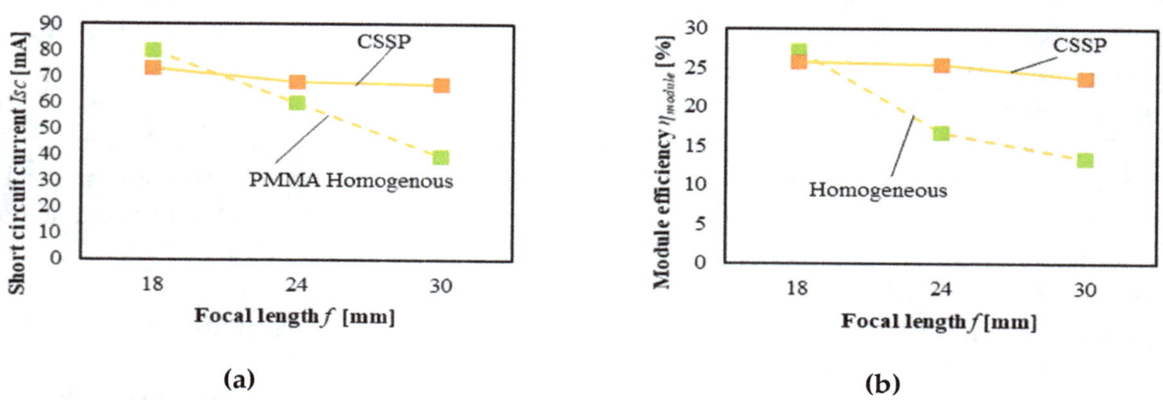

Figure 16. Experimental results of the lens-cell module for the homogenous and CSSP lenses regarding the (**a**) short circuit current and (**b**) module efficiency.

It was confirmed that the lens-cell module with the CSSP lens exhibited greater I_{sc} and η_{module} than that with the homogenous lens for a longer f. This trend was more emphasized in the η_{module}. This is because, with the homogenous lens, not only I_{sc} but also the fill factor decreases as f increases. The fill factor decreased because of non-uniformity of concentrated sunlight at the solar cell surface, as shown in the left side of Figure 15b. The resulting I_{sc} and η_{module} of the homogenous lens at f = 18 mm were high: I_{sc} = 79.5 mA and η_{module} = 27.1%. However, as f increased, I_{sc} and η_{module} decreased drastically. On the other hand, for the CSSP lens, the results for f = 18, 24, and 30 mm were I_{sc} = 73.3 mA and η_{module} = 25.7%, I_{sc} = 68.0 mA and η_{module} = 25.4%, and I_{sc} = 67.1 mA and η_{module} = 23.6%, respectively. Even when f increased further, I_{sc} and η_{module} decreased slightly.

A previous study [23] presented the relationship between the normalized focal length, f/r_{shell}, and the mechanical limit of the tracking angle, θ_{in}. The results for f = 18, 24, and 30 mm were θ_{in} = 33.6°, 51.3°, and 60°, respectively. A high module efficiency at a long focal length leads to an increase in θ_{in} in the SMTCPV system.

Figure 17 shows a comparison between the measured I_{sc} values from Figure 16 and the simulated η_{opt} of Figure 8a. The I_{sc} of the solar cell is generally proportional to irradiation at the solar cell surface. Thus, I_{sc} is considered a measure of the optical efficiency of the concentrator. The red line indicates the simulated η_{opt}, while the orange line indicates the measured I_{sc}. The dashed line indicates the homogenous spherical lens, whereas the solid line indicates the CSSP lens. The orange points indicate the actual measuring points. The horizontal axis is the normalized focal length f/r_{shell}. This result shows that the tendency of the measured I_{sc} is qualitatively consistent with the simulated η_{opt}. The trend of I_{sc} of the CSSP lens was lower compared to that of the simulated η_{opt}; this is likely caused by the occurrence of stray light at the bonding part of the upper and bottom parts of the shell. Otherwise, lens shape error may have occurred in the lens fabrication process. A higher concentration experiment using a smaller triple-junction solar cell was also attempted; however, sharp focusing and the concentration performances were very poor. This may also be attributed to lens shape error during fabrication and/or deformation during experimentation. By further identifying reasons for the lower I_{sc} of the CSSP lens compared to that of the simulated η_{opt}, I_{sc} and η_{module} can be increased further. Usage of a solar cell designed for a proper concentration ratio nearer to the experimental condition can also increase η_{module}.

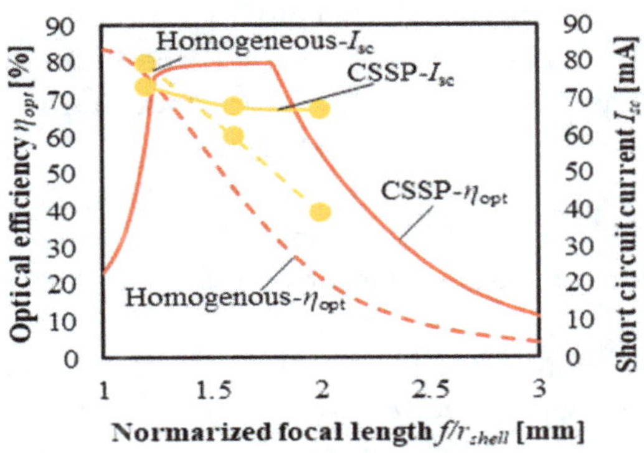

Figure 17. Comparison of the measured short circuit current and the simulated optical efficiency.

5. Conclusions

The optical characteristics of a radially symmetrical CSSP lens is analyzed regarding its suitability to microtracking CPV applications. The CSSP lens, with a core composed of a material with a lower refractive index, exhibits a higher optical efficiency than the homogenous spherical lens for longer focal lengths. The CSSP lens with the appropriate core radius can reduce the spherical and chromatic aberrations for a long focal length. This characteristic is preferable for microtracking CPVs with a spherical lens array to achieve a higher module efficiency for a wider incidence angle. In the outdoor experiment, the lens-cell CPV module with the prototype CSSP lens made of commercially available materials showed superior optical efficiency and module efficiency compared to the module with the homogenous spherical lens, as expected by the optical simulation. Because the simulated and experimental conditions were limited to the on-axis condition, i.e., an incidence angle of zero, and a single lens, further investigations are necessary to determine the practical characteristics and performance of the microtracking CPV system with the CSSP lens array. Additionally, the lower optical performance of the fabricated CSSP lens compared to the simulated one should be further investigated. Thus, a manufacturing method to fabricate the CSSP lens without optical defects is also necessary, as well as the selection of appropriate core and shell materials.

Author Contributions: Conceptualization, M.N.; formal analysis, M.N.; experimentation, M.N.; writing—original draft preparation, M.N.; writing—review and editing, N.Y.; supervision, N.Y.

References

1. Dimroth, D.F. New World Record for Solar Cell Efficiency at 46%—French-German Cooperation Confirms Competitive Advantage of European Photovoltaic Industry. Available online: https://www.ise.fraunhofer.de/en/press-media/press-releases/2014/new-world-record-for-solar-cell-efficiency-at-46-percent.html (accessed on 11 September 2019).
2. Steiner, M.; Bösch, A.; Dilger, A.; Dimroth, F.; Dörsam, T.; Muller, M.; Hornung, T.; Siefer, G.; Wiesenfarth, M.; Bett, A.W. FLATCON® CPV module with 36.7% efficiency equipped with four-junction solar cells. *Prog. Photovolt. Res. Appl.* **2015**, *23*, 1323–1329. [CrossRef]
3. Apostoleris, H.; Stefancich, M.; Chiesa, M. Tracking-integrated systems for concentrating photovoltaics. *Nat. Energy* **2016**, *1*, 16018. [CrossRef]
4. Duerr, F.; Meuret, Y.; Thienpont, H. Tailored free-form optics with movement to integrate tracking in concentrating photovoltaics. *Opt. Express* **2013**, *21*, A401–A411. [CrossRef]
5. Zagolla, V.; Dominé, D.; Tremblay, E.; Moser, C. Self-tracking solar concentrator with an acceptance angle of 32°. *Opt. Express* **2014**, *22*, A1880–A1894. [CrossRef]
6. Ma, H.; Wu, L. Horizontally staggered lightguide solar concentrator with lateral displacement tracking for high concentration applications. *Appl. Opt.* **2015**, *54*, 6217–6223. [CrossRef]
7. Michel, C.; Blain, P.; Clermont, L.; Languy, F.; Lenaerts, C.; Fleury-Frenette, K.; Décultot, M.; Habraken, S.; Vandormael, D.; Cloots, R.; et al. Waveguide solar concentrator design with spectrally separated light. *Sol. Energy* **2017**, *157*, 1005–1016. [CrossRef]
8. Hallas, J.M.; Baker, K.A.; Karp, J.H.; Tremblay, E.J.; Ford, J.E. Two-axis solar tracking accomplished through small lateral translations. *Appl. Opt.* **2012**, *51*, 6117–6124. [CrossRef]
9. Dhakal, R.; Lee, J.; Kim, J. Bio-inspired thin and flat solar concentrator for efficient, wide acceptance angle light collection. *Appl. Opt.* **2014**, *53*, 306–315. [CrossRef]
10. Karp, J.H.; Tremblay, E.J.; Hallas, J.M.; Ford, J.E. Orthogonal and secondary concentration in planar micro-optic solar collectors. *Opt. Express* **2011**, *19*, A673–A685. [CrossRef]
11. Bouchard, S.; Thibault, S. Planar waveguide concentrator used with a single axis tracker. *Opt. Express* **2014**, *51*, 6848–6854. [CrossRef]
12. Price, J.S.; Sheng, X.; Meulblok, B.M.; Rogers, J.A.; Giebink, N.C. Wide-angle planar microtracking for quasi-static microcell concentrating photovoltaics. *Nat. Commun.* **2015**, *6*, 6223. [CrossRef]
13. Lloyd, J.; Pavilonis, M.; Gladden, C.; Casper, C.; Schneider, K.; McMahon, W. Performance of a prototype stationary catadioptric concentrating photovoltaic module. *Opt. Express* **2018**, *26*, A413–A419. [CrossRef]
14. Vu, N.; Shin, S. Flat Concentrator Photovoltaic System with Lateral Displacement Tracking for Residential Rooftops. *Energies* **2018**, *11*, 114. [CrossRef]
15. Ito, A.; Sato, D.; Yamada, N. Optical design and demonstration of microtracking CPV module with bi-convex aspheric lens array. *Opt. Express* **2018**, *26*, A879–A891. [CrossRef]
16. Schrank, H.; Sanford, J. A Luneberg-Lens Update. *IEEE Antennas Propag. Mag.* **1995**, *37*, 76–79.
17. Morgan, S.P. General solution of the Luneberg lens problem. *J. Appl. Phys.* **1958**, *29*, 1358–1368. [CrossRef]
18. Kotsidas, P.; Modi, V.; Gordon, J.M. Realizable planar gradient-index solar lenses. *Opt. Lett.* **2012**, *37*, 1235–1237. [CrossRef]
19. Dodge, M.J. Refractive properties of magnesium fluoride. *Appl. Opt.* **1984**, *23*, 1980–1985. [CrossRef]

20. Liu, Z.; Yang, S.; Nie, Z. A dielectric lens antenna design by using the effective medium theories. In Proceedings of the 2010 International Symposium on Intelligent Signal Processing Communication Systems, Chengdu, China, 6–8 December 2010; pp. 1–4.
21. Xi, J.Q.; Schubert, M.F.; Kim, J.K.; Schubert, E.F.; Chen, M.; Lin, S.Y.; Liu, W.; Smart, J.A. Optical thin-film materials with low refractive index for broadband elimination of Fresnel reflection. *Nat. Photonics* **2007**, *1*, 176–179. [CrossRef]
22. Tabata, M.; Adachi, I.; Ishii, Y.; Kawai, H.; Sumiyoshi, T.; Yokogawa, H. Development of transparent silica aerogel over a wide range of densities. *Nucl. Instrum. Methods Phys. Res. A* **2010**, *623*, 339–341. [CrossRef]
23. Nakatani, M.; Yamada, N. Optical Simulation of Two-Shell Spherical Lens for Microtracking CPV System. In Proceedings of the IEEE 7th World Conference on Photovoltaic Energy Conversion, Waikoloa, HI, USA, 10–15 June 2018; pp. 927–930.
24. Rhys, T.L. The Design of Radially Symmetric Lenses. *IEEE Trans. Antennas Propag.* **1970**, *18*, 497–506. [CrossRef]
25. Kotsidas, P.; Modi, V.; Gordon, J.M. Nominally stationary high-concentration solar optics by gradient-index lenses. *Opt. Express* **2011**, *19*, 2325–2334. [CrossRef]
26. Shin-Etsu-Chemical Silicone Fluid Catalog Shin-Etsu Chemical Catalog. Available online: http://www.shinetsusilicone-global.com/catalog/pdf/fluid_e.pdf (accessed on 11 September 2019).
27. Sultanova, N.; Kasarova, S.; Nikolov, I. Dispersion properties of optical polymers. *Acta Phys. Pol. A* **2009**, *116*, 585–587. [CrossRef]
28. Cooper, P.R. Refractive-index measurements of paraffin, a silicone elastomer, and an epoxy resin over the 500–1500-nm spectral range. *Appl. Opt.* **1982**, *21*, 3413–3415. [CrossRef]
29. New Energy and Industrial Technology Development (NEDO), Solar Spectrol Database VER-3 2018. Available online: https://www.nedo.go.jp/library/nissharyou.html (accessed on 11 September 2019).

12

Experimental Study on Vacuum Performance of Parabolic Trough Receivers based on a Novel Non-Destructive Testing Method

Fangyuan Yao [1,2,3,4], Dongqiang Lei [1,2,3,4,*], Ke Yu [5], Yingying Han [5], Pan Yao [1,2,3,4], Zhifeng Wang [1,2,3,4], Quanxi Fang [5] and Qiao Hu [5]

1 Key Laboratory of Solar Thermal Energy and Photovoltaic System, Chinese Academy of Sciences, Beijing 100190, China; yaofangyuan@mail.iee.ac.cn (F.Y.); yaopan@mail.iee.ac.cn (P.Y.); zhifeng@vip.sina.com (Z.W.)
2 Institute of Electrical Engineering, Chinese Academy of Sciences, Beijing 100190, China
3 University of Chinese Academy of Sciences, Beijing 100049, China
4 Beijing Engineering Research Center of Solar Thermal Power, Beijing 100190, China
5 Royal Tech CSP Limited, Changzhou 213163, Jiangsu, China; operating@royalcsp.com (K.Y.); han.tech@royalcsp.com (Y.H.); royaltech-fang@royalcsp.com (Q.F.); joey.hu@royalcsp.com (Q.H.)
* Correspondence: ldqljmei@126.com.

Abstract: The loss of vacuum in the parabolic trough receivers is one of the most common problems in the parabolic trough solar power plants. The vacuum level and gas species in the annulus of the receiver determine the heat loss and have an important influence on the thermal efficient of the solar system. If hydrogen is inside the annulus, it can cause heat losses to be almost four times that of a receiver with good vacuum. However, it is hard to non-destructively measure the gas species and partial pressure in the annulus of the receiver. In this paper, a novel non-destructive method was presented to evaluate the vacuum performance by using combined dielectric barrier discharge and the spectral analysis technology. The discharge characteristics and spectrometric properties of four kinds of gases, which are the most likely gases to be found in the receivers, were studied in the experiments. The test results of the non-destructive vacuum evaluation method agree well with the results of the residual gas analysis. The feasibility and accuracy of the non-destructive test method was verified. The relationship between the vacuum performance of receiver and the spectral characteristics of dielectric barrier discharge were obtained by a series of experiments.

Keywords: parabolic trough receiver; vacuum performance; partial vacuum pressure; optical fiber spectrometer; gas discharge

1. Introduction

Concentrating solar power (CSP) is renewable energy technology and offers important advantages as it has the ability of thermal storage. In CSP systems, parabolic trough technology is the most mature and widely used solar thermal power technologies worldwide [1,2]. Parabolic trough receiver tubes are the core components, which convert solar energy into thermal energy. As shown in Figure 1, the tubes are comprised of glass envelope coated in a layer of anti-reflective coating, and absorber tubes covered in a solar selective coating. The glass envelope and the metal tube are connected on both ends by a glass-metal seal and a metal bellows to achieve the necessary vacuum-tight annulus space and to accommodate for thermal expansion difference between the absorber tube and the glass envelope [3,4]. The annulus space must be in a vacuum state to significantly reduce the gas heat convection or conduction and to prevent the solar selective coating oxidizing in high temperatures, which can lead to heat losses from the parabolic trough receiver tubes. A getter is also emplaced within the annulus space to absorb hydrogen and other gases to maintain its vacuum state [5,6].

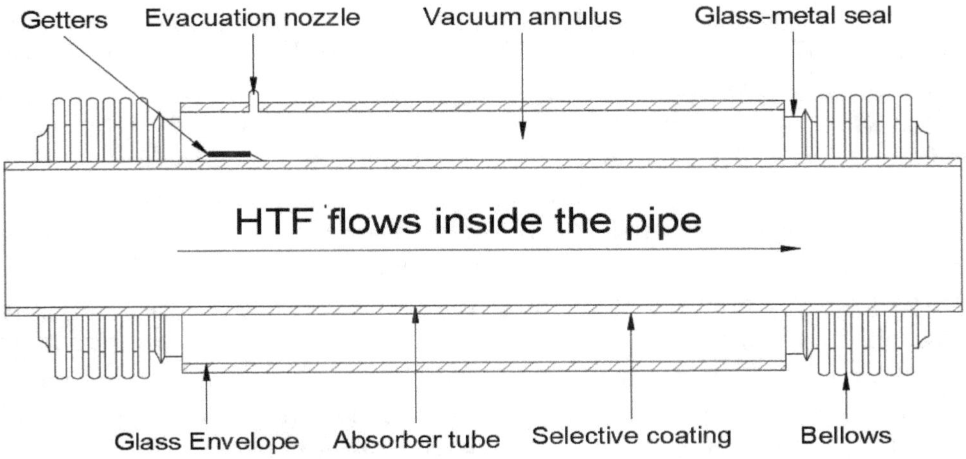

Figure 1. Typical parabolic trough receiver.

The failures of the receiver tubes have been a problem in parabolic trough solar power plants. The data on the Solar Energy Generating Systems (SEGS) plants that have been operating for a number of years show that receiver tubes have an accumulated failure rate of over 4% [7]. Survey by the U.S. National Renewable Energy Laboratory (NREL) found that the most recent data for the Solar Energy Generating Systems (SEGS) plants indicated that receiver tube failures had decreased to 3.37% of the total field receivers per year [8]. Failures involved vacuum loss, hydrogen infiltration, glass envelope breakage, and degradation of the coating. Of these failures, 55% were reported to involve broken glass and 29% involved loss of vacuum [7,8]. Changes in vacuum performance within the annulus space are inevitable, factors contributing towards the loss of vacuum include materials outgassing, gases permeation, vacuum leakage at seals and welds, getter failure, or saturation [9,10]. Liu et al. [11] determined experimentally that there were H_2O, CO_2, H_2, N_2, Ar, and He in the annulus space. Moens [10] presented that the long-term exposure of the heat transfer fluid (diphenyl oxide/biphenyl) to 400 °C temperatures could lead to hydrogen pressures that exceed the gas-absorbing capacity of the hydrogen getters, thereby causing the significant heat losses that were currently observed in several solar trough plants.

The vacuum performance of receiver tubes affects the efficiency and operating economy of the whole solar power system. The heat losses mainly result from the radiation between the solar selective coating and the glass envelope and the convection in the annulus space. In particular, once the annulus space in receiver tubes experiences a loss of vacuum, its heat losses would significantly increase, and its solar selective coating would also become less efficient due to oxidation in high temperatures, significantly reducing the thermal efficiency of the trough solar thermal power plants. In recent years, it has been discovered that hydrogen from degradation of heat transfer oils could permeate through the absorber tube into the vacuum annulus in many parabolic trough stations, which caused the increasing glass temperature and showed the so-called "hot tube phenomena" [12]. Especially, if the partial pressure of hydrogen is more than 10 Pa in the annulus, the heat loss at the level is approximately a factor of 4 higher than the loss for a receiver with good vacuum [3]. The annual plant revenue can then be reduced by as much as 20% by receivers infiltrated with hydrogen [12,13]. Meanwhile, vacuum life is the main factor that decides the life of parabolic trough receiver tubes. It is not only the key index that the manufacturers of receiver tubes focus on, but also the index that power plants operators and maintenance teams care most about. Therefore, it is necessary to identify the vacuum failure among all receiver tubes installed in a large solar field in order to maintain a higher level of thermal efficiency of the system. However, how to non-destructively test and evaluate the vacuum performance of the receiver installed in the solar filed is a technical problem at present.

Price et al. [3] used an infrared camera to measure the surface temperature of the parabolic trough receiver in the solar field. This method makes it possible to roughly judge the heat loss according to the surface temperature of the receiver tube, but can not identify whether the heat loss is caused by vacuum failure or the solar selective coating degradation. It is also hard to judge the type of residual gas and its partial pressure. Pagola et al. [14] developed a fluorescence sensor based on a single fiber measurement system that can monitor degradation of heat transfer fluid (HTF) and fluorescent substances generated by H_2 generation, and proposed a method to detect hydrogen molecules in heat transfer fluid as the hydrogen in residual gases comes from permeation of hydrogen in HTF. This method is helpful to define the quantity of hydrogen in the HTF and analyze how much hydrogen can permeate into the vacuum according to the hydrogen forming mechanism in the annulus. To analyze residual gases, Ren et al. [15] presented a residual gas analysis (RGA) method for parabolic trough receiver tubes. The RGA system comprised of a receiver tube opening device, a high vacuum system and a quadrupole mass spectrometer (QMS), which is able to analyze the components and the partial pressure of the residual gases in the annulus space. The RGA method is destructive and needs to be tested indoor. Espinosa-Rueda et al. [16] presented a procedure to evaluate the annulus gas of receivers in the solar field based on the plasma generation and spectral analysis in the annulus space. The method can be used in limits of working pressure range from 1 to 1000 Pa for some gases and mixtures. Setien et al. [17] presented a surface temperature method to determine partial vacuum pressure of receiver tubes not completely damaged by a portable IR camera. It can distinguish partial vacuum pressure among 10^{-2}–10^{5} Pa when the temperature of the receiver is higher than 300 °C and there should be air in the annulus. It does require accurate and careful measurement.

In this paper, a nondestructive testing method is developed to evaluate the vacuum performance of parabolic trough receiver tubes based on the principles of dielectric barrier discharge and spectrometric analysis. A novel test device, including two elements: a gas ionization system and a spectrometric analysis system, was developed to identify the gas composition and partial pressure and evaluate the state of vacuum in the annulus space. The feasibility and accuracy of the testing method were verified by using the RGA system [15]. The different types of gases and partial pressures in the annulus space have been analyzed by using the novel device for the receivers in a series of experiments.

2. Methodology and Test Device

The non-destructive vacuum evaluation of the receiver tubes is based on the principles of dielectric barrier discharge and spectrometric analysis.

2.1. *Principle of Dielectric Barrier Discharge*

The dielectric barrier discharge (DBD) is a form of electrical discharge in low-pressure and atmospheric pressure. By placing an insulating dielectric barrier between two electrodes, plasma is generated between the electrodes. This is a typical form of high voltage alternating current discharge. Usually, DBD takes three forms according to the working pressure (P) and the discharge gap (D): (1) filamentous discharge; (2) dispersion of discharge; and (3) glow discharge [18,19]. For the given particular annulus space of the receiver tube, the plasma generation depends on the pressure and gas composition.

For filamentous discharge, the current in DBD is formed by the movement of charge in the micro-discharge channel. As the PD value is large, the breakdown process in DBD filamentous discharge belongs to the streamer breakdown. The generation process of discharge is the process from electron to initial electron avalanche and then to secondary electron avalanche, and finally produces discharge channel and forms self-sustaining discharge [20]. Figure 2 shows the filamentous discharge in the annulus space of the receiver tube.

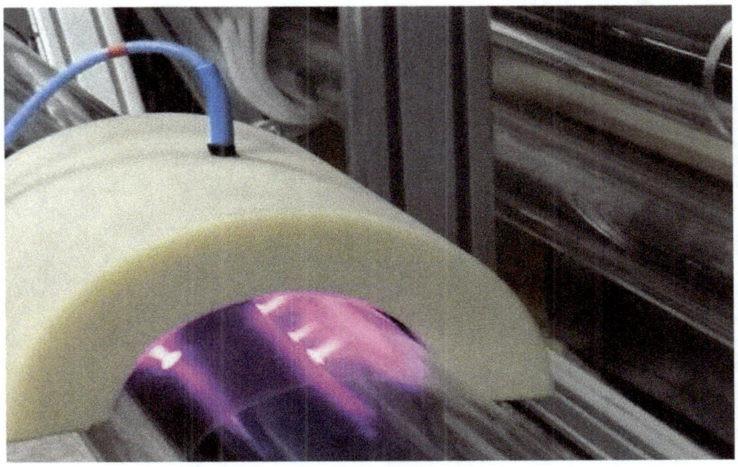

Figure 2. Emission image of air filamentous discharge at 7000 Pa.

For diffuse discharge, it is an intermediate transition stage between filamentous discharge and glow discharge. It is known as a spot pattern discharge, which is a type of macro-structure displaying certain patterns in both time and space and is a typical non-linear spontaneous phenomenon. Figure 3 shows the emission image of air diffuse discharge in the experiment.

Figure 3. Emission image of air diffuse discharge at 700 Pa.

For glow discharge, the breakdown process of DBD glow discharge belongs to Townsend breakdown as the PD value is small. Under the condition of low PD value, the positive ion is accelerated by the electric field and bombards the cathode, thus emitting secondary electrons. Then the electron avalanche is generated by the movement of the secondary electrons and the continuous collision of ionized atoms and molecules. Many electron avalanches generated by cathode emission overlap and influence each other in the discharge space, thus forming uniform discharge in the discharge space. Figure 4 shows a luminescent image of DBD glow discharge of helium gas in the experiment.

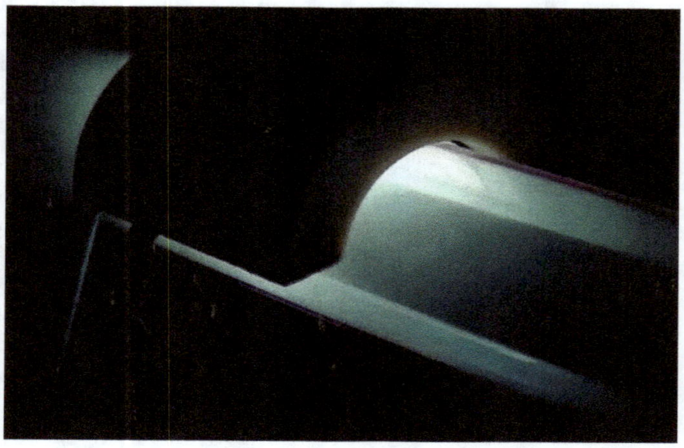

Figure 4. Emission image of glow discharge of helium gas in the receiver.

2.2. Principles of Spectrometric Analysis

Light spectral emission is a spectrum formed by a large number of photons released by plasma gases from a high energy state to a low energy state. Under the action of electric energy, the outer electrons of atoms and molecules will get a certain amount of energy, causing them to be excited from the lower energy level E_1 to the higher energy level E_2. At this time, the atoms or molecules are in the excited state. Atoms gain energy $\Delta E = E_2 - E_1$ referred to as the excitation energy. An electron moving in a high energy state is in an unstable state. When it directly returns to its original energy state, it emits light at a characteristic wavelength of the atom composing the plasma [21]:

$$\lambda = \frac{c}{v} = \frac{ch}{E_2 - E_1}, \tag{1}$$

where c is the speed of light; h represents Planck's constant; and E_1 and E_2 represents the energy of an electron in a lower-level state and a higher-level state, respectively.

An electron in an excited state may also pass through several intermediate energy levels before it returns to its original energy level. At this time, light of several different wavelengths will be generated, forming several spectral lines in the spectrum, with wavelengths of respectively

$$\lambda_1 = \frac{ch}{E_2 - E_a}; \ \lambda_2 = \frac{ch}{E_a - E_b}; \ \lambda_3 = \frac{ch}{E_b - E_c}; \ \lambda_n = \frac{ch}{E_{n-1} - E_1}, \tag{2}$$

where, E_a, E_b, E_c, $\cdots E_{n-1}$ are the energies of the intermediate stages.

Through the measurement and analysis of the profile spectrums, the type of gas compositions in the discharge channel formed during gas discharge can be obtained. Generally, the characteristic wavelengths of the light emitted by plasma are in the range from 200 to 1100 nm.

2.3. Test Method and Device

A non-destructive test method and device were developed for the non-destructive evaluation of the vacuum state and gas compositions in the annular space of the receiver tube in this study. The method included the dielectric barrier discharge to generate plasma and the spectrometric analysis to obtain the characteristic wavelength and the spectrum intensity of the light spectral emission. Meanwhile, when the dielectric barrier discharge cannot happen in the annulus space of the receiver tube, the voltage between the two electrodes depends on the pressure of the annulus space that can be measured. Thus, the pressure of the annulus space could be obtained through the relationship between the pressure and the voltage built by experiments. The flow chart of the test method is shown in Figure 5.

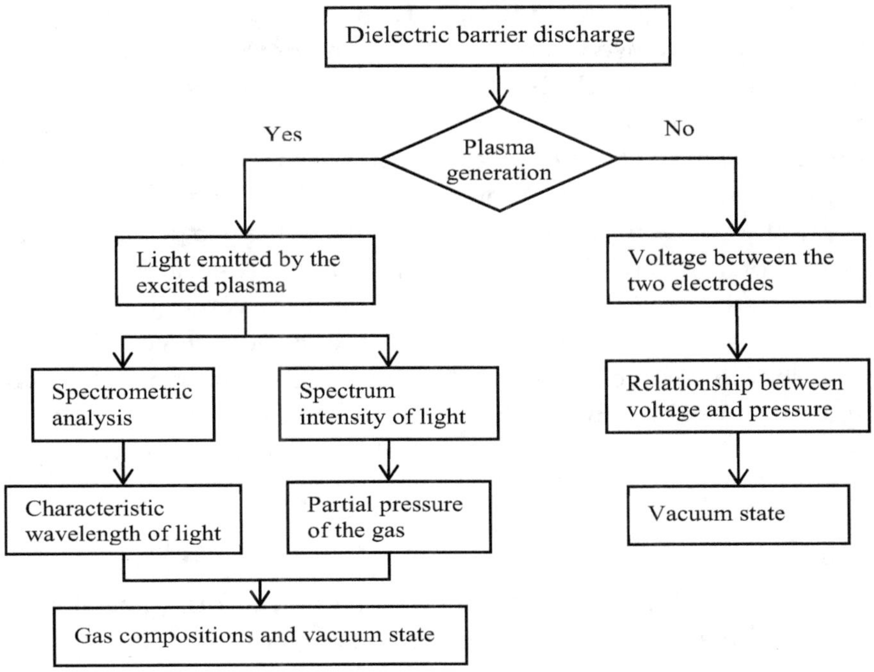

Figure 5. The flow chart of the test method.

The test device included two elements: a dielectric barrier discharge system and a spectral measurement system. The schematic of the test device is as shown in Figure 6. The dielectric barrier discharge system included an HMV-1 high frequency power source (HFPS), which was made by Yuantuo technology Co. Ltd and a semicircular gas discharge applicator composed of an insulated fixture and a copper electrode and covered on the outer surface of the glass envelope. The HFPS with high voltage had one end connected to the copper electrode and the other end connected to the absorber tube (as the other electrode) of the parabolic trough receiver. The field of the dielectric barrier discharge was formed between the copper electrode covered on the glass envelope and the absorber tube so that the gas in the annular space could realize the dielectric barrier discharge and generate plasma. The absorber tube must be connected to the ground (GND) for safety. The HFPS had a constant current and working frequency mode so that it had the ability to measure the voltage between the two electrodes.

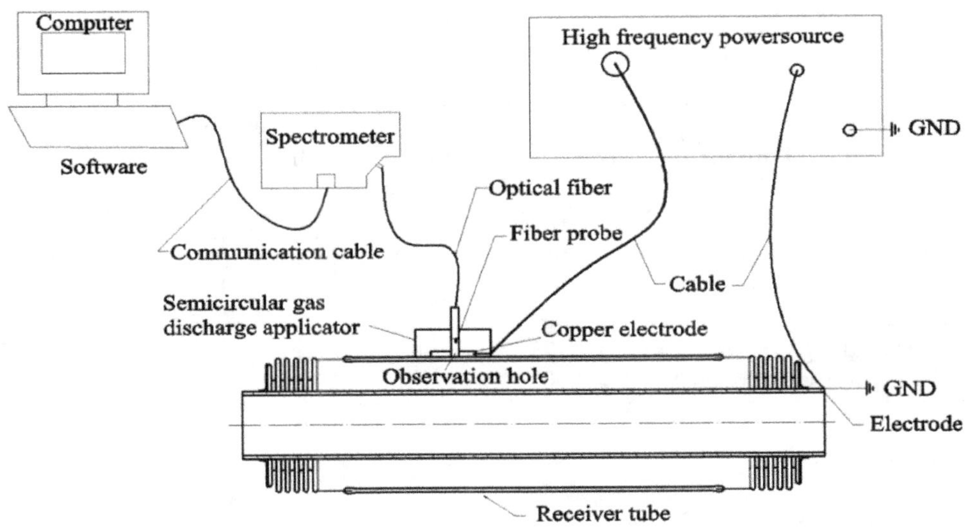

Figure 6. Schematic of the test device for vacuum evaluation.

The spectral measurement system included a Maya-Pro spectrometer (trademark of Ocean optics), optical fiber and analysis software used to control the measurement and record the emission spectra of 200–1100 nm. One end of the optical fiber was connected to the spectrometer, and the other end was the fiber probe, which was inserted into the observation hole of the fixture. The fiber probe was fixed and kept the same place in the hole for each measurement. The fiber probe transmitted the light emitted in the annulus space to the spectrometer, and then the optical signal was converted into an electrical signal. Finally, the characteristic wavelength and the spectrum intensity could be analyzed through the Ocean View software (1.6.7, Ocean Optics, Florida, America).

For the solar field test, the device was easy to carry and convenient to operate. The semicircular gas discharge applicator must be fixed on the outer surface of the receiver. Then the copper electrode and the other electrode should connect to the HFPS. The absorber tube must be connected to the GND. In order to avoid the device from burning by the solar flux, the test should be taken when the parabolic trough collectors were temporarily defocused. Finally, the spectrometer could obtain and analyze the test data.

2.4. Experimental Design

In order to explore the relationship between the vacuum performance of receiver tubes and the spectral characteristics of dielectric barrier discharge, a series of experiments including pure argon, nitrogen, helium, and air under various pressures, which are the main gases in annulus space of the receiver, were carried out firstly. Table 1 shows the types and the pressure range of the gases in the experiments. After filling various gases at different pressure into the annular space, the corresponding light spectral emissions of various gases at different pressure were recorded. Through these the light spectral emission, the corresponding characteristic spectral lines of each gas can be obtained. Then the non-destructive vacuum evaluation method was compared with the method by using the RGA system. Finally, the feasibility and accuracy of the non-destructive method were proved.

When the dielectric barrier charge reaches uniform discharge, the spectrometric measurement should be finished in 10 s to avoid the thermal effect of copper electrode.

The dimensions and materials of the parabolic trough receiver used in the experiment were shown in Table 2.

Table 1. Types and the pressure range of the gases in the experiments.

No	Gas Type	Pressure (Pa)
1	Argon	2–900
2	Helium	7–100000
3	Nitrogen	3–80000
4	Air	3–80000

Table 2. Dimensions and materials of the parabolic trough receiver.

Structural Parameters	Value (m)	Components	Materials
Outer diameter of the absorber tube	0.07	The absorber tube	316L stainless steel
Thickness of the absorber tube	0.003	The glass envelope	Pyrex glass
Length of the absorber tube	4.06		
Outer diameter of the glass envelope	0.125		
Thickness of the glass envelope	0.003		

3. Results and Discussions

3.1. Spectral Characteristics of Gases Discharge

Four kinds of gases, which are the most likely gases to be found in the receivers, were filled into the annulus space under various pressures [12,16]. The experiments of the dielectric barrier discharge of gases were carried out. The spectral measurement system was used to define the relationship between the vacuum performance of receiver tubes and the spectral characteristics of the gases. The non-destructive vacuum evaluation system is shown in Figure 7.

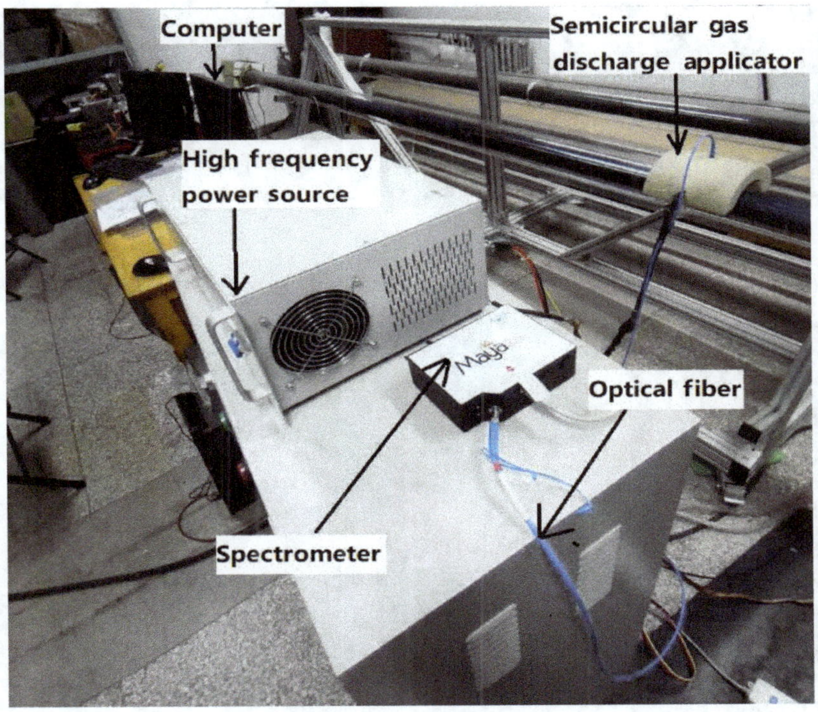

Figure 7. Non-destructive vacuum evaluation system.

For each gas, 15–25 sets of data were tested within its discharge range of pressure. Electrical parameters and spectra of different gas types and pressure were obtained in the experiment. Among them, the characteristic wavelengths of each gas were obtained. Each group of data was measured for several times, and each spectrum had good repeatability. When it was out of the discharge range for each gas, the relationship of the voltage between the copper electrode and the absorber tube and the pressure in the annulus space were obtained.

3.1.1. Argon Analysis

The working pressure range of argon discharge was 2–900 Pa in the experiments. The discharge mode was glow discharge with purple color when the pressure was 2–80 Pa. The axial length of the glow discharge decreased with the pressure increase in the annulus space, as shown in Figure 8a, which shows a large range discharge in the axial direction along the receiver tube while the pressure was 2 Pa. Figure 8b shows that the discharge happened only near the semicircular gas discharge applicator and was circumferentially distributed. The intensity of the light emitted by argon also decreased with the pressure increase.

When the pressure range was between 80 and 500 pa, the discharge mode gradually changed from glow discharge to diffusion discharge, which distributed only near the semicircular gas discharge applicator. As the pressure increased, the color gradually changed from purple to fuchsia and the patterns in the center of the discharge slowly appeared and increased, as shown in Figure 9a,b. At 500–900 Pa, the discharge mode gradually changed from diffusion discharge to filamentous discharge, as shown in Figure 9c. The discharge was filamentous and ran through the annular space. With the increase of argon pressure, the color of plasma changes from purple to bright purple, and several filamentous discharge channels gradually formed in the annular space, and the number of discharge channels gradually decreased.

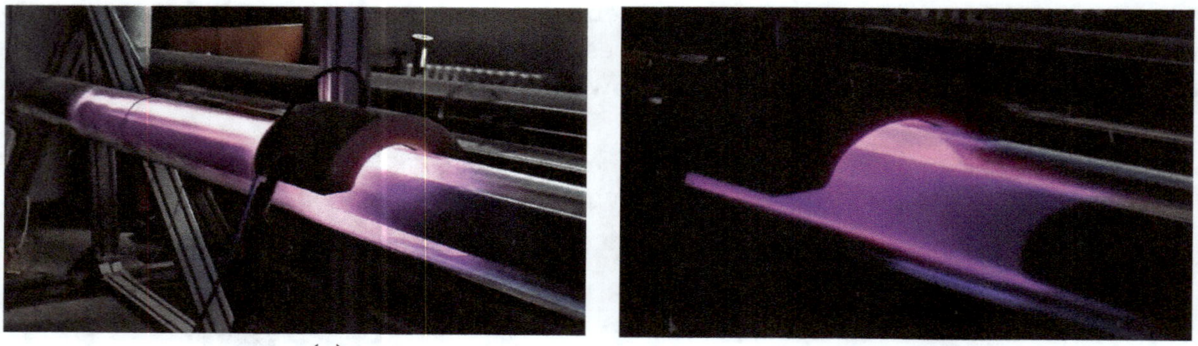

Figure 8. (**a**) Argon glow discharge at 2 Pa and (**b**) argon glow discharge at 80 Pa.

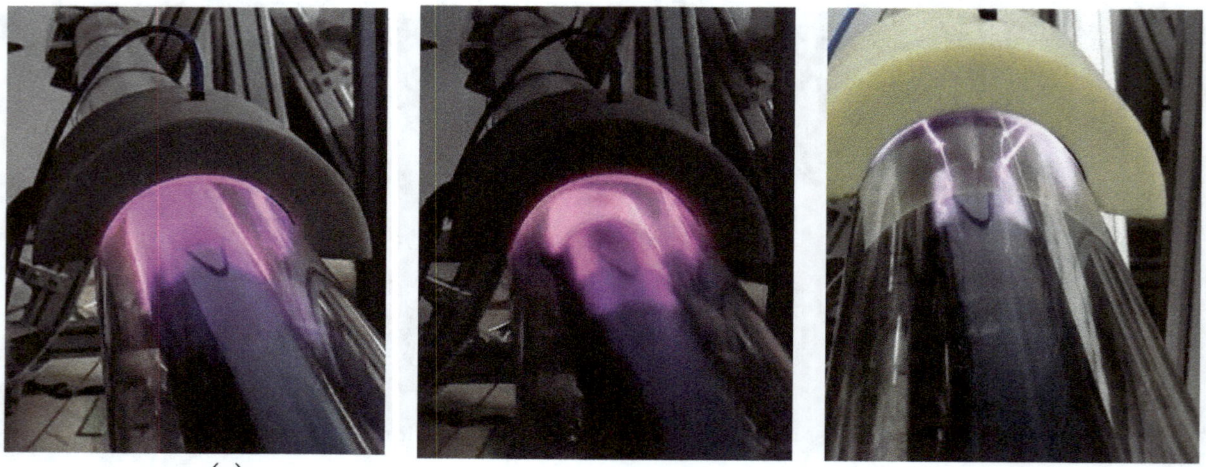

Figure 9. (**a**) Argon diffusion charge at 160 Pa; (**b**) argon diffusion charge at 500 Pa; and (**c**) argon filamentous charge at 900 Pa.

Emission spectra of the tested argon under different pressure are shown in Figure 10. The characteristic spectral lines at the wavelength of 750 nm, 763 nm, 810 nm, 841 nm, and 911 nm were obtained for argon gas. The wavelength of the spectral line with the highest intensity measured in the experiment was 810 nm, indicating that there were many argon ions involved in this electron transition in the argon discharge. The second higher intensity measured was the wave peak of 763 nm, which the electron transition was also easily generated and was also the characteristic wave peak for argon discharge. When the pressure increased from 2–700 Pa, more electrons participated in the transition, which caused the increase of both the photon quantity and the intensity of the spectral peak, as shown in Figure 10.

In order to distinguish the argon discharge, the characteristic peaks of 763 nm and 810 nm were selected, and the intensity ratio of the spectral peaks at 763 nm and 810 nm (I763/I810) was used in the vacuum evaluation. Figure 11 shows the curves of the intensity of the characteristic spectral peaks and ratio of I763/I810 with the pressure. The fitted function of intensity ratio of (I763/I810) and the pressure was obtained to analyze the partial pressure of the argon in the receiver:

$$I_{ar} = 1.647 \times 10^{-6} P^2 + 8.779 \times 10^{-4} P + 0.590, \quad (3)$$

where I_{ar} is I763/I810 of argon and P is the argon partial pressure.

It can be seen that the fitting curve was in good agreement with the experimental data and the intensity ratio changed roughly in a quadric curve with the pressure.

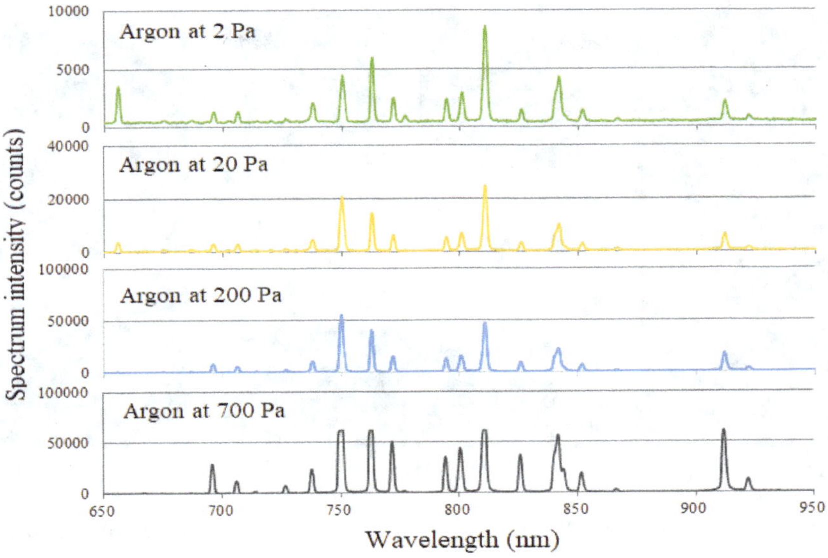

Figure 10. Emission spectra of the argon under different pressure.

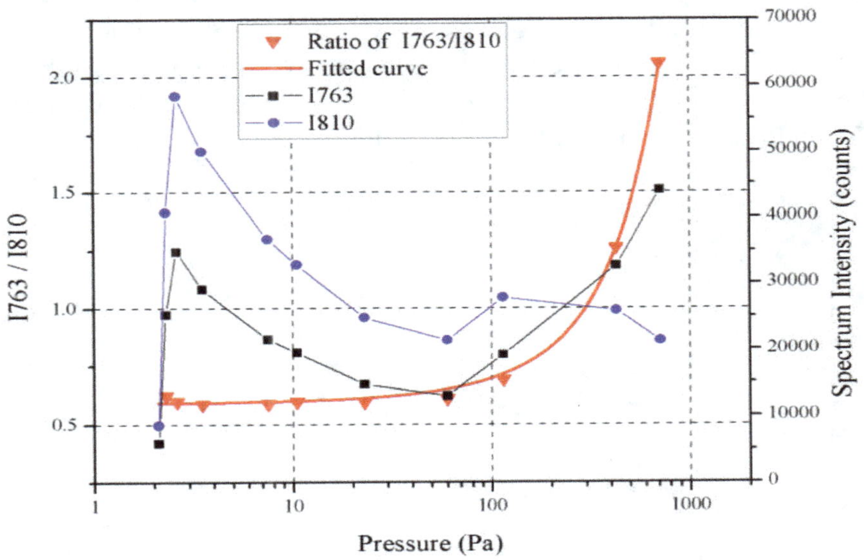

Figure 11. Intensity of the characteristic spectral peaks and ratio of I763/I810 with the pressure.

3.1.2. Helium Analysis

The working pressure range of helium discharge was from 7 to 100000 Pa in the experiments. The glow discharge mode happened in a long range of the pressure, which was from 7–40000 Pa. When the pressure was less than 300 Pa, the discharge was cylindrical and distributed in annular space, as shown in Figure 12a. The color of plasma slowly changed from green to light yellow with the increase of pressure. It gradually changed to pink in the range of 300–40000 Pa. At the range of 40000–80000 Pa, the discharge mode was diffusion discharge, which indicated the stability of the helium discharge, as shown in Figure 12b. The discharge was curved and distributed on the surface of the absorber tube and the glass envelope. The center of the discharge gradually formed several patterns and the number of patterns increased with the pressure increase. The discharge mode was filamentous discharge as the pressure was above 80000 Pa, as shown in Figure 12c. It can be seen that the color of plasma changed from pink to bright red with the increase of helium pressure and a large number of filamentous discharge channels were gradually formed in the annulus space.

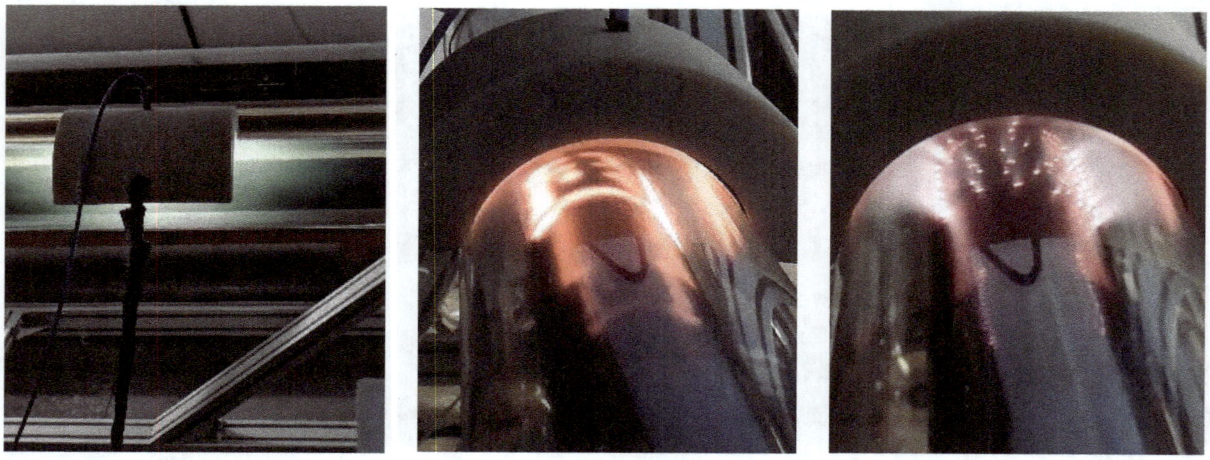

Figure 12. (**a**) Helium glow discharge at 7 Pa; (**b**) helium diffusion discharge at 80000 Pa; and (**c**) helium filamentous discharge at 100000 Pa.

The emission spectra of tested helium gas at different pressures are shown in Figure 13. The characteristic spectral peaks at the wavelength of 501 nm, 587 nm, 667 nm, and 727 nm were obtained for helium gas. From Figure 13, it can be seen that the intensity of the spectral peak at 501 nm was relatively the highest at 7 Pa and then the spectral peak at 667 nm was the highest with the pressure increase for the helium discharge. The intensity of the spectral line remarkably increased in the helium pressure range of 7–300 Pa, and then slightly increased with the pressure increase.

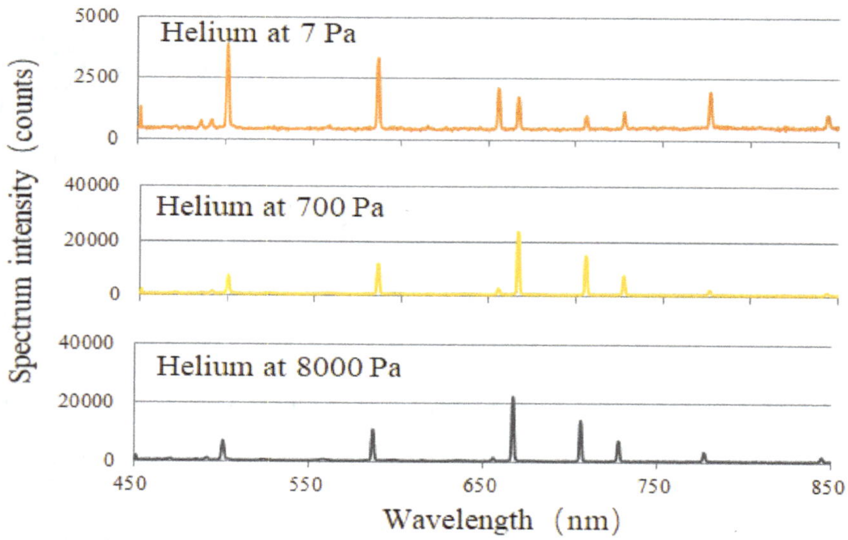

Figure 13. Emission spectra of the helium under different pressure.

The characteristic peaks of 501 nm and 667 nm were selected to distinguish the helium discharge. The intensity ratio of the spectral peaks at 667 nm and 501 nm (I667/I501) was used in the vacuum evaluation. Figure 14 shows the curves of the intensity of the characteristic spectral peaks and ratio of I667/I501 with the pressure. The fitted function of intensity ratio of (I667/I501) and the pressure was obtained to analyze the partial pressure of the helium in the receiver:

$$I_{he} = -5.567 \times 10^{-6} P^2 + 0.007P + 0.407, \quad (4)$$

where, I_{he} is I667/I501 of helium.

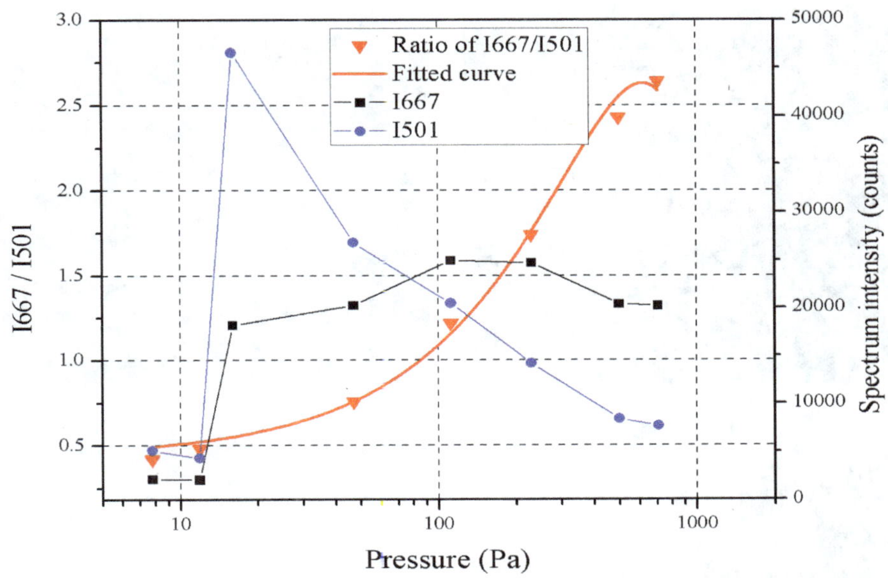

Figure 14. Intensity of the characteristic spectral peaks and ratio of I667/I501 with the pressure.

3.1.3. Nitrogen Analysis

For the nitrogen, the glow discharge mode happened in the range of 3–180 Pa, as shown in Figure 15a,b. The discharge was cylindrical and distributed in the whole annulus space and its color gradually changed from fuchsia to purple with the pressure increase. When the range of pressure was from 180 to 600 Pa, the discharge mode changed to diffusion discharge, which was a curved shape and distributed on the surface of the absorber tube and glass envelope. Several patterns were formed between the two electrodes and the number of patterns increased with the pressure increase. When the pressure was above 600 Pa, the discharge mode was filamentous discharge as shown in Figure 15c. With the increase of nitrogen pressure, the color of plasma in the annular space changed from pink to red and yellow, and filamentous discharge channels began to form in the annular space as shown in Figure 15d.

The corresponding characteristic spectral lines of nitrogen at different pressure were obtained. There were characteristic spectral peaks at 336 nm, 356 nm, 390 nm, 451 nm, 656 nm, and 777 nm for nitrogen, as shown in Figure 16. The intensity of the spectral peak at 390 nm was the highest in the range of 3–100 Pa, indicating that more nitrogen ions participating in the transition. The intensity of the spectral line remarkably increased with the pressure increase, which resulted in an increase in photon quantity. While the highest intensity of the spectral peak was at 336 nm as the pressure was higher than 100 Pa, as shown in Figure 16. The intensity of the peak at 390 nm was relatively lower than that at lower pressure, indicating that there were relatively few nitrogen ions involved in the transition. The intensity of the spectral line slowly increased when the pressure was higher than 100 Pa.

For nitrogen, the characteristic peaks of 336 nm and 390 nm were chosen to determine the nitrogen gas. The intensity ratio of the spectral lines at 336 nm and 390 nm was calculated (I336/I390). Figure 17 shows the curves of the intensity of the characteristic spectral peaks and ratio of (I336/I390) with the pressure. The fitted function of intensity ratio of (I336/I390) and the pressure was obtained to analyze the partial pressure of the nitrogen in the receiver:

$$I_n = -4.053 \times 10^{-6} P^2 + 0.009P + 0.037, \quad (5)$$

where, I_n is I336/I390 of nitrogen.

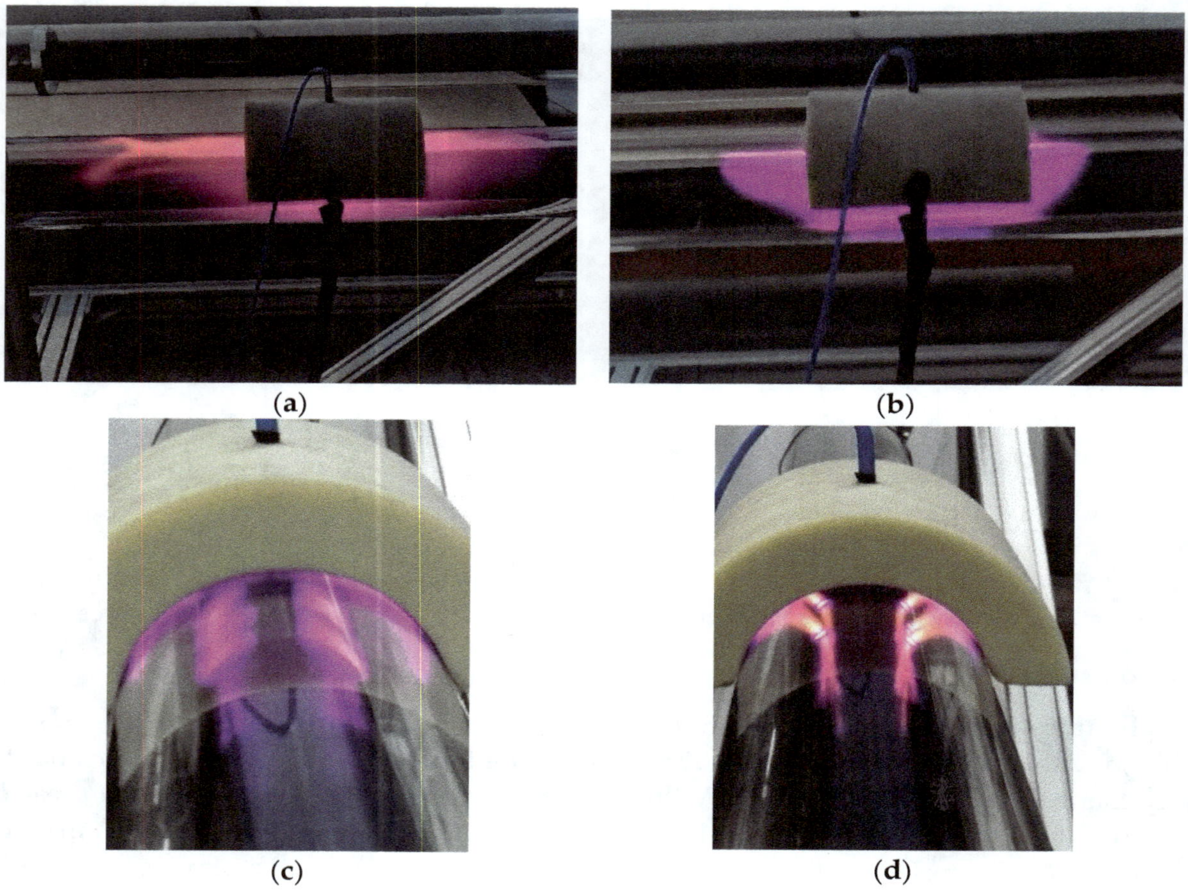

Figure 15. (**a**) Nitrogen glow discharge at 3 Pa; (**b**) nitrogen glow discharge at 180 Pa; (**c**) nitrogen diffusion discharge at 600 Pa; and (**d**) nitrogen filamentous discharge at 3000 Pa.

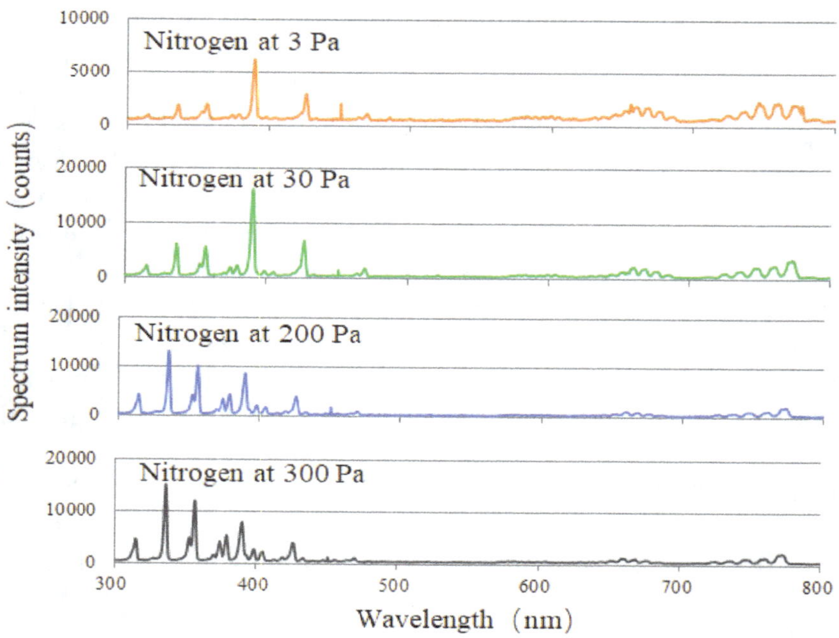

Figure 16. Emission spectra of the nitrogen under different pressure.

3.1.4. Air Analysis

The pressure range of the air in the annulus space of the receiver was from 3 to 80000 Pa in the experiment. When the pressure was between 3 and 200 Pa, the air discharge mode was glow discharge, as shown in Figure 18a,b. The discharge was cylindrical and distributed in whole annular space. The color of the discharge was purple. When the pressure was from 200 to 700 Pa, the discharge mode was diffusion discharge. The discharge distributed on the surface of absorber tube and glass envelope and the patterns appeared in the annulus space, as shown in Figure 3. With the increase of air pressure, the patterns became larger and more between the glass envelope and the absorber tube. The filamentous discharge mode happened in the annulus space when the air pressure was in the range of 700–80000 Pa, as shown in Figure 2. With the increase of air pressure, the number of filamentous discharge channels increased.

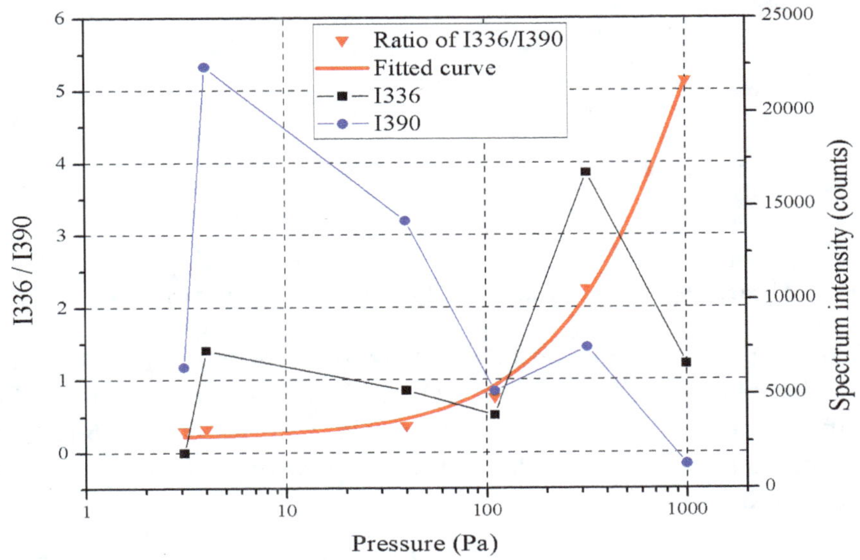

Figure 17. Intensity of characteristic spectral lines and their ratio (I336/I390) at different air pressure for nitrogen.

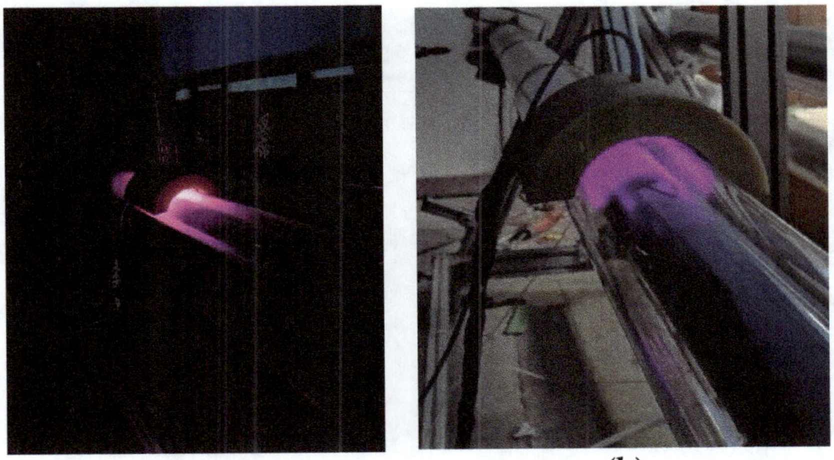

(**a**) (**b**)

Figure 18. (**a**) Air glow discharge at 7 Pa and (**b**) air glow discharge at 70 Pa.

The discharge spectra of the tested air at different pressure were shown in Figure 19. The comparison between the discharge spectra of air and nitrogen at the same pressure showed that the intensity of the characteristic spectral line of air is lower than that of nitrogen. It is reasonable because the percentage of the nitrogen in the air at a certain pressure was lower than that of nitrogen at the same pressure. However, the spectral peaks at 451 nm, 655 nm, and 776 nm in the air spectrum were stronger than those of nitrogen at the same pressure, indicating that these spectral peaks were superimposed by the spectral peaks formed by the transition of other gases in the air, as shown in Figure 19.

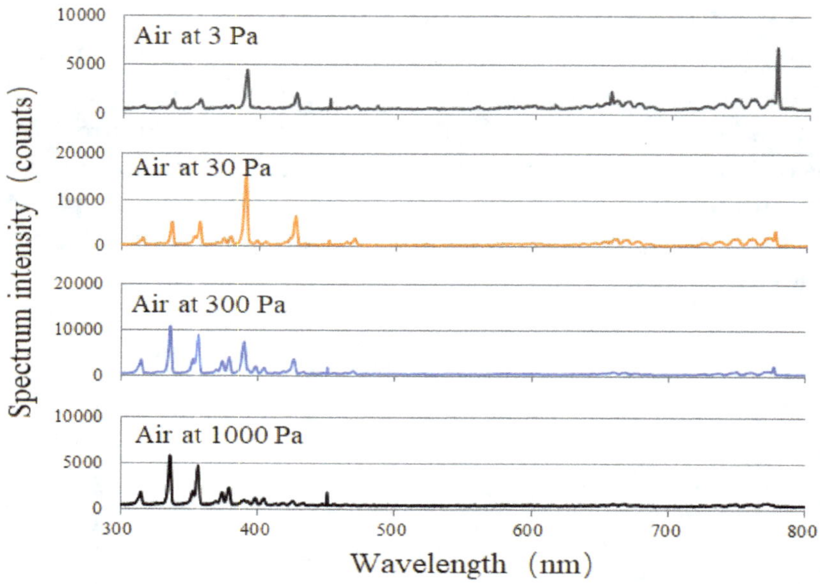

Figure 19. Emission spectra of air under different pressure.

The characteristic peaks of 336 nm and 390 nm were chosen to analyze the air discharge. The intensity ratio of the spectral lines at 336 nm and 390 nm (I336/I390) was obtained. Figure 20 shows the curves of the intensity of the characteristic spectral peaks and ratio of (I336/I390) with the pressure. The fitted function of intensity ratio of (I336/I390) and the pressure was obtained to analyze the partial pressure of the air in the annulus:

$$I_{air} = -5.654 \times 10^{-6} P^2 + 0.019P + 0.033 \quad (6)$$

where, I_{air} is I336/I390 of air.

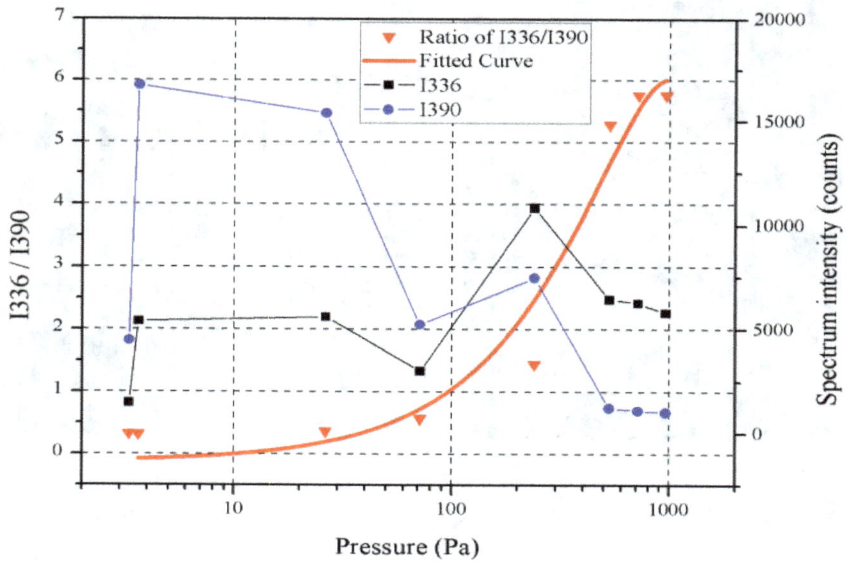

Figure 20. Intensity of characteristic spectral lines and their ratio (I336/I390).

3.2. Verification

In order to verify the accuracy and feasibility of the nondestructive testing method for the vacuum performance of receiver tubes, an RGA system was used to analyze the components and the partial pressure of the gases in the annulus space. The RGA system introduced in Ref. [15] comprises of a receiver tube opening device, a high vacuum system, a mass flow controller system, and a quadrupole mass spectrometer (QMS), as shown in Figure 21. For the RGA system, the QMS can directly analyze the gas composition and the gas partial pressure with a high accuracy. The mass flow controller system can accurately control the quantity of various gases into the annulus space of the receiver. The non-destructive test device and the RGA system can simultaneously measure the vacuum performance for the same receiver tube.

Figure 21. Residual gas analysis (RGA) system of the receiver.

The comparative experiments were carried out by using argon, nitrogen, and helium, respectively. According to the RGA test results, the annular space was pure argon at 7.5 Pa, pure helium at 11.3 Pa, and pure nitrogen at 39 Pa. For the nondestructive test method, the pressure value was obtained by the fitted function of the intensity ratio of the two characteristic peaks. The non-destructive test device showed that the characteristic spectral lines were consistent with the pure argon, pure helium, and pure nitrogen, respectively. Among them, the ratio of (I762/I810), (I667/I501) and (I336/I390) was 0.59, 0.48, and 0.33, respectively. According to Equations (3)–(5), the compared results between the RGA test and the non-destructive test were shown in Figure 22. It shows that the test results of the non-destructive vacuum evaluation method proposed were in good agreement with the RGA test results. The test error was ±10 Pa, which verified the feasibility and accuracy of this method. Therefore, the non-destructive test method combining DBD and spectrometric analysis could obtain ideal results for vacuum performance of the parabolic trough receiver.

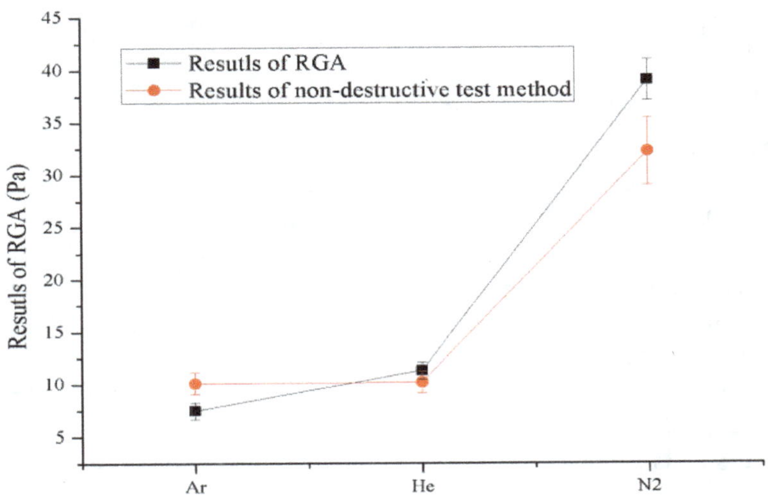

Figure 22. Compared results between the RGA test and the non-destructive test.

4. Conclusions and Future Work

This paper proposed a non-destructive test method to evaluate the vacuum performance of the parabolic trough receiver by combining the DBD method and spectrometric analysis. The non-destructive test device including a gas ionization system and a spectrometric analysis system was developed. The discharge characteristics and spectrometric properties of four kinds of gases, which are the most likely gases to be found in the receivers, were studied in the experiments. The test results of the non-destructive vacuum evaluation method proposed were in good agreement with the RGA test results, which verified the feasibility and accuracy of this non-destructive test method. The relationship between the vacuum performance of receiver and the spectral characteristics of dielectric barrier discharge were obtained by a series of experiments. Through this study, the following conclusions could be obtained:

1. The working pressure range of each gas discharge was obtained by the experiments, which was 2–900 Pa for argon, 7–100000 Pa for helium, 3–80000 Pa for nitrogen, and 3–80000 Pa for air. The discharge characteristics of each gas were studied. The common features of each gas discharge were: when the gas pressure increase in the annulus space, the gas discharge mode was glow discharge and the discharge was circumferential distribution near the semicircular gas discharge applicator. Then the discharge mode turned to diffuse discharge, and the distribution of the discharge was only on the surfaces of the glass envelope and the absorber tube near the semicircular gas discharge applicator, and the patterns appeared. Finally, the gas discharge mode turned to filamentous discharge.
2. The emission spectra and characteristic peaks of each gas under different pressure were obtained to distinguish the gas composition in the receiver. According to the fitted function of the intensity ratio of two characteristic peaks and the pressure proposed in the paper, the intensity ratio of (I763/I810), (I667/I501), (I336/I390), and (I336/I390) were used to distinguish the partial pressure of argon, helium, nitrogen, and air, respectively.

Through experiment, the relationship between the vacuum performance of the receiver and the spectral characteristics of dielectric barrier discharge were obtained. This information is of great importance for the vacuum evaluation of the parabolic trough receivers. In the future, the hydrogen discharge characteristics and spectrometric properties will be tested when the safety of hydrogen discharge can be solved in the experiments. In addition, the on-site measurement of the vacuum performance of the receivers will be carried out in the solar field.

Author Contributions: Conceptualization, F.Y. and D.L.; methodology, D.L. and F.Y.; software, F.Y.; validation, F.Y.; formal analysis, F.Y.; investigation, F.Y., P.Y., and Y.H.; resources, D.L. and K.Y.; data curation, F.Y. and P.Y.; writing—original draft preparation, F.Y.; writing—review and editing, D.L.; visualization, F.Y. and Y.H.; supervision, Z.W. and K.Y.; project administration, D.L. and F.X.; funding acquisition, D.L. and Q.H.

References

1. Yılmaz, İ.H.; Mwesigye, A. Modeling, simulation and performance analysis of parabolic trough solar collectors: A comprehensive review. *Appl. Energy* **2018**, *225*, 135–174. [CrossRef]
2. Ma, L.; Wang, Z.; Lei, D.; Xu, L. Establishment, Validation, and Application of a Comprehensive Thermal Hydraulic Model for a Parabolic Trough Solar Field. *Energies* **2019**, *12*, 3161. [CrossRef]
3. Price, H.; Forristall, R.; Wendelin, T.; Lewandowski, A.; Moss, T.; Gummo, C. Field survey of parabolic trough receiver thermal performance. In Proceedings of the ISEC2006: ASME International Solar Energy Conference, Denver, CO, USA, 8–13 July 2006.
4. Lei, D.; Fu, X.; Ren, Y.; Yao, F.; Wang, Z. Temperature and thermal stress analysis of parabolic trough receivers. *Renew. Energy* **2019**, *136*, 403–413. [CrossRef]
5. Wang, J.; Huang, X.; Gong, G.; Hao, M.; Yin, F. A systematic study of the residual gas effect on vacuum solar receiver. *Energy Convers. Manag.* **2011**, *52*, 2367–2372. [CrossRef]
6. Tang, Z.; Zhao, X.-P.; Li, Z.-Y.; Tao, W.-Q. Multi-scale numerical analysis of flow and heat transfer for a parabolic trough collector. *Int. J. Heat Mass Transf.* **2017**, *106*, 526–538. [CrossRef]

7. Wu, Z.; Lei, D.; Yuan, G.; Shao, J.; Zhang, Y.; Wang, Z. Structural reliability analysis of parabolic trough receivers. *Appl. Energy* **2014**, *123*, 232–241. [CrossRef]
8. Kutscher, C.; Mehos, M.; Turchi, C.; Glatzmaier, G.; Moss, T. *Line-Focus Solar Power Plant Cost Reduction Plan*; National Renewable Energy Laboratory NREL/TP-5500-48175; National Renewable Energy Lab. (NREL): Golden, CO, USA, 2010.
9. Li, J.; Wang, Z.; Li, J.; Lei, D. Vacuum reliability analysis of parabolic trough receiver. *Sol. Energy Mater. Sol. Cells* **2012**, *105*, 302–308. [CrossRef]
10. Moens, L.; Blake, D.M. Mechanism of Hydrogen Formation in Solar Parabolic Trough Receivers. *J. Sol. Energy Eng.* **2010**, *132*, 031006. [CrossRef]
11. Liu, J.; Lei, D.; Li, Q. Vacuum lifetime and residual gas analysis of parabolic trough receiver. *Renew. Energy* **2016**, *86*, 949–954. [CrossRef]
12. Möllenhoff, M.; Albert, C.; Werner, M.; Sohr, O.; Kuckelkorn, T. A new approach for lifetime extension of parabolic trough receivers. In Proceedings of the SolarPACES 2011, Granada, Spain, 20–23 September 2011; pp. 1–6.
13. Burkholder, F.; Brandemuehl, M.; Kutscher, C.; Wolfrum, E. Heat conduction of inert gas-hydrogen mixtures in parabolic trough receivers. In Proceedings of the Energy Sustainability 2008, Jacksonville, FL, USA, 10–14 August 2008; Volume 2, pp. 449–458.
14. Pagola, I.; Funcia, I.; Sánchez, M.; Gil, J.; González-Vallejo, V.; Bedoya, M.; Orellana, G. On-line monitoring of H2 generation and the HTF degradation in parabolic trough solar thermal power plants: Development of an optical sensor based on an innovative approach. In *AIP Conference Proceedings*; AIP Publishing: Melville, NY, USA, 2017; Volume 1850, pp. 1–8.
15. Ren, Y.; Lei, D.; Wang, Z. Experimental Analysis of Residual Gas of Vacuum Annulus in Parabolic Trough Solar Receivers. In *AIP Conference Proceedings*; AIP Publishing: Melville, NY, USA, 2019; Volume 2126, pp. 1–10.
16. Espinosa-Rueda, G.; Navarro Hermoso, J.L.; Martínez-Sanz, N.; Gallas-Torreira, M. Vacuum evaluation of parabolic trough receiver tubes in a 50MW concentrated solar power plant. *Sol. Energy* **2016**, *139*, 36–46. [CrossRef]
17. Setien, E.; López-Martín, R.; Valenzuela, L. Methodology for partial vacuum pressure and heat losses analysis of parabolic troughs receivers by infrared radiometry. *Infrared Phys. Technol.* **2019**, *98*, 341–353. [CrossRef]
18. Kogelschatz, U. Filamentary, Patterned and Diffuse Barrier Discharges. *IEEE Trans. Plasma Sci.* **2002**, *30*, 1400–1410. [CrossRef]
19. Brandt, S.; Schütz, A.; Klute, F.; Kratzer, J.; Franzke, J. Dielectric barrier discharges applied for optical spectrometry. *Spectrochim. Acta Part B At. Spectrosc.* **2016**, *123*, 6–32. [CrossRef]
20. Li, M.; Han, C.; Liu, W. Influence of gas pressure on plasma parameters in a dielectric barrier discharge plasma actuator. *Optik* **2017**, *145*, 672–677. [CrossRef]
21. Xin, R. *Plasma Atomic Emission Spectrometry*, 3rd ed.; Chemical Industry Press: Beijing, China, 2018; pp. 114–119.

13

Study on Optimized Dispatch and Operation Strategies for Heliostat Fields in a Concentrated Solar Power Tower Plant

Dongchang You [1,2,3,4], Qiang Yu [1,2,3,4,*], Zhifeng Wang [1,2,3,4,5] and Feihu Sun [1,2,3,4]

1 Key Laboratory of Solar Thermal Energy and Photovoltaic System, Chinese Academy of Sciences, Beijing 100190, China
2 Institute of Electrical Engineering, Chinese Academy of Sciences, Beijing 100190, China
3 School of Electronic, Electrical and Communication Engineering, University of Chinese Academy of Sciences, Beijing 100049, China
4 Beijing Engineering Research Center of Solar Thermal Power, Beijing 100190, China
5 Joint Laboratory, Institute of Electrical Engineering, Chinese Academy of Sciences and Guangdong Five Star Solar Energy Co., Ltd., Dongguan 523051, China
* Correspondence: yuqiang_21st@sina.com.

Abstract: Concerning solar flux densities during the operation of a concentrated solar power tower plant, their uneven distribution on a central receiver not only leads to abrupt variations of thermal gradient on the receiver surface but also makes it possible for the receiver to break down. Specific to such problems, a "concentrating-receiver" coupling system of a 1 MWe concentrated solar power tower plant in Yanqing was selected as the research object. On this basis, a spliced heliostat model was firstly established in this paper. The model was used to investigate solar flux distribution on the receiver surface. Considering that heliostats in different positions make diverse contributions to receiver surface energy and the incidence cosines of adjacent heliostats are similar to each other, a new grouping method for heliostat fields was subsequently proposed; moreover, focal point selection criteria were designed for the receiver surface according to solar spot sizes. Finally, an optimized dispatch and operation strategy was established based on the genetic algorithm for the heliostat field. Therefore, a standard deviation of solar flux distribution can be minimized. To verify the reliability of the established model and the proposed strategy, a small-scale heliostat field was adopted to check the simulation results by means of experiments. It has been demonstrated that a heliostat field subjected to optimized dispatch makes solar flux densities distribute more uniformly on the receiver surface. Hence, the safe and steady operation of the receiver is guaranteed.

Keywords: heliostat field; solar flux; genetic algorithm; optimized dispatch

---

1. Introduction

Energy is the fundamental economic support in a country and the basis for human survival. With constant social development, environmental pollution and energy crisis become critical factors restricting growth [1]. As a more economical, more efficient and high-capacity renewable energy technology, concentrated solar power (CSP) technology has been recognized and accepted in an increasing number of countries [2]. It has been predicted by the Resources for the Future (RFF) organization that CSP plants may provide electricity of 4100 TWh in 2040, which accounts for 10–11% of global gross generation [3]. Up to 2018, the cumulative installed capacity of CSP plants throughout the world reached 5.5 GW, indicating an accelerating increase rate [4]. CSP transforms solar irradiation to heat power during the day and stores the redundant energy in a thermal energy storage (TES) system. The stored energy can be used to generate electricity if no solar irradiation is available at night or little is available on a cloudy day. Incorporation with thermal energy storage TES system allows CSP plants to shift electricity production to meet load demands. Hence, CSP is a completely clean renewable energy technology. Compared with traditional coal-fired plants, CSP uses solar energy to generate electricity instead of traditional coal-fired plants that produce steam by burning coal or natural gas. Therefore, almost no greenhouse gases are emitted into the atmosphere [5]. According to the latest energy report from World Resources Institute (WRI) [6], the carbon dioxide emissions from coal-fired plants account for 72% of global emissions. Though the latest technologies have been used to reduce carbon dioxide emissions, the amount of carbon dioxide emitted into the atmosphere is still significant. The average carbon dioxide emission intensity of coal-fired and gas-fired plants are around 890 g/kWh. The average coal consumption of gas-fired units is 247 g/kWh, and the average carbon dioxide emission intensity is 390 g/kWh [6]. In economic terms, the levelized cost of renewable power generation technology is currently still not competitive with conventional thermal power technology. For instance, the levelized costs of electricity (LCOE) of concentrated solar power is ¢7.3/kWh, wind power is ¢0.087/kWh, and Photovoltaics (PV) is ¢0.15/kWh, while conventional thermal power is ¢0.05/kWh [7]. Though the cost of CSP plants is significantly higher than those at this time, many countries are making efforts to promote the development of CSP from the view of enhancing power system flexibility by using renewable energy. It should be noted that the LCOE of CSP plants is estimated to drop sharply to around the cost of a base-load thermal unit.

Thus far, many CSP plants have been successively constructed in lots of countries. As for a CSP plant, highly efficient solar thermal power tower plants that feature mature techniques and large-scale commercial applications hold a dominant position, such as the 110 MWe Crescent Dunes molten salt solar tower power plant in Nevada, USA, and the 150 MWe Noor 3 molten salt solar tower power plant in Morocco [2]. During the operation of a solar tower power plant, energy concentrated on a central receiver may dramatically fluctuate due to defects (e.g., intermittency and instability) in solar radiation resources. Consequently, the thermal stress on the receiver surface rapidly changes. In addition, excessive thermal gradients may cut the service life of the receiver and exert an adverse influence on power generation stability [8]. One practical and feasible method to prevent the occurrence of such occasions is to carry out optimized dispatch of solar flux distribution on the receiver surface in a certain way [9]. The so-called optimized dispatch of a heliostat field refers to adjusting the focusing situations of a heliostat field to make the concentrated energy meet the system running demands to the greatest extent. This is of great significance to extend the service life of the receiver and ensure the safe and stable operation of the receiver.

Many scholars have conducted relevant studies on heliostat field dispatch issues. For the purpose of minimizing the difference of peak solar flux in a cavity receiver, a TABU heuristic algorithm was utilized by Qiang Yu et al. [10] to establish a linear multi-point focusing model on an aperture of a cavity receiver. In order to minimize the total energy consumed by a heliostat to track the sun, a mathematical model was proposed by Nicole Fernandez et al. [11] based on the energy consumption of electrical machine and solar spot center offset, and the maximum control cycle of heliostats at different positions was successfully obtained. Specific to the thermal gradient changes of the receiver, a meta-heuristic algorithm was selected by Salomé A et al. [12] to seek an optimal focal point on the receiver surface according to solar spot size of every heliostat, and a final optimal distribution of focal points was acquired. Considering that a receiver might undergo local overheating during operation, Guo Tiezheng et al. [13] proposed an overheating evaluation model to study influencing factors that were used to judge whether a receiver was overheated, and an optimal strategy of adjusting target points of a heliostat was finally presented. In order to realize a uniform temperature distribution on the receiver surface, a tabu heuristic algorithm was utilized by F. J. Garcia-Martin et al. [14] to define an appropriate number of heliostats for focal points on a receiver surface. In this way, the temperature distribution curve of a receiver surface was obtained C. Maffezzoni et al. [15] developed a static aim processing system (SAPS) and a dynamic aim processing system (DASP) for the safety protection of the Solar Two Plant's receiver, and the operation strategies of heliostat field were finally obtained.

For the purposes of maximizing solar energy received by a receiver and realizing uniform distribution of solar flux as much as possible, a convolution optimized dispatch algorithm for the heliostat field was raised by Gallego A J et al. [16]. As a result, the optimal focusing positions were acquired during the operation of a heliostat field. Belhomme B et al. [17] first studied the optimized dispatch strategies of a heliostat field from an economical aspect. With the aim to minimize the operation cost of the heliostat field, they not only developed a ray tracing tool to study solar flux distribution on a receiver but also accordingly raised a time-segmented optimized dispatch strategy. The influence of a heliostat's control cycle on the position drifting of a solar spot on a receiver was explored by Xu Ming et al. [18], whose target was to find the maximum control cycle. In addition, they also transformed the problem of optimal control cycle into a non-linear optimization issue with equality and inequality constraints. On this basis, an optimized control strategy was put forward for a large-scale heliostat field with an optimal time-segmented control cycle. With the goal of acquiring optimal system efficiency, the optimized dispatch of heliostat fields was converted into a 0–1 knapsack problem based on an intelligence algorithm by Ding Tingting et al. [19]. Hence, both the quantity and the distribution of heliostats that should be put into service in various periods of optimal system performance were obtained. The above studies investigated the optimized dispatch of a heliostat field for diverse purposes and ultimately obtained optimal operation strategies for respective heliostat fields. For the majority of scholars, the surface of a heliostat has usually been considered an ideal curved surface and the solar flux distribution of receiver surface has been directly regarded as a Gaussian distribution, so the actual reflecting surface has rarely been considered. For this paper, on the basis of the existing literature, a reflection model was first established in conformity with the real heliostat surface which was composed of multiple small plane mirrors, and the solar flux distribution of actual solar spot on the receiver surface was finally acquired. Then, a novel grouping method for a heliostat field was raised, and this shortened the calculating time of optimization process. Eventually, an optimized dispatch strategy was presented for a heliostat field based on the genetic algorithm (GA). In this strategy, a uniformly-distributed solar flux on the receiver surface was selected as its optimization objective. Additionally, the spillage and maximum allowable solar flux of the receiver were deemed as constraints.

In this study, a genetic algorithm was adopted to improve the heliostat's focusing positions on the receiver. The aim of this optimized dispatch is to make the solar flux distribution on the receiver as uniform as possible while reducing the spillage loss to the greatest possible extent. In comparison with integer optimization and other algorithms in previous literatures, a self-adaptive genetic algorithm has the advantages of rapid optimal-searching and invulnerability to local optimization; therefore, it was chosen as the solver of the optimization strategy in this paper. The results show that the peak solar flux on the receiver surface declined to half of that subjected to central point focusing after an optimization strategy. Additionally, the solar flux densities in most regions were significantly similar to each other. Through the application of optimized dispatch strategy, in addition to enhancing the system efficiency of the whole plant, the safe and steady operation of the receiver was guaranteed.

2. System Introduction of 1 MWe Solar Thermal Power Plant in Yanqing

A 1 MWe CSP plant in Yanqing is the first solar thermal tower power plant at a megawatt level in Asia. This plant was constructed to study the system integration techniques for CSP tower plants [20]. It is primarily composed of a concentrating system, a receiver system, a storage system, and a power generation system. To be specific, the concentrating system is laid out in a north-south direction. Located at the north of a central tower, this system contains 100 heliostats that are in a staggered arrangement (as shown in Figures 1 and 2). Figure 1 shows a perspective view of the heliostat field in the Yanqing CSP plant, and Figure 2 is a layout drawing of the heliostat field's coordinates in the Yanqing CSP plant. Therefore, the whole heliostat field has 15 rinds in total, and each heliostat has its own inherent coordinate. The receiver system is constituted by a cavity receiver installed at a location 78 m height from the tower. In this way, solar energy concentrated by the concentrating system can be converted into the thermal energy of the working fluid. In this study, research contents were only limited to the "concentrating-receiver" coupling system.

Figure 1. A perspective view of the heliostat field in the Yanqing concentrated solar power (CSP) plant.

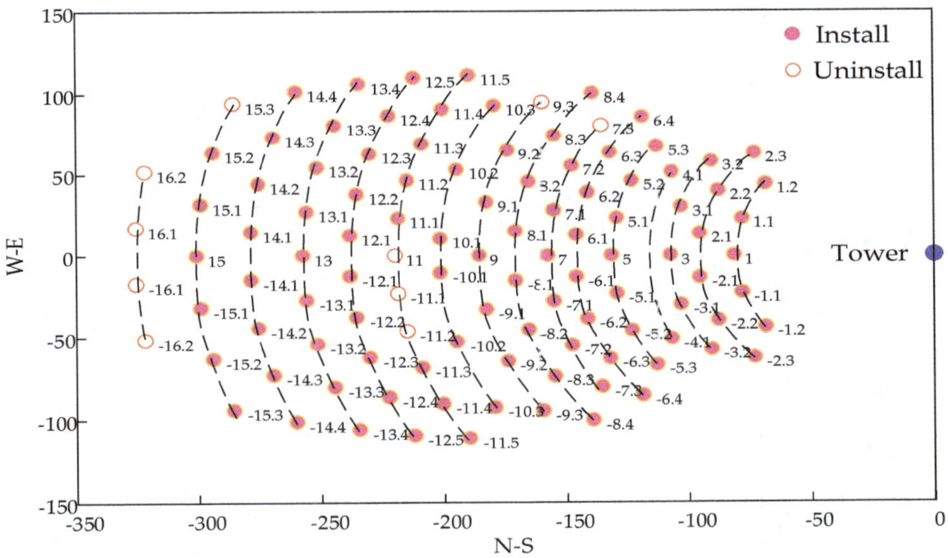

Figure 2. A layout drawing of the heliostat field's coordinates in the Yanqing CSP plant.

During the operation of a power plant, the control precision of a concentrating system largely affects the tracking precision of a heliostat. The control system of the heliostat field in the Yanqing CSP Plant consists of an host computer, a beam characterization system (BCS), a central control interchanger, and a lower computer. Among them, a BCS provides a technical guarantee for a heliostat to accurately focus on the receiver. It is comprised of a high-resolution industrial CCD camera (effective number of pixels 1600 × 1200, cell size 4.4 × 4.4 μm). Each pixel has eight bits or, equivalently, 256 levels. The selected lens was the Nikon® (Nikon Corporation, Tokyo, Japan) AF-S Ø95 (zoom range: 200–500 mm, F5.6), and an image processing software and its working diagram is presented in Figure 3. The fundamental service principles of a BCS are described as follows: Based on a track angle bias strategy, the concentrating characteristics analysis system is utilized to gain the corrected values of two reference tracking axis positions of the heliostat, and the corrected values are also saved in the bias correction database [21]. During operation, the reference positions of heliostat's azimuth and altitude axes are calculated in line with solar moving orbit. While a heliostat tracks the sun, the corrected values of reference positions for two tracking axes are incorporated to reduce the time-varying tracking errors of the heliostat. Under the action of a BCS, the heliostat's tracking error of the Yanqing CSP plant can be controlled below 0.4 mrad. During the experiment, a white Lambert target was selected as the focal target. As a perfect diffusing surface, the Lambert target presented the same apparent brightness from any angle. This provided a theoretical basis for our experiments [22]. Once the heliostat focused on the Lambert target, a CCD camera was utilized to take photographs of solar spots (the camera was situated at a middle position of the 11th ring in the heliostat field). The acquired solar spot information, subsequent to grey processing, was converted into a percentage based chart of flux distribution. Currently, the heliostat field of this plant is mostly controlled by single-point focusing, which is located at the center of the receiver. In this case, the uneven solar flux distribution on the receiver was likely to affect the secure operation of the receiver system and shorten the service life of the receiver.

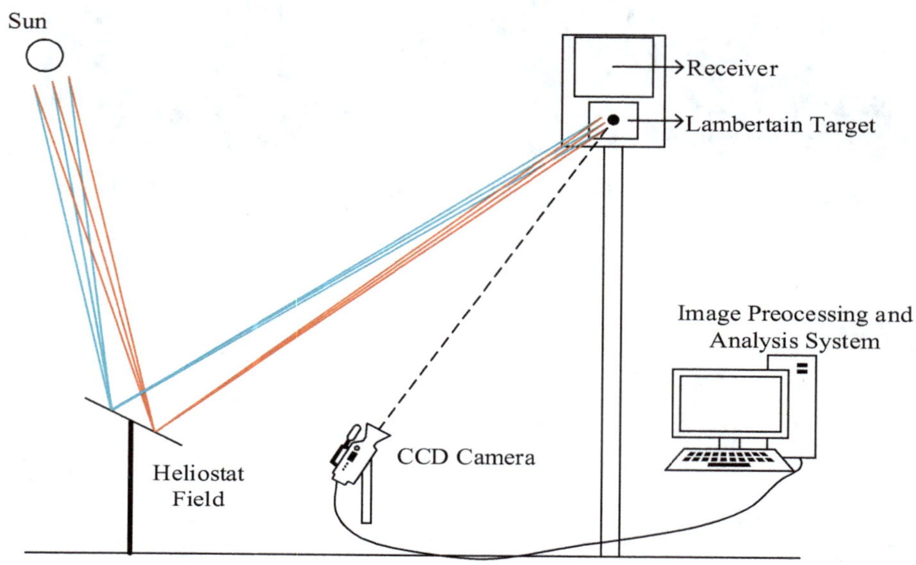

Figure 3. Working diagram of the beam characterization system (BCS).

3. Modeling of Heliostat Field

In the heliostat field of the Yanqing CSP plant, each heliostat as high as 6.6 m covers an area of 100 m^2 (10×10 m). Moreover, it is formed by 64 small square plane mirrors (each 1.25×1.25 m) by means of stitching; this is shown in Figure 4. To explore the solar flux distribution on the receiver surface, modeling must be conducted for the entire heliostat field. The only differences in heliostats are their positions. Specific to the whole concentrating system, an individual heliostat was thus selected as the modeling object to set up a mathematical model based on the Monte Carlo ray tracing method. Since there has been much research which has adopted the Monte Carlo method for the modeling of a heliostat field by previous scholars [23,24], the detailed modeling process is not elaborated here. In this study, only investigations that are different from those of former researchers are described. In preceding literature, heliostat modeling has assumed the reflector as an ideal continuous curved surface in most cases, or the direct fitting of solar flux distribution is fulfilled on the receiver in accordance with the law of Gaussian distribution. Additionally, not only does a tilted angle error exist when the plane mirror approaches the curved surface, the influence of heliostat support structure on solar flux distribution has hardly been considered by preceding scholars in the course of modeling. To carry out concentrating system modeling in this paper, both the stitching structures of the actual curved surface and the gaps of individual plane mirrors were taken into account. Meanwhile, it was assumed that heliostat errors complied with normal distribution. A schematic diagram of the heliostat model is presented in Figure 5.

Figure 4. Surface shape of heliostat in the Yanqing CSP plant.

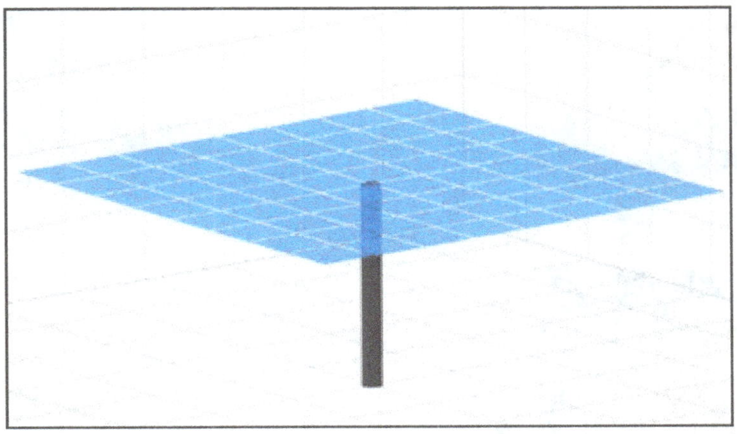

Figure 5. A schematic diagram of the heliostat model.

To study the effects of different mirror surface processing modes on solar flux distribution during modeling, a heliostat with coordinates of #8.1 was randomly selected to carry out the simulation of solar flux at noon of the vernal equinox day. Such simulation was targeted at both the ideal and the actual mirror surfaces, and the simulation results are given in Figure 6.

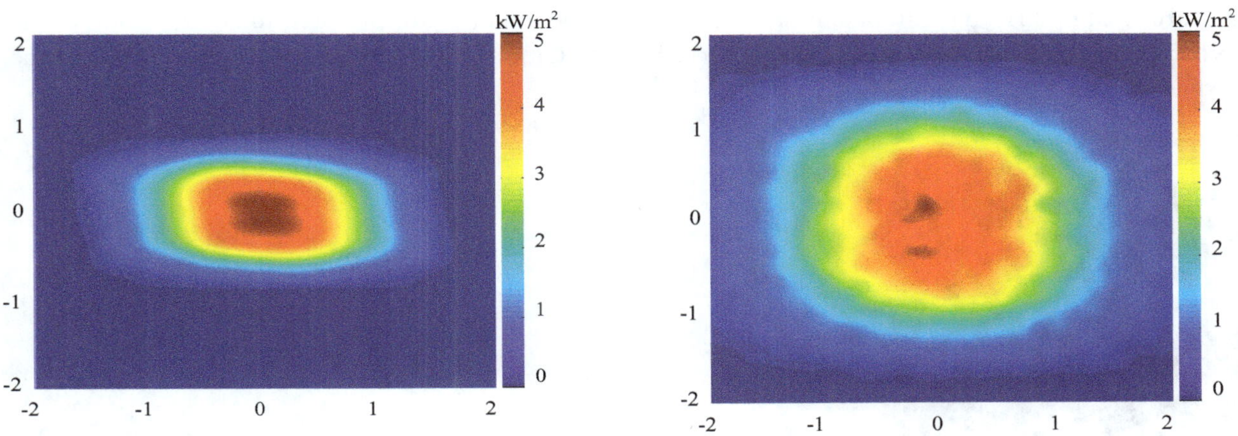

(a) (b)

Figure 6. Solar spot shape by the ideal and the spliced mirror surfaces. (**a**) Solar spot of the ideal surface. (**b**) Solar spot of the spliced surface.

As shown in this figure, it can be clearly observed that the solar flux distribution of the ideal mirror surface on the receiver perfectly conformed to the Gaussian distribution, which was embodied in regular shapes and small spot diameters. In regard to the spliced mirror surface, although the solar flux distribution of its focus spot presented a Gaussian distribution on the whole, peak energy could be found in local areas. Additionally, the shape of solar spot was relatively large. To verify the validity of the established model, the focusing points of the #8.1 heliostat were set at No. 4 and No. 5 (the set of focusing points is shown in Section 4.1 in this paper) at noon on 4 September 2019 (clear weather). The comparison results between simulation and experiment were given in Figures 7 and 8. To better perform the comparative analysis, the solar flux distribution was normalized.

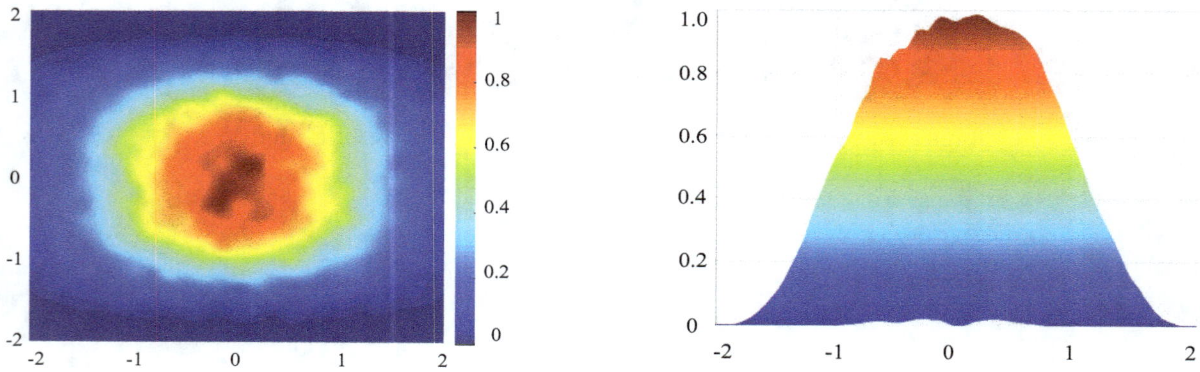

(**a**)

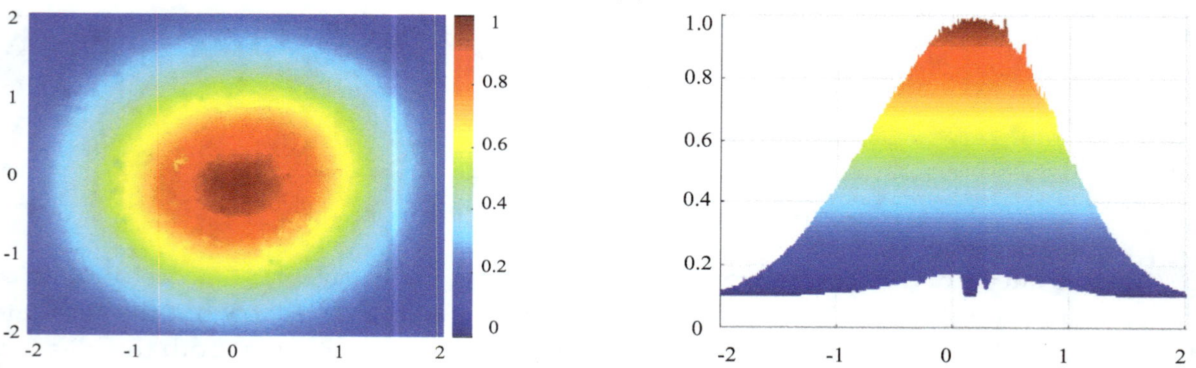

(**b**)

Figure 7. Comparison result of simulation and experimental data for the solar flux distribution of the No. 5 focal point. (**a**) Solar flux distribution of the simulation result. (**b**) Solar flux distribution of the experimental result.

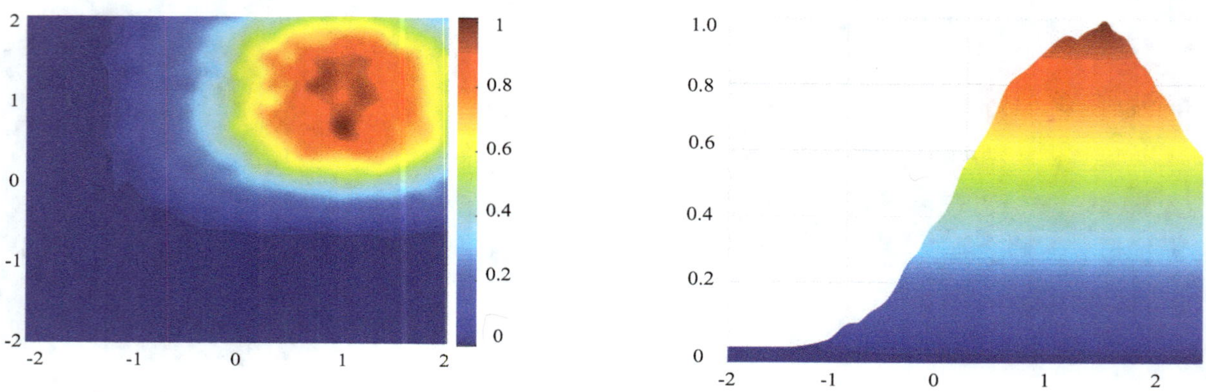

(**a**)

Figure 8. *Cont.*

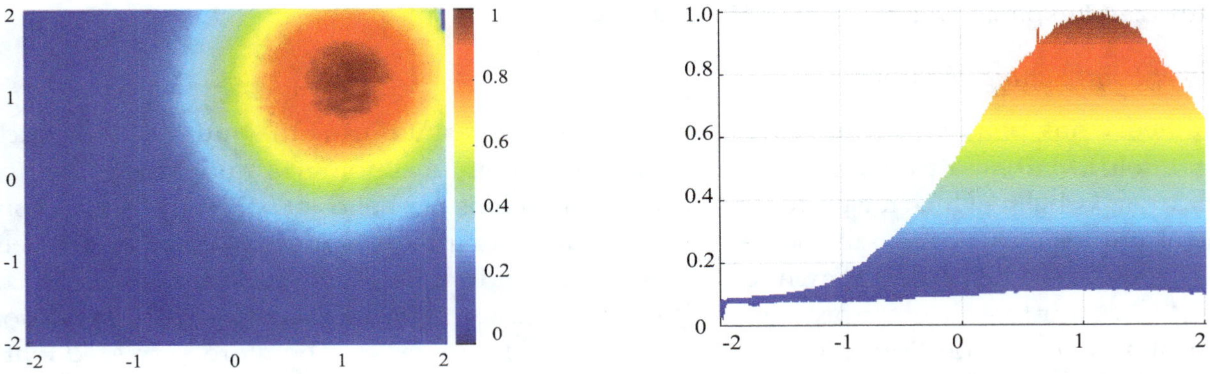

(b)

Figure 8. Comparison result of simulation and experimental data for the solar flux distribution of the No. 4 focal point. (**a**) Solar flux distribution of the simulation result. (**b**) Solar flux distribution of the experimental result.

To observe solar spot similarities in a more intuitive manner, a curve chart (as shown in Figure 9) was portrayed according to the information presented in Figures 7 and 8; thus, variations in proportions taken by the spot energy and their radius could be clearly presented.

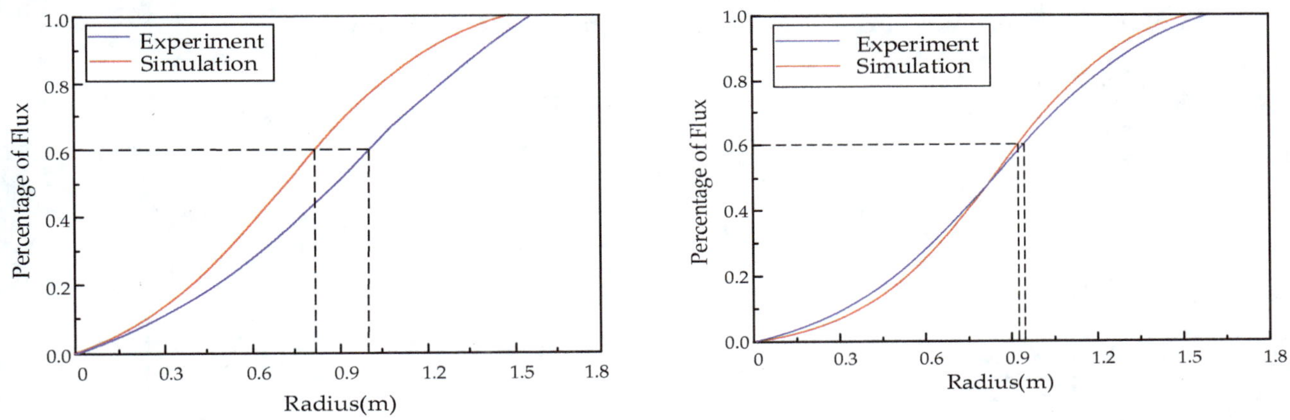

(a) (b)

Figure 9. Relation curves of solar spot energy percentage and radius. (**a**) Spot energy percentage of the No. 5 focal point. (**b**) Spot energy percentage of the No. 4 focal point.

Figure 9a presents the variation trend of the energy percentage as the spot radius changed when the heliostat focused on the #5 focal point. To be specific, the solar spot radius of the simulation and the experiment were, respectively, 1.48 and 1.55 m. In this case, the relative error was less than 4.8%. Moreover, the radii of the simulation and experimental focus spot containing 60% energy were 0.87 and 0.98 m, respectively, and the difference seemed very small. Figure 9b shows the relation curve when heliostat focused on the #4 focal point. Here, the simulated and experimental spot radii were 1.53 and 1.58 m, respectively, and the relative error was less than 3.2%. Different from Figure 9a, the simulation and experimental radii for areas containing 60% of the spot energy turned out to be 0.91 and 0.93 m, respectively, and the difference was only 0.02 m, which is fully acceptable.

As shown in these figures above, it is noted that the solar flux distribution of simulated solar spot was rather close to experimental result. Moreover, the spot sizes also agreed well with the experimental data. According to comparison results, the spliced curved surface could reflect the actual effect of the heliostat better, which verified the validity of the model established here.

4. Optimized Dispatch Design for the Heliostat Field

4.1. Target Point Settings for Concentration

In theory, any point on a receiver can be seen as a focal point during the optimized dispatch of a heliostat field. Mathematically, however, the distribution of focal points is a non-linear continuous optimization problem, which means that the focusing situations of a heliostat field may become complex beyond example. If a heliostat only focuses on several target points, the optimized dispatch of a heliostat field is deemed as a target assignment optimization issue. In this case, the continuous optimization problem turned into a discrete problem that was associated with target assignment. Additionally, the complexity of the solution space was also substantially lowered. Therefore, provided that both quantities and locations of focal points on the receiver are confirmed, a focusing strategy optimization model can be established for a heliostat field. Here, the solar flux distribution of the receiver's opening surface was adopted as the research object, and the actual spot sizes and the intercept efficiency were selected to define the focal point locations. Hence, the following selecting criteria were formulated. Based on these criteria, a schematic diagram for a heliostat's concentrating process and focal point selecting is presented in Figure 10.

(1) The edges of the receiver should be avoided to reduce the spillage of the solar spot.
(2) The distance between two adjacent focal points should be no less than the minimum distance that plays a role in distinguishing the spot centers of heliostats in diverse regions.
(3) The center of the receiver surface should be selected as its focal point.

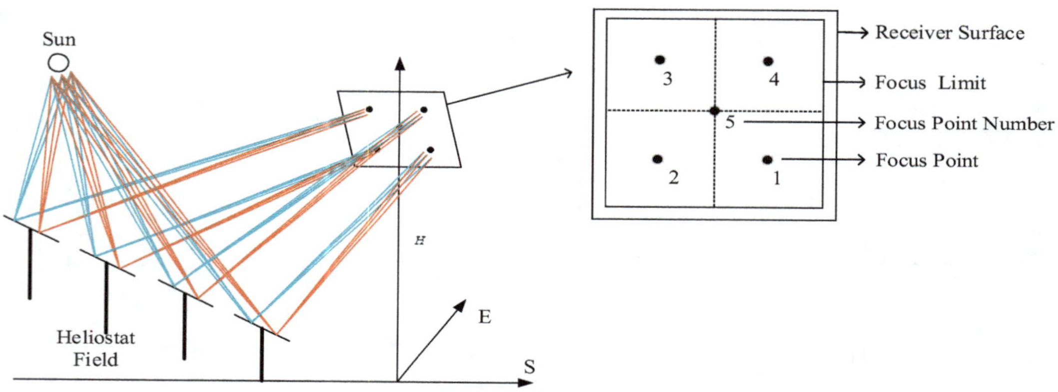

Figure 10. A schematic diagram of heliostat concentrating and focal point selecting.

Regarding the selection of focal points, it should be noted that the distance between adjacent focal points should be determined by spot size. As seen in the comparative analysis shown in Figure 9, the radius of a spot containing 60% of the energy was about 1 m. During focusing, adjacent focal points should be firstly capable of differentiating the centers of various spots. Secondly, within a region where the spots overlap with each other, solar flux density is close to the peak solar flux of an individual solar spot. In other words, the part where the spot energy is below 50% should fall into the category of an overlap area. Therefore, the optimal gap of focal points is defined at 2 m in this paper.

In the context of multi-points focusing of a heliostat, the most important factor that affects the solar flux distribution on a receiver is the tracking precision of heliostats. Under the actions of a mechanical error and the external environment, a tracking error is necessarily inherent in the heliostat. This enables the central point of the focusing spot to deviate from the actual focusing point. In the heliostat field, a heliostat that is most far away from the receiver is located at the 15th ring, and the distance from this heliostat to the receiver is 270 m. In this case, the maximum offset between the actual central position of the heliostat's focusing point on the receiver and the target point was 0.1 m. The least distance was 79 m, and an offset of the spot center was 0.05 m. With regard to the distances of focal points, the spot center offset led to a rather low distribution error of solar flux. For this reason, the influence caused by the tracking error may have been neglected during analysis. During the multi-points experiment, the center of the focus point was deemed to coincide with the target point.

4.2. Heliostat Regionalization Methods

In a commercialized CSP plant, a heliostat field is large and contains thousands of heliostats. During operation, the actual manipulations may be rather complicated, and the implementation cost can be even higher if their focal points are controlled one by one. In order to better serve the actual power plants, dividing the heliostat field into several groups is considered to be a comparatively economical, highly feasible, and effective method according to the similar characteristics of their focused spots. For this purpose, a novel heliostat grouping approach was proposed for this paper. By regarding heliostats' contribution degrees and the incidence cosine similarity as grouping standards, the entire heliostat field was divided into multiple sub-regions. Here, two groups of heliostats in the symmetrical position (#9.2 vs. #−9.2 and #8.2 vs. #−8.2 in Figure 2) were randomly selected to study the incidence cosines on the vernal equinox day. The corresponding comparison results are given in Figure 11.

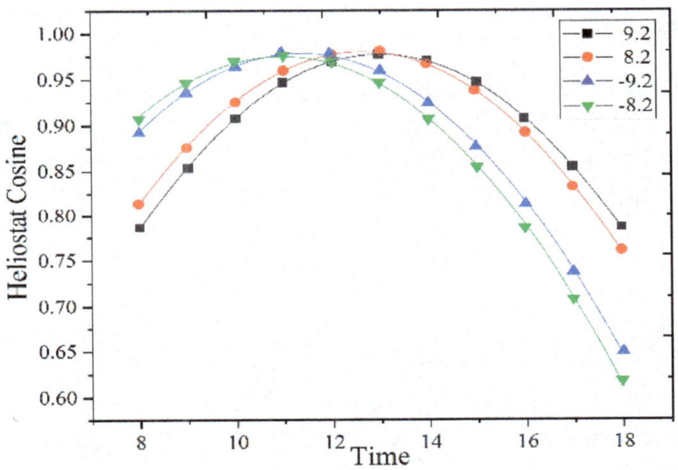

Figure 11. Incidence cosines of heliostats on different positions of the heliostat field.

Apparently, incidence cosines of adjacent heliostats (#9.2 vs. #8.2 or #−9.2 vs. #−8.2) had an identical variation tendency. Within the same ring, the symmetrical heliostats (e.g., #9.2 vs. #−9.2 or #8.2 vs. #−8.2) had significantly different incidence cosines in a day. For example, the incidence cosine of heliostats in the east (#9.2 and #8.2) reached its maximal value at 1:00 PM, while the maximum incidence cosine of heliostats in the west (#−9.2 and #−8.2) was generated at 11:00 AM. This means that the heliostats in different locations contributed different energies to the receiver under the same solar radiation. Based on the above analysis, the grouping method of the heliostat field should abide by the following fundamental principles:

(1) The number of heliostats in a group should be similar to that in another group.
(2) Heliostats in the same group should not be excessively distant from each other.
    (a) Spot shapes reflected by heliostats close to each other are similar, and no great discrepancies exist in their solar flux distribution.
    (b) Their incidence cosines are similar, and no great discrepancies exist in the total energy provided by each heliostat.

On account of the above principles, the heliostat field in the Yanqing CSP plant was divided into 14 sub-regions, and the result is shown in Figure 12. During focusing, heliostats in the same group focused on the same target point as a whole.

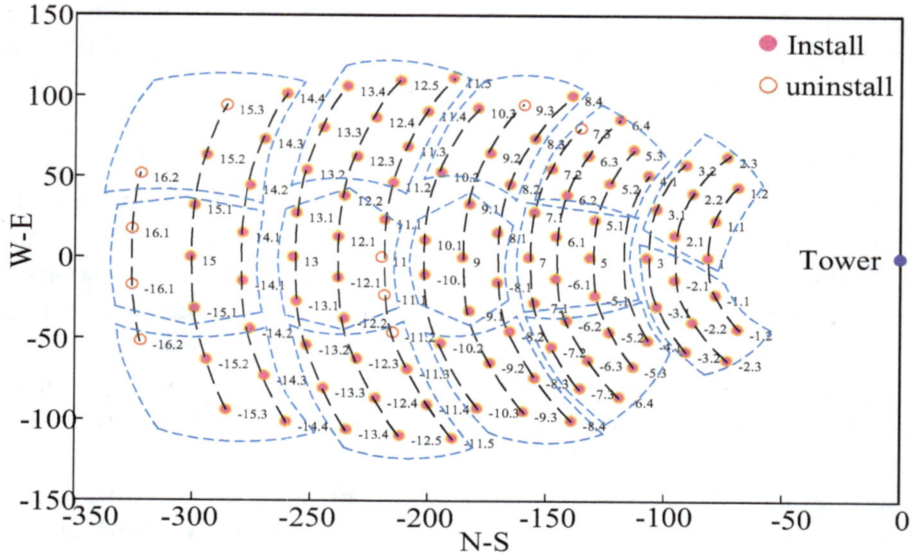

Figure 12. A schematic diagram of heliostat field regionalization.

4.3. Optimization Goal Settings

Once the number of target focal points and the size of heliostat field sub-regions were defined, the optimized dispatch of a heliostat field became a discrete optimization combination problem in a confined space. As a result, the complexity of the solution space greatly dropped. The sample space of heliostat field optimization is expressed in Equation (1):

$$\max = T^h \tag{1}$$

where T refers to the number of target points and h refers to that of the sub-regions in the heliostat field.

The so-called optimal focusing strategy of a heliostat field should be formulated from two perspectives. One is the safety of the receiver, and the other is the energy collected by receiver. On the premise of ensuring the receiver's safety, the solar flux densities on the receiver should be uniformly distributed to the greatest extent. At the same time, the spillage loss of the solar spot should be minimized. A mathematical expression for optimization goals and boundary constraint conditions of the model is given below:

$$\begin{cases} \min \delta \\ F \leq F_0 \\ \eta_{\text{int}} \leq 15\% \end{cases} \tag{2}$$

In Equation (2), δ stands for a standard deviation of solar flux density, F is the actual average solar flux density on the receiver, F_0 is the maximum solar flux density that is allowed by the receiver, and η_{int} is the spillage loss of receiver. Moreover, both the optimization goal and the constraint condition are established to meet solar flux distribution and receiver safety requirements.

4.4. Optimization Algorithms

Nowadays, many algorithms have been applied in heliostat field optimization, including the heuristic tabu method [10], mixed-integer programming [11], the greedy algorithm [25], the genetic algorithm (GA) [8] and particle swarm optimization (PSO) [19]. Here, the GA was selected as a solver of relevant problems. Historically, the GA is a stochastic algorithm derived by professor Holland from the law of nature known as "it is not the strongest of the species that survive, but the one most responsive to change" proposed by Darwin [26]. Among stochastic optimization algorithms, the GA is the most widely applied approach. The basic thought of this algorithm is to simulate biological evolutionism and genetic mechanism and, thus, see an optimization process as a natural evolutionary process. In this manner, an optimal solution [27] can be found. During iterative optimization, the GA starts with population of problem solutions to search, instead of individual solutions to the problem like other optimization algorithms. Considering this, it has a wide coverage and benefits to global optimization. Additionally, the corresponding solving process is independent of information related to search space. On the contrary, only fitness function values are used to evaluate individuals and implement genetic manipulations. Moreover, the GA may undergo self-optimization during iteration, indicating that such an algorithm has the features of self-adaptability and self-learning [28]. Concerning the optimization problems in this study, the solar flux distribution on the receiver cannot be accurately expressed by a mathematical equation. In addition, the solar spot formed by a single heliostat is also a discrete datum. Therefore, the traditional optimization algorithm based on calculus theory could be applied to this situation, as it also requires that the objective function in the optimization problem to be differentiable or at least continuous. Though the heliostat grouping method and database were designed to reduce the running time of the optimization program in this study, an efficient search program was still needed to completed the optimization strategy in the relatively shorter control period of a heliostat field. In search algorithms, a traversal method is usually used to find the optimal solution of the target in the process of searching, such as the meta-heuristic and particle swarm optimization. As a result, the calculation process takes a long time, which did not meet the application requirements of this study. In contrast, the genetic algorithm has great advantages in search speed by means of population evolution, and it can search for multiple possible optimal solutions in one iteration. At the same time, it has a low requirement for the mathematical analytical properties of the optimization objective, and it is especially suitable for the optimization of discrete problems. In such case, it is rather appropriate to solve these problems by means of the genetic algorithm.

In regard to the application to heliostat fields, at the very beginning of the iteration, a random target point should be confirmed for each group of heliostats. Alternatively, no target points may be defined—that is, no focused spot on the receiver. On the corresponding focus point, the spot information of heliostats was retrieved from a database in consistency with the confirmed target point. Then, fitness was figured out for every individual in the population of solutions, and an individual with the lowest fitness was removed. As for elite individuals in the retained population, they were directly reproduced to a new-generation population and also selected as evolved individuals. Hence, crossover and mutation operations could be executed to further generate a new generation of individuals that could be incorporated in the population. During the successive population update, new individuals produced from the new-generation population were able to evolve to those of higher fitness. For the purpose of ensuring convergence to a globally optimal solution, the probabilities of both crossover and mutation may have automatically varied during optimization based on fitness values. In this way, the diversity of population is maintained and locally optimal solutions could be avoided. Computational formulas for crossover probability p_c and mutation probability p_m are given below.

$$p_c = \begin{cases} \frac{p_{c1}(f_{\max} - f_c)}{f_{\max} - f_{avg}} & f_c \geq f_{avg} \\ p_{c2} & f_c < f_{avg} \end{cases} \tag{3}$$

$$p_m = \begin{cases} \frac{p_{m1}(f_{\max} - f_m)}{f_{\max} - f_{avg}} & f_m \geq f_{avg} \\ p_{m2} & f_m < f_{avg} \end{cases} \tag{4}$$

where $f_{\max}$ and f_{avg}, respectively, represent the maximum fitness and the average fitness of population; f_c is the higher fitness among two individuals on which crossover is performed; f_m is fitness of individuals subjecting to mutations; p_{c1}, p_{c2}, p_{m1} and $p_{m2} \in [0,1]$, with $p_{c1} < p_{c2}$ and $p_{m1} < p_{m2}$. In this context, a fitness function can be expressed in the following Equation (5).

$$f = \delta \tag{5}$$

Upon the termination of the algorithm, an optimal individual was acquired, and it was the optimal focusing strategy of the current heliostat field. On this basis, the respective focus points of subregions in the heliostat field were determined. Under the premise of ensuring receiver safety, such a focusing strategy has the capability to make solar flux densities that are uniformly distributed on the receiver. A flow chart of optimization is presented in Figure 13.

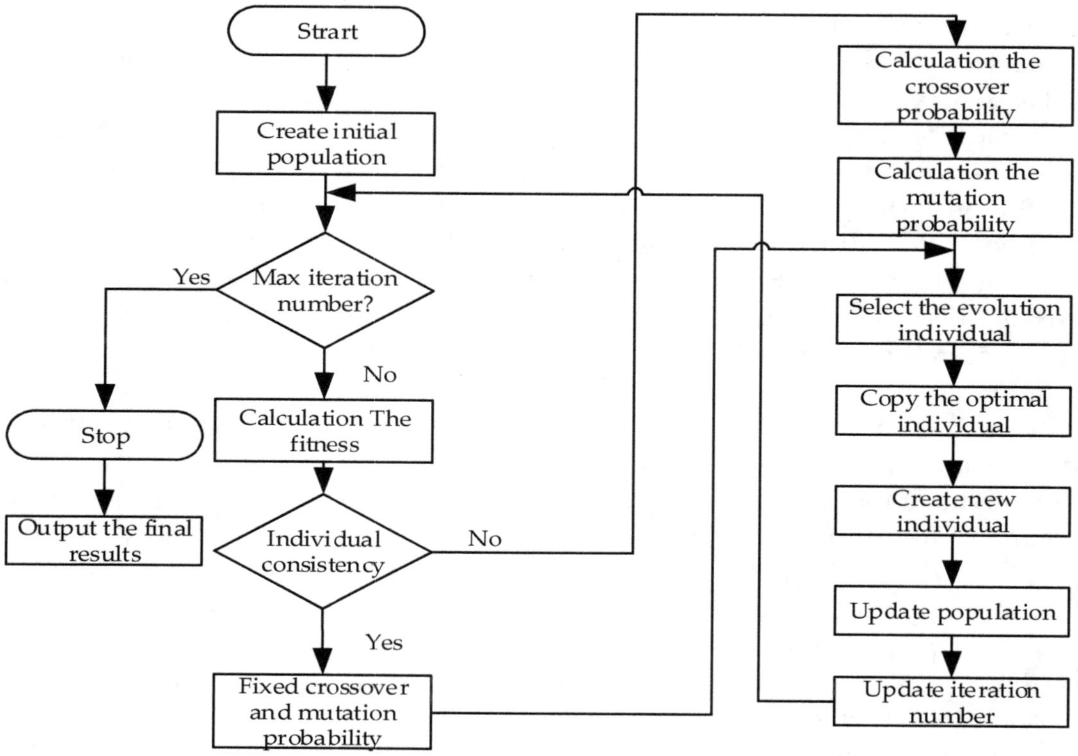

Figure 13. Flow chart of heliostat field optimization.

5. Results and Discussions

5.1. Optimization Strategy Verification

To verify the thought that the proposed optimized dispatch strategy could realize uniform flux distribution on the receiver and check both validity and uniqueness of such a strategy, two groups of focal points in different quantities were designed on a 4 × 4 m receiver surface, as shown in Figure 14. Their biggest difference lied in the addition of a central point of the receiver.

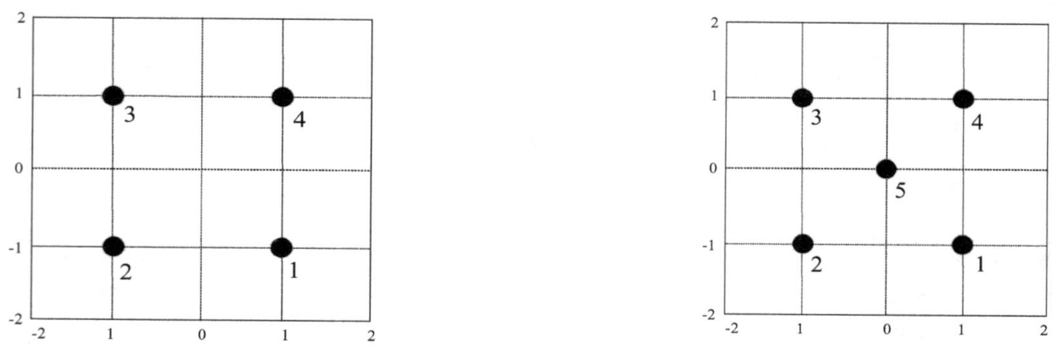

(**a**) 4 Target points. (**b**) 5 Target points.

Figure 14. Distribution of focus points on a 4 × 4 m receiver surface.

The distribution and quantity of focal points on the receiver exert significant influences on optimization space. Under the circumstance that the number of heliostats remains unchanged, one more focal point corresponds to a multiple growth of solution space, which is expressed in Equation (6). This dramatically increases the calculated amount of optimization.

$$M = \frac{(T+1)^h}{T^h} \tag{6}$$

According to focal point distribution in Figure 14, the #−4.1, #5.1, #6.2 and #8.1 heliostats (see Figure 12) in the heliostat field were randomly selected to perform a simulation study. As the heliostat field was regionalized above, the selected heliostats were located at different sub-regions and distributed on the eastern and western sides of the heliostat field. At different times on the vernal equinox day, both the quantity and positions of focal points focused on the receiver after an optimized dispatch was obtained. For details, please see Table 1.

Table 1. Focus positions of heliostats (four or five focal points).

Time \ Focal Point	No. 1	No. 2	No. 3	No. 4	No. 5
9:00 AM	#5.1/#5.1	#−4.1/#−4.1	#8.1/#8.1	#6.2/#6.2	NO/NO
12:00 AM	#5.1/#5.1	#−4.1/#−4.1	#6.2/#6.2	#8.1/#8.1	NO/NO
15:00 PM	#8.1/#8.1	#−4.1/#−4.1	#6.2/#6.2	#5.1/#5.1	NO/NO

It can be observed from Table 1 that distribution optimization results of the two groups of focal points were completely identical at different times on the vernal equinox day. In other words, the addition of the central point did not lead to a change in the optimization result. This means that the optimization algorithm has the potential to not only effectively define the positions of focal points for heliostats but also make solar flux densities uniformly distributed on the receiver. Meanwhile, the uniqueness of solutions acquired through such an algorithm was also validated.

In order to verify the feasibility and validity of optimization results, the above four heliostats were selected to perform experimental verification at different times on 4 September. The focal point positions after optimized dispatch is shown in Table 2.

Table 2. Focal points of heliostats at different times on 4 September.

Time \ Focal Point	No. 1	No. 2	No. 3	No. 4
9:00 AM	#5.1	#−4.1	#8.1	#6.2
12:00 AM	#5.1	#−4.1	#6.2	#8.1
15:00 PM	#8.1	#−4.1	#5.1	#6.2

During the experiment, the heliostat focal points were set as corresponding coordinates according to their positions at different times, as presented in Table 2. The solar flux distribution on the receiver surface at diverse moments are reflected in Figures 15–17.

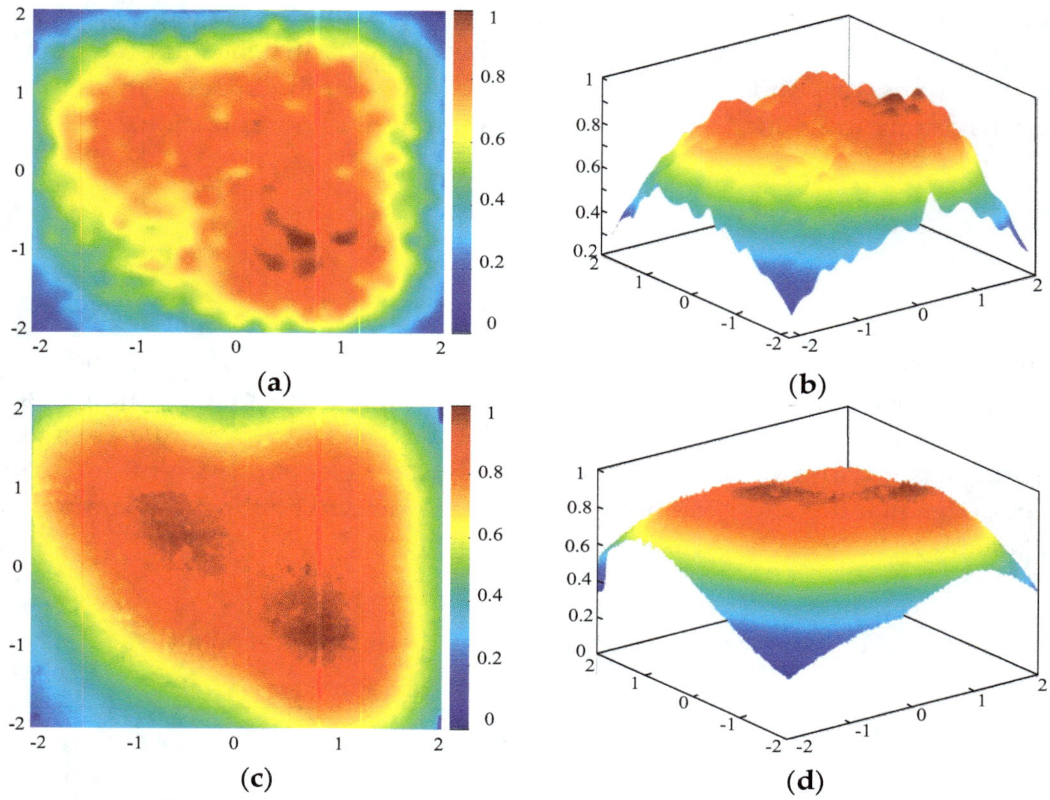

Figure 15. Comparative results of solar flux distribution on receiver surface at 9:00 AM.(**a**) Flux distribution for simulation (plane graph). (**b**) Flux distribution for simulation (perspective view). (**c**) Flux distribution for experiment (plane graph). (**d**) Flux distribution for experiment (perspective view).

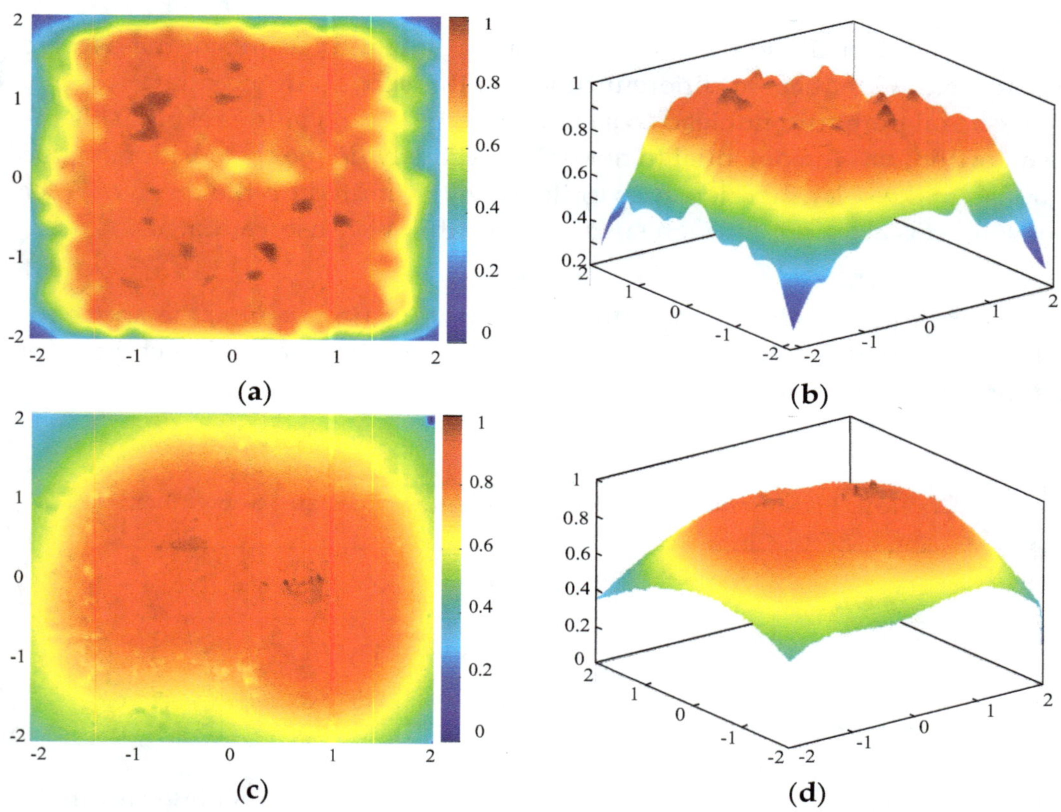

Figure 16. Comparative results of solar flux distribution on receiver surface at 12:00 AM. (**a**) Flux distribution for simulation (plane graph). (**b**) Flux distribution for simulation (perspective view). (**c**) Flux distribution for experiment (plane graph). (**d**) Flux distribution for experiment (perspective view).

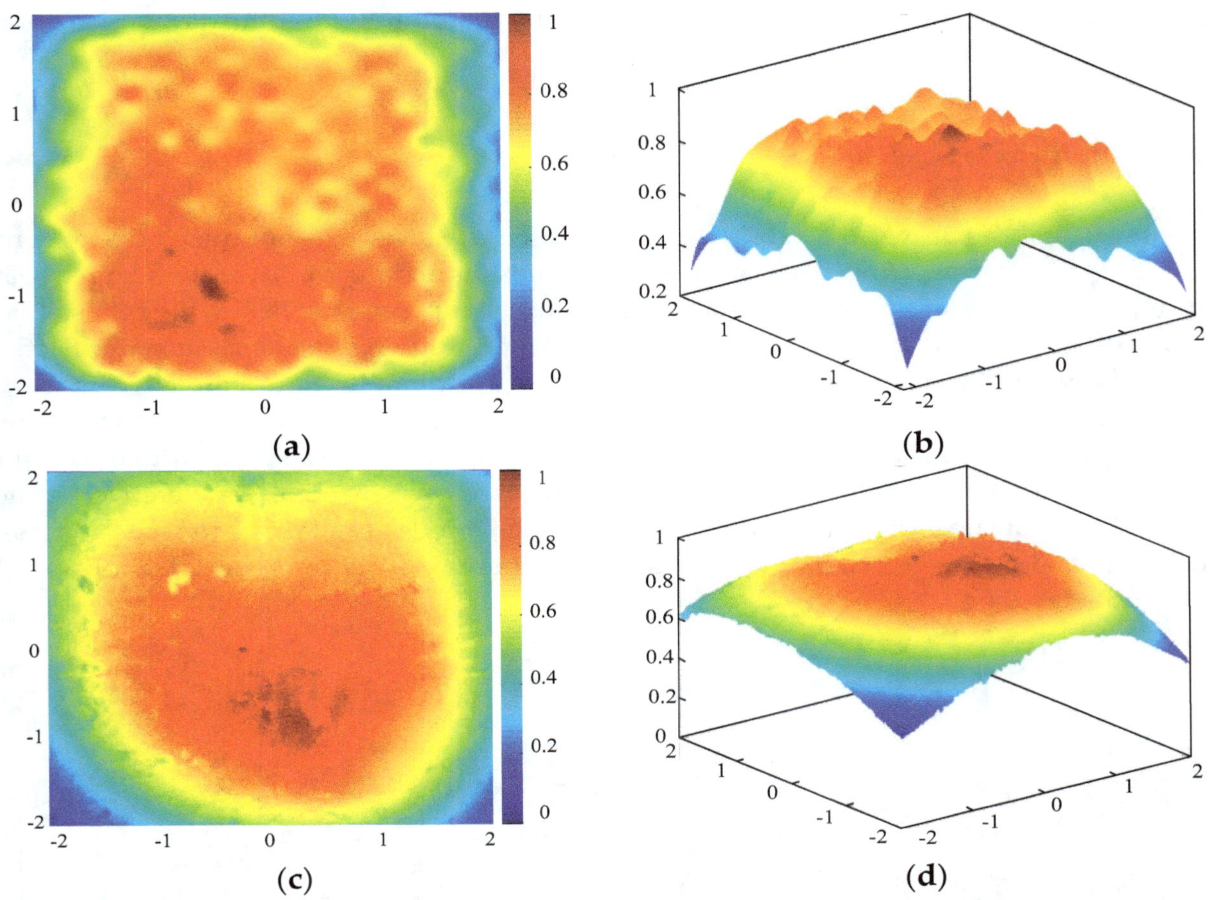

Figure 17. Comparative results of solar flux distribution on receiver surface at 15:00 PM. (**a**) Flux distribution for simulation (plane graph). (**b**) Flux distribution for simulation (perspective view). (**c**) Flux distribution for experiment (plane graph). (**d**) Flux distribution for experiment (perspective view).

In Figure 15, the solar flux densities of the No. 2 focal point can be seen to be significantly lower than those in another sub-region. The major reason for this was that the sun in the morning was closer to the east, and the solar altitude was accordingly smaller. The No. 2 focal point was formed by the #−4.1 heliostat at the westernmost periphery of the heliostat field. Subjected to such an external condition, the incidence cosine loss of the heliostat was the largest, and the energy that it contributed to the receiver was comparatively less. Consequently, the solar flux density at such a focal point was below that of other focal points. As shown in Figure 16, the sun was due south at 12:00 at noon, and the solar altitude reached its highest level at the same time. Clearly, heliostat positions in the heliostat field placed the least influence on spot formation, and the heliostats provided the receiver with balanced energy. Therefore, the most uniform flux distribution phenomenon could be found on the receiver at high noon. The case shown in Figure 17 was just opposite to that revealed in the morning. In the afternoon when the sun moved toward the west, the energy provided by heliostats in the west of the heliostat field was greater than that in other sub-regions. Hence, the solar flux density of the No. 2 focal point at 3:00 PM was above that of remaining focal points.

As shown in Figures 15–17, there were slight differences between the simulation and experimental results on the plane graph. The reasons are as follows: First, the existence of the tracking error of the heliostat caused the center deviation of the focused solar spot from the actual focal point. Secondly, the slight surface error resulted in a difference in the spot shape. However, it seems that the difference was relatively acceptable.

Through analysis, it could be found that, at different times selected in the morning, noon and the afternoon in a day, the experimental and simulation results of solar flux distribution on the receiver surface are generally coincided with each other. This indicates that the optimization algorithm was capable of producing an optimal dispatch strategy at any time throughout the day, which got rid of temporal influence.

5.2. Analysis of the Whole Heliostat Field

The above analysis proves the validity of the heliostat model and its optimized dispatch model from multiple perspectives. However, the number of heliostats and quantities of their target points in practical applications are far greater than those defined for the experiments. Below, the heliostat field as our research object in the Yanqing CSP plant was divided into 14 sub-regions as discussed in Section 4.2. In addition, multi-point focusing and central point focusing were employed to simulatively investigate solar flux distribution on a receiver in different periods on 4 September 2019. Corresponding results are presented in Figure 18. The heliostat count corresponding to each focal point after the optimized dispatch was conducted for sub-regions, and the result is shown in Table 3. There were different quantities of heliostats on each focal point in different time periods. This means the focusing position of heliostats may also change in a day, as the position of the sun and Direct Normal Irradiance (DNI) both change all day long.

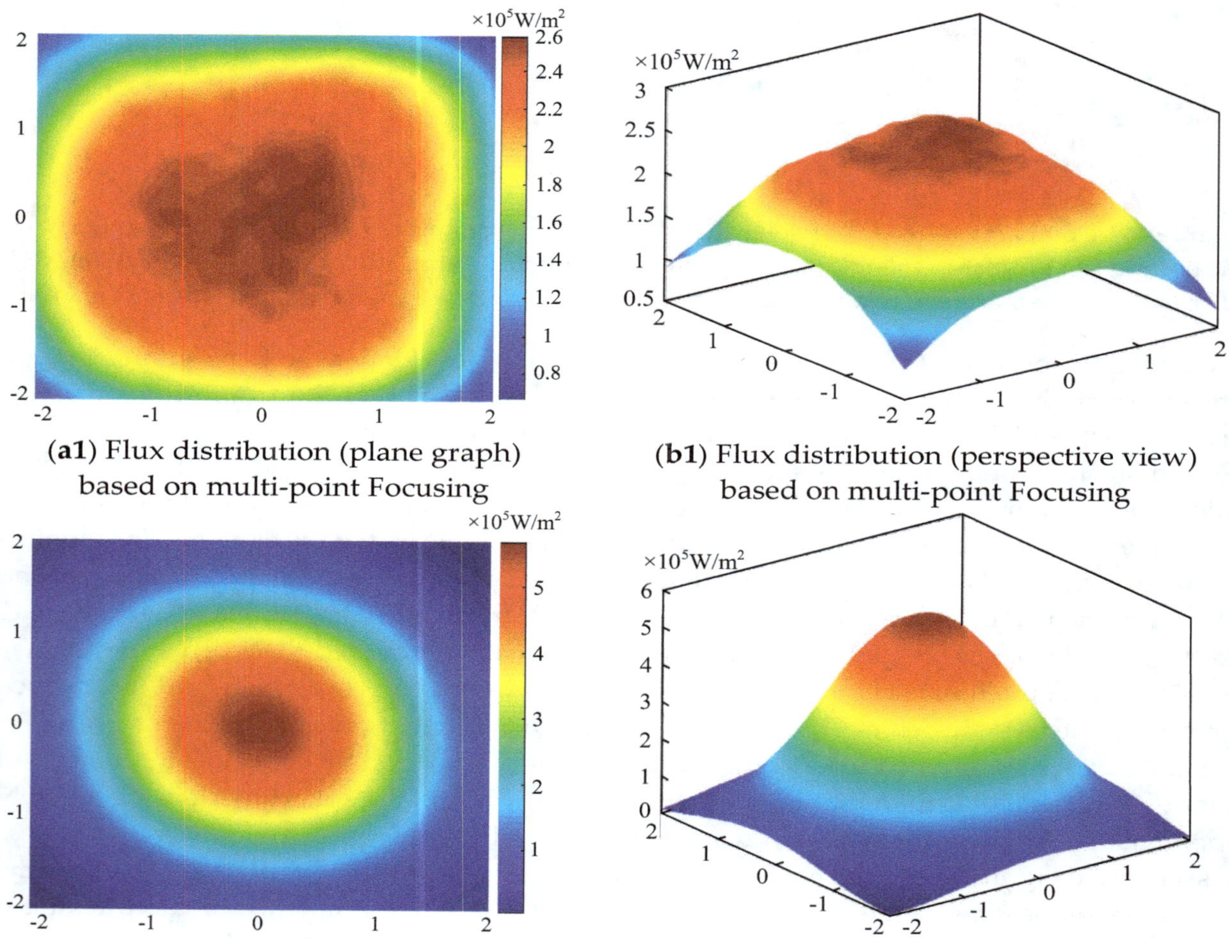

(**a1**) Flux distribution (plane graph) based on multi-point Focusing

(**b1**) Flux distribution (perspective view) based on multi-point Focusing

(**c1**) Flux distribution (plane graph) based on central point focusing

(**d1**) Flux distribution (perspective view) based on central point focusing

(1) Solar flux distribution at 9:00 AM (DNI = 700 W/m^2)

Figure 18. *Cont.*

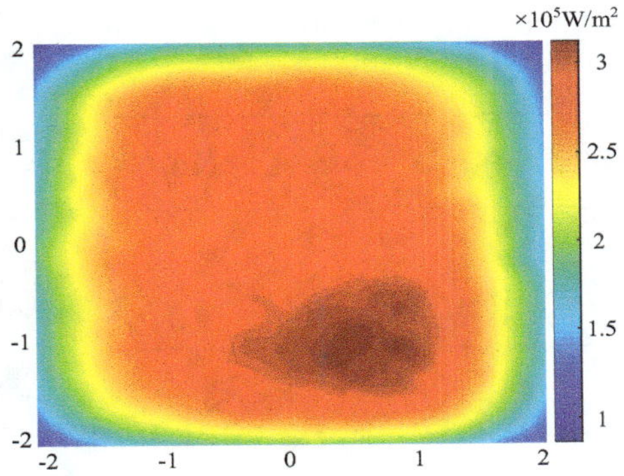

(**a2**) Flux distribution (plane graph) based on multi-point focusing

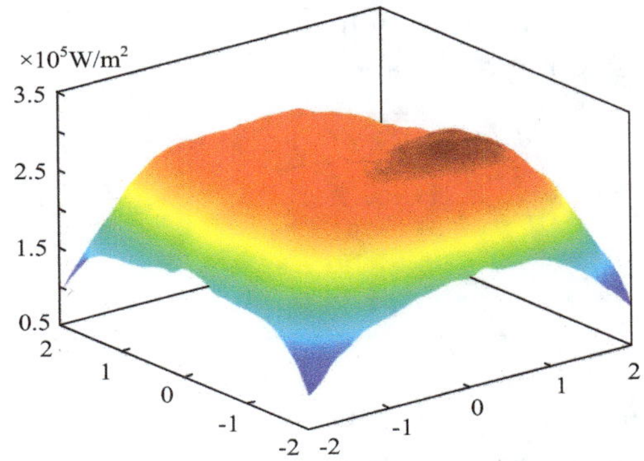

(**b2**) Flux distribution (perspective view) based on multi-point focusing

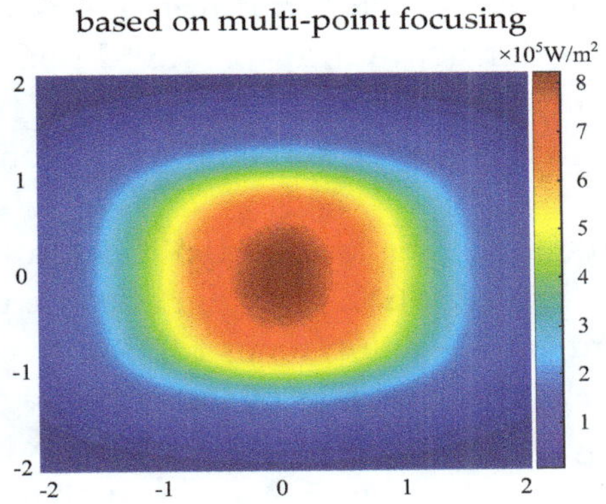

(**c2**) Flux distribution (plane graph) based on central point focusing

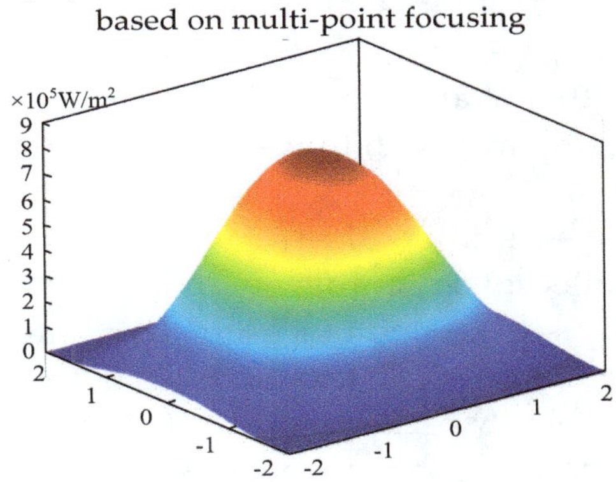

(**d2**) Flux distribution (perspective view) based on central point focusing

(2) Solar flux distribution at 12:00 AM (DNI = 800 W/m^2)

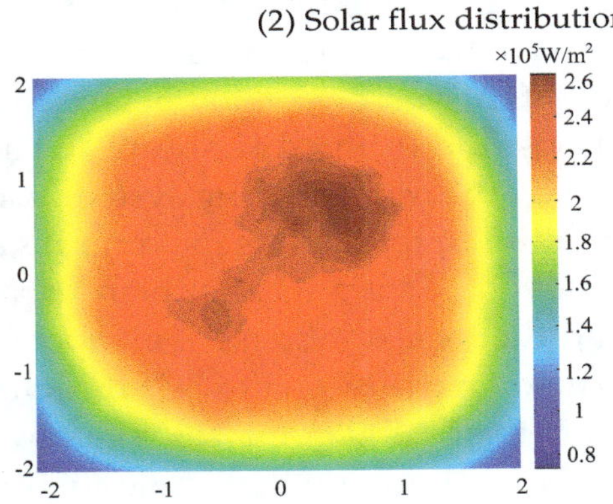

(**a3**) Flux distribution (plane graph) based on multi-point focusing

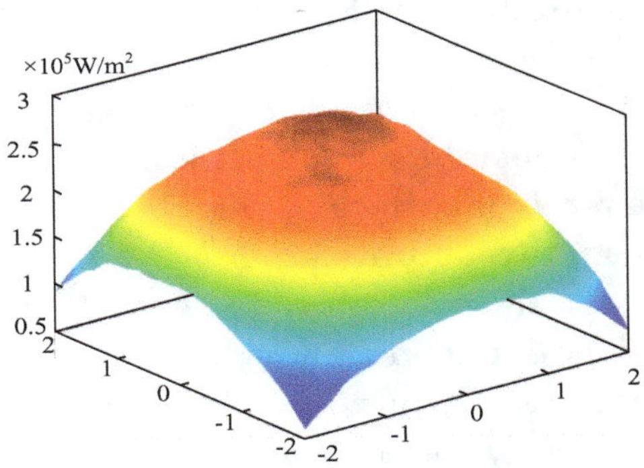

(**b3**) Flux distribution (perspective view) based on multi-point focusing

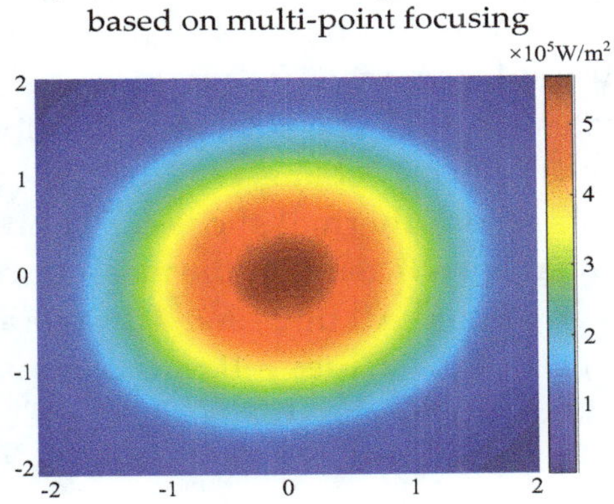

(**c3**) Flux distribution (plane graph) basedon central point focusing

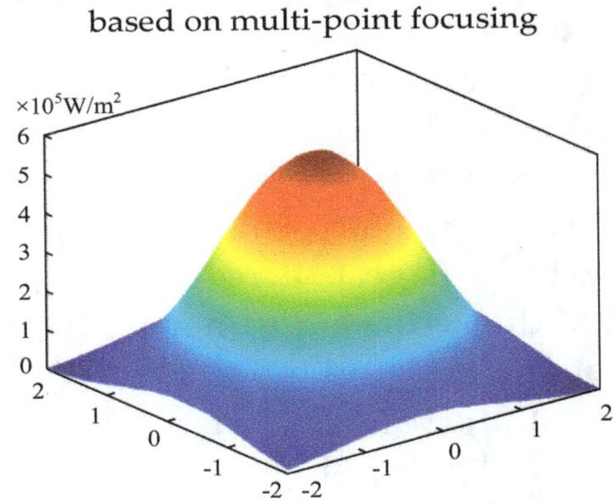

(**d3**) Flux distribution (perspective view) based on central point focusing

(3) Solar flux distribution at 15:00 PM (DNI = 700 W/m^2)

Figure 18. Flux distribution on the receiver surface after heliostat field regionalization. (**a1–a3**) Flux distribution (plane graph) based on multi-point Focusing. (**b1–b3**) Flux distribution (perspective view) based on multi-point Focusing. (**c1–c3**) Flux distribution (plane graph) based on central point focusing. (**d1–d3**) Flux distribution (perspective view) based on central point focusing.

Table 3. The number of heliostats corresponding to focal points based on multi-point focusing after heliostat field regionalization.

Time \ Focal Point	No. 1	No. 2	No. 3	No. 4	No. 5
9:00 AM	3	4	4	3	0
12:00 AM	4	3	4	3	0
15:00 PM	4	2	5	3	0

As can be clearly observed from Table 3, no group focused on the central point (the No. 5 focal point) when multi-point focusing was adopted due to the two following major aspects. First, the solar spots of other focal points all covered the central point. As a result, multiple spots within such a sub-region overlapped with each other, and the solar flux density of the No. 5 focal point was slightly different from that of another focal point. Second, the solar spot formed by a group of heliostats was larger when compared with those formed by an individual heliostat, which led to an enlargement of the overlap area.

As can be seen in Figure 18, when multi-point focusing was applied, although the solar flux distribution failed to be completely uniform, the solar flux variation gradient became much gentler. This differentiates multi-point focusing from central point (single-point) focusing. Moreover, this figure reveals that the solar flux distribution was not very ideal as the solar flux of some sub-regions was still greater than others. Especially in periods of the morning or the afternoon, the solar flux density of the central sub-region was above those in several other focal points. The reasons for such a phenomenon are as follows: Firstly, the heliostat field of the Yanqing CSP plant belonged to a small-scale field in the current CSP plants, and the number of groups after grouping was relatively small. Thus, the addition of an additional group of heliostats in focal points may have had an enormous influence on the solar flux distribution. Secondly, the influence of the position of the sun on solar spot size changes cannot be ignored. As the sun slowly rose and arrived at a position in the due south, the solar altitude became increasingly large, while the focal spot generated by the concentrating system got smaller and smaller on the receiver surface. At high noon, the solar altitude reached its maximum value, and the smallest focal spot was formed there. When the sun moved beyond its zenith and gradually went down to the west, the solar altitude began to drop, and the focal spot described above enlarged as time goes on. Under circumstances that solar spots formed on the receiver were the smallest at high noon, an overlap area of multiple overlapping spots within the central point sub-region was also much smaller. Consequently, the solar flux distribution was rather uniform on the whole. In time periods of the morning and the afternoon, the decline in solar altitude not only led to the enlargement of the focal spot, it also led to a large overlap area in the central sub-region. For this reason, the solar flux density of such a sub-region was greater than those of remaining sub-regions.

A comparative analysis based on Figure 18 indicates that: (1) The solar flux distribution on the receiver surface was rather uniform; (2) the solar flux density gently changed at different time periods of a day when multi-point focusing was adopted. In this case, a thermal shock effect on the receiver could be lowered to the largest degree. In regard to the intercept efficiency of the receiver, no significant changes occurred due to the adjustment of focal point positions. All day long, the average intercept efficiency was about 89%, and the spillage was about 11%. Both values fell into the designed value range. By contrast, the solar flux distribution perfectly conformed to Gaussian distribution when the traditional central point (single-point) focusing was implemented. In this scenario, the intercept efficiency of the receiver throughout the day averaged 96.07%. However, the gradient variations of solar flux densities were rather fierce, and highly non-uniform heating occurred on the entire receiver, which may have caused considerable damages to it.

Furthermore, through comparative analysis, it was also noted that the peak flux on the receiver may have been significantly reduced by means of multi-point focusing to about half of that generated by conventional central point focusing. In addition, the temperature distribution on the entire receiver surface could have been more uniform, and the thermal shock effect on the receiver could have been much insignificant. Relative to damages caused by non-uniform heating on the receiver, the thus incurred energy loss was completely within the acceptable range, although its intercept efficiency slightly declined. Such a phenomenon was in accordance with objectives of the optimization

In order to quantitatively observe the solar flux densities on the receiver surface, the horizontal and vertical profile were deeply analyzed based on a central focusing point [29], and the result of this analysis is shown in Figure 19. As shown in this figure, for the horizontal and vertical profiles subjected to multi-point focusing, the peak solar flux densities were far below those generated in a central point focusing mode in different moments of a day. According to the traditional single-point focusing, the maximum solar flux density was 8.13×10^5 W/m^2 at high noon, which was 2.6 times greater than that generated by multi-point focusing (3.12×10^5 W/m^2). From 9:00 AM to 3:00 PM, the DNI was equal to 700 W/m^2, the peak solar flux densities for the multi-point focusing and central point focusing turned out to be 2.6×10^5 and 5.65×10^5 W/m^2, respectively, and it was clearly shown that the latter was 2.17 larger than the former. In order to more intuitively learn about optimization results, the relationship between the maximum and minimum solar flux densities was analyzed on the horizontal profile in a condition of different focusing modes at the same time. For the central point focusing, the peak flux densities at 9:00 AM., 3:00 PM. and by 12:00 noon, respectively, were 11.3, 11.3 and 12.3 times higher than their minimum flux densities. However, these became 2.3, 2.1 and 1.9, respectively, when multi-point focusing was adopted. For this reason, the optimized focusing strategy was deemed to have the potential to effectively decrease the peak solar flux on the receiver and achieve an uniform solar flux distribution. From Figure 19, it can be seen that the solar flux densities on the receiver surface were very close to each other in most sub-regions after optimization, regardless of the vertical or horizontal profile.

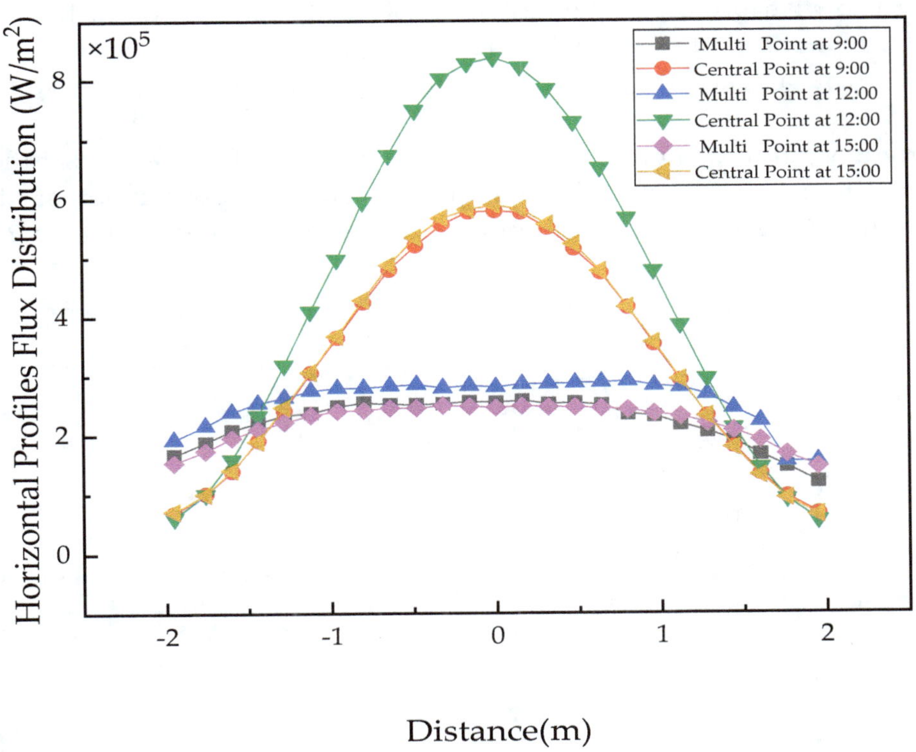

(a)

Figure 19. *Cont.*

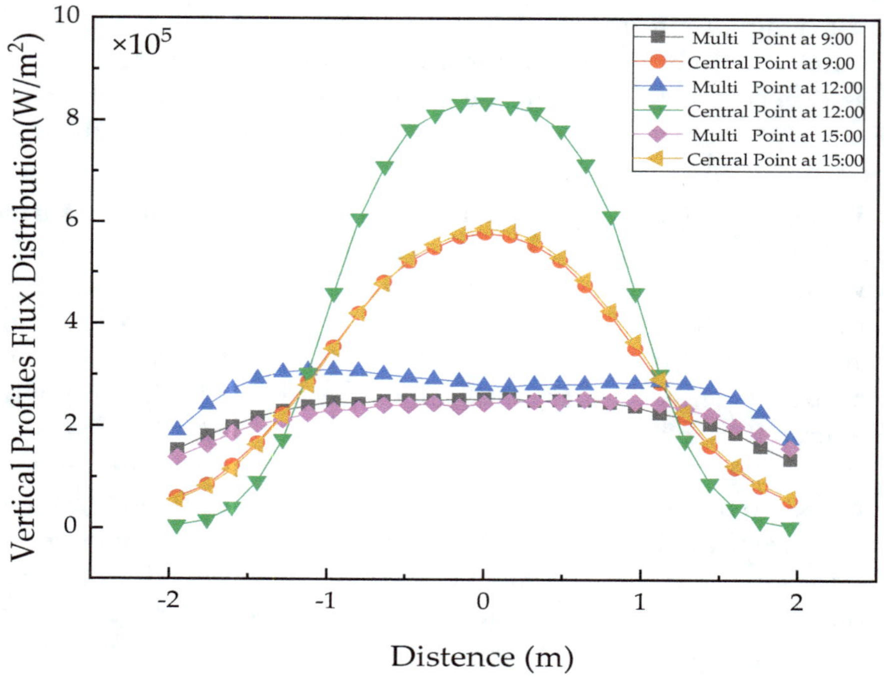

(**b**)

Figure 19. Flux distribution on the receiver surface central profiles. (**a**) Flux distribution on horizontal profiles. (**b**) Flux distribution on vertical profiles.

5.3. Compared with Published Data

In order to further verify that the optimization strategy adopted in this paper can be effectively applied to other CSP plants, the Themis plant in France was selected as the research objective to simulate the solar flux distribution on the receiver surface at 11:00 AM on 21 June. The heliostat of the Themis plant is composed of nine small mirrors (eight mirrors with the size 3.63 × 1.794 m and one mirror of 2.46 × 0.828 m), and the total area is 53.85 m^2. The heliostat field layout can be found in reference [12].

Before comparison, all the parameters (which are shown in Table 4) are set the same with the published data in reference [12], the only difference is that the optimization strategy adopted in reference [12] was the TABU algorithm. The comparative result is shown in Figure 20.

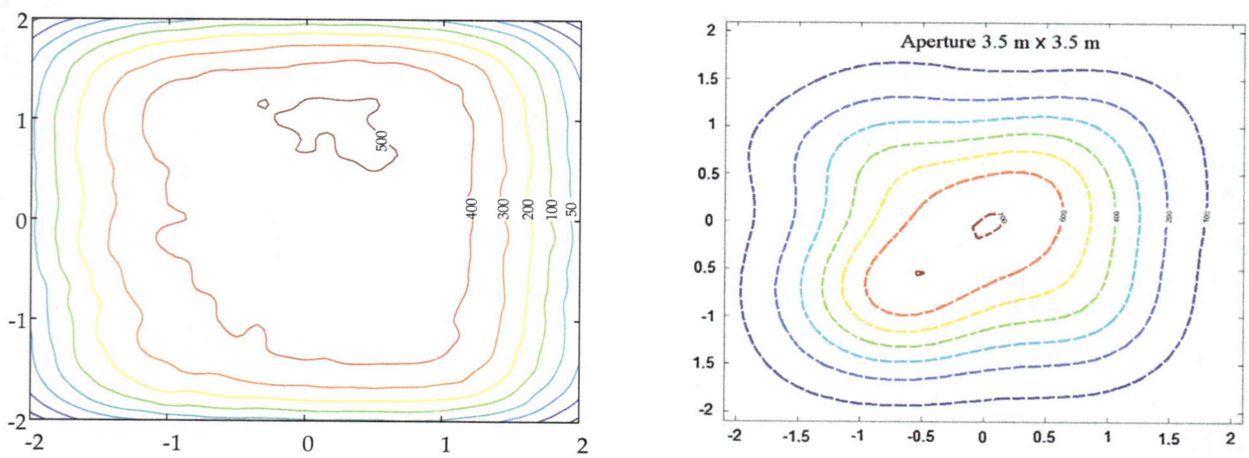

(**a**) (**b**)

Figure 20. Comparative results of solar flux distribution on the receiver surface of the Themis plant. (**a**) The result with the GA. (**b**) The result with TABU.

Table 4. System parameters setting during the optimization.

Time	Receiver Size	Receiver Height	Threshold Spillage	Number of Focal Points	DNI
11:00 AM	3.5 × 3.5 m	100 m	10%	25	1000 W/m^2

6. Conclusions

In order to more intuitively observe the comparative results, the unit in our paper was converted into the same unit of that in reference [12]. It could be seen that, after optimization, the peak solar flux density obtained in our paper was 500 suns, which was 200 suns lower than that in reference [12]. At the same time, the distribution area of solar flux in our paper seemed to be more extensive with the same incident energy. The radius of 400 suns was about 1.2 m in our paper and 1 m in reference [12]. In regard the spillage loss, the value after optimization in our paper was 8.67%, which was only increased by 0.84% (the value in reference [12] was 7.83%). This difference is fully acceptable. From the comparative result, it can be noted that the optimization strategy adopted in this paper can more effectively reduce the peak solar flux density on the receiver surface. Meanwhile, the solar flux distribution seems to be more uniform. Ultimately, it seems that the optimized dispatch proposed in this paper can meet the requirements for the safe operation of a central receiver.

A heliostat field plays a critical role in an entire solar thermal power tower plant and serves as its energy source. In this study, a "concentrating-receiver system" of 1 a MWe CSP Plant in Yanqing was selected as the research object to establish a spliced reflection model in the first place. Then, a grouping approach of heliostat fields was proposed according to the facts that heliostats in different positions have different contribution degrees and the incidence cosines of adjacent heliostats share resemblances. The selection principles of focal points are also designed based on spot sizes. Finally, an optimized dispatch and operation strategy have been confirmed for the heliostat field to realize an optimization goal of minimizing the standard deviation of solar flux distribution. It has been proven that the optimized dispatch strategy can be used to tremendously lower the peak values of solar flux densities on the receiver and to achieve uniform flux distribution. This is of great significance to extend the service life of a receiver and ensure that the receiver runs in a safe and stable state.

Author Contributions: Conception, D.Y.; methodology, D.Y. and. Q.Y.; software, D.Y.; validation, D.Y. and F.S. formal analysis, D.Y.; investigation, D.Y.; resources, D.Y.; data curation, D.Y.; handwriting—original draft preparation, D.Y.; handwriting—review and editing, D.Y. and. Q.Y.; visualization, D.Y.; supervision, Q.Y.; project administration, Z.W. and Q.Y.; funding acquisition, Z.W. and Q.Y.

References

1. Desideri, U.; Campana, P.E. Analysis and comparison between a concentrating solar and a photovoltaic power plant. *Appl. Energy* **2014**, *113*, 422–433. [CrossRef]
2. Behar, O.; Khellaf, A.; Mohammedi, K. A review of studies on central receiver solar thermal power plants. *Renew. Sustain. Energy Rev.* **2013**, *23*, 12–39. [CrossRef]
3. Richardg.newell Daniel-Raimi-Gloria-Aldana. Global Energr Outlook 2019: The Next Genengeneration of Energy. Available online: https://www.rff.org/publications/reports/global-energy-outlook-2019/ (accessed on 1 July 2019).
4. Llamas, J.M.; Bullejos, D.; Ruiz de Adana, M. Optimization of 100 MWe Parabolic-Trough Solar-Thermal Power Plants under Regulated and Deregulated Electricity Market Conditions. *Energies* **2019**, *12*, 3973. [CrossRef]
5. Zhan, J.; Wang, Z. Solar Thermal Power Generation Technology in a New Generation of Energy System Positioning. *Adv. Energy Power Eng.* **2018**, *6*, 1–9. [CrossRef]
6. Dunnan, L.; Xiaodan, G.; Bowen, X. What causes growth of global greenhouse gas emissions? Evidence from 40 countries. *Sci. Total Environ.* **2019**, *661*, 750–766.
7. Du, E.; Zhang, N.; Hodge, B.-M.; Wang, Q.; Xia, Q. The Role of Concentrating Solar Power Towards High Renewable Energy Penetrated Power Systems. *IEEE Trans. Power Syst.* **2018**, *33*, 6630–6641. [CrossRef]
8. Besarati, S.M.; Yogi Goswami, D.; Stefanakos, E.K. Optimal heliostat aiming strategy for uniform distribution of heat flux on the receiver of a solar power tower plant. *Energy Convers. Manag.* **2014**, *84*, 234–243. [CrossRef]

9. Li, W.; Ma, J. Researches and simulation on heliostat concentrating strategy in solar power tower system. *Renew. Energy Resour.* **2016**, *34*, 475–480.
10. Yu, Q.; Wang, Z.; Xu, E. Analysis and improvement of solar flux distribution inside a cavity receiver based on multi-focal points of heliostat field. *Appl. Energy* **2014**, *136*, 417–430. [CrossRef]
11. Fernandez, N.; Street, A.; Montenon, A.; Maussion, P. A Scheduling Optimization Model for Sun Tracking of an Autonomous Heliostat. Available online: https://ieeexplore.ieee.org/stamp/stamp.jsp?tp=&arnumber=6652249 (accessed on 1 July 2019).
12. Salomé, A.; Chhel, F.; Flamant, G.; Ferrière, A.; Thiery, F. Control of the flux distribution on a solar tower receiver using an optimized aiming point strategy: Application to THEMIS solar tower. *Sol. Energy* **2013**, *94*, 352–366. [CrossRef]
13. Guo, T.; Liu, G.; Liu, D.; Guo, S. Research on Supreheated Evaluation Mode of Receiver on Solar Power Tower Plant and Dispatching Methods of heliostats Field. *Acta Energ. Sol. Sin.* **2014**, *35*, 166–170.
14. GarcÍa-MartÍn, F.J.; Berenguel, M.; Valverde, A.; Camacho, E.F. Heuristic knowledge-based heliostat field control for the optimization of the temperature distribution in a volumetric receiver. *Sol. Energy* **1999**, *66*, 355–369. [CrossRef]
15. Maffezzoni, C.; Parigi, F. Dynamic analysis and control of a solar power plant—II. Control system design and simulation. *Sol. Energy* **1982**, *28*, 117–128. [CrossRef]
16. Gallego, A.J.; Fele, F.; Camacho, E.F. On the optimization of irradiance distribution in solar tower plants with flat receivers. In Proceedings of the 2014 European Control Conference (ECC), Strasbourg, France, 24–27 June 2014.
17. Pitz-Paal, R.; Schwarzbozl, P.; Ulmer, S.; Belhomme, B. A New Fast Ray Tracing Tool for High-Precision Simulation of Heliostat Fields. *J. Sol. Energy Eng.* **2009**, *131*, 031002.
18. Xu, M. Study on Modeling and Control Strategy of Solar Thermal Power Condensing System. Ph.D. Thesis, Nan Jing Normal University, Nanjing, China, 2012.
19. Ding, T.; Hu, X. Application Research of Heliostats Field Scheduling with Intelligent Algorithm. *Acta Energ. Sol. Sin.* **2014**, *35*, 2139–2144.
20. Xu, E.; Yu, Q.; Wang, Z.; Yang, C. Modeling and simulation of 1 MW DAHAN solar thermal power tower plant. *Renew. Energy* **2010**, *36*, 848–857. [CrossRef]
21. Sun, F. Study on the Heliostat Tracking Correction System in a Solar Tower Power Plant. Ph.D. Thesis, The University of Chinese Academy of Sciences, Beijing, China, 2014.
22. Ballestrín, J.; Carra, M.E.; Enrique, R.; Monterreal, R.; Fernández-Reche, J.; Polo, J.; Casanova, M.; Barbero, F.J.; Marzo, A. Diagnosis of a Lambertian target in solar context. *Measurement* **2018**, *119*, 265–269. [CrossRef]
23. Liu, Y. Theoretical and Experimental Result of Focal Spot Flux Density Distribution of Solar Energy Concentrations. Ph.D. Thesis, HaRBin Institute of Technology, Harbin, China, 2008.
24. Zhou, Y. Research on Flux Distribution Simulation and Focus Strategy of TSP Plant. Master's Thesis, ZheJiang University, Zhejiang, China, 2014.
25. Zhao, Q. Optimization on the Schedualing of the Heliostat Field in a Solar Tower Power Plant. Master's Thesis, ZheJiang University, Zhejiang, China, 2017.
26. Koza, J.R. *An Introduction to Genetic Algorithms*; MIT Press: Boston, MA, USA, 1996.
27. Lei, Y.; Zhang, S.; Li, X.; Zhou, C.M. *MALAB Genetic Algorithm Toolbox and Applications*; XiDian University Press: Xian, China, 2014.
28. Mokshin, A.V.; Mokshin, V.V.; Sharnin, L.M. Adaptive Genetic Algorithms Used to Analyze Behavior of Complex System. *Commun. Nonlinear Sci. Numer. Simul.* **2018**, *2019*, 174–186. [CrossRef]
29. Ballestrín, J.; Casanova, M.; Monterreal, R.; Fernández-Reche, J.; Setien, E.; Rodríguez, J.; Galindo, J.; Barbero, F.J.; Batlles, F.J. Simplifying the measurement of high solar irradiance on receivers. Application to solar tower plants. *Renew. Energy* **2019**, *138*, 551–561. [CrossRef]

PERMISSIONS

All chapters in this book were first published in MDPI; hereby published with permission under the Creative Commons Attribution License or equivalent. Every chapter published in this book has been scrutinized by our experts. Their significance has been extensively debated. The topics covered herein carry significant findings which will fuel the growth of the discipline. They may even be implemented as practical applications or may be referred to as a beginning point for another development.

The contributors of this book come from diverse backgrounds, making this book a truly international effort. This book will bring forth new frontiers with its revolutionizing research information and detailed analysis of the nascent developments around the world.

We would like to thank all the contributing authors for lending their expertise to make the book truly unique. They have played a crucial role in the development of this book. Without their invaluable contributions this book wouldn't have been possible. They have made vital efforts to compile up to date information on the varied aspects of this subject to make this book a valuable addition to the collection of many professionals and students.

This book was conceptualized with the vision of imparting up-to-date information and advanced data in this field. To ensure the same, a matchless editorial board was set up. Every individual on the board went through rigorous rounds of assessment to prove their worth. After which they invested a large part of their time researching and compiling the most relevant data for our readers.

The editorial board has been involved in producing this book since its inception. They have spent rigorous hours researching and exploring the diverse topics which have resulted in the successful publishing of this book. They have passed on their knowledge of decades through this book. To expedite this challenging task, the publisher supported the team at every step. A small team of assistant editors was also appointed to further simplify the editing procedure and attain best results for the readers.

Apart from the editorial board, the designing team has also invested a significant amount of their time in understanding the subject and creating the most relevant covers. They scrutinized every image to scout for the most suitable representation of the subject and create an appropriate cover for the book.

The publishing team has been an ardent support to the editorial, designing and production team. Their endless efforts to recruit the best for this project, has resulted in the accomplishment of this book. They are a veteran in the field of academics and their pool of knowledge is as vast as their experience in printing. Their expertise and guidance has proved useful at every step. Their uncompromising quality standards have made this book an exceptional effort. Their encouragement from time to time has been an inspiration for everyone.

The publisher and the editorial board hope that this book will prove to be a valuable piece of knowledge for researchers, students, practitioners and scholars across the globe.

LIST OF CONTRIBUTORS

Paweł Obstawski, Tomasz Bakoń and Dariusz Czekalski
Institute of Mechanical Engineering, Department of Fundamental Engineering and Energetic, Warsaw University of Life Sciences, 02-787 Warsaw, Poland

Joanna Aleksiejuk-Gawron and Andrzej Chochowski
Department of Fundamentals in Engineering and Energy, Institute of Mechanical Engineering, Warsaw University of Life Sciences, 02-787 Warsaw, Poland

Cristóbal Villasante, Saioa Herrero and David Olasolo
Fundacion Tekniker, Polo Tecnológico de Eibar, Calle Iñaki Goenaga 5, 20600 Gipuzkoa, Spain

Marcelino Sánchez, Iñigo Pagola, Adrian Peña and Ana Bernardos
CENER (National Renewable Energy Centre of Spain), Ciudad de la Innovación 7, 31621 Navarra, Spain

Seoyong Shin
Department of Information and Communication Engineering, Myongji University, Gyeonggi-do 17058, Korea

Thanh Tuan Pham
Department of Information and Communication Engineering, Myongji University, Gyeonggi-do 17058, Korea
Institute of Applied Material Science, Vietnam Academy of Science and Technology, Ho Chi Minh City 70072, Vietnam

Ngoc Hai Vu
Electrical and Electronics Engineering, Phenikaa University, Hanoi 1000, Vietnam

Jing Liu and Yongqing He
School of Chemical Engineering, Kunming University of Science and Technology, Kunming 650500, China

Xianliang Lei
State Key Laboratory of Multiphase Flow in Power Engineering, Xi'an Jiaotong University, Xi'an 710049, China

José Galindo
Plataforma Solar de Almería, Crta. Senes, km. 4.5, E04200 Tabernas, Spain

Diego Pulido-Iparraguirre
Plataforma Solar de Almería, Crta. Senes, km. 4.5, E04200 Tabernas, Spain
Universidad de Antofagasta, Avenida Angamos, Antofagasta 601, Chile
Universidad de Almería, Crta. De Sacramento s/n, E04120, La Cañada de San Urbano, E04120 Almería, Spain

Loreto Valenzuela, Jesús Fernández-Reche and José Rodríguez
CIEMAT-Plataforma Solar de Almería, Crta. Senes, km. 4.5, E04200 Tabernas, Spain

Kristina Kilikevičienė, Mindaugas Jurevičius and Vytautas Makarskas
Department of Mechanical and Material Engineering, Vilnius Gediminas Technical University, J. Basanaviciaus St. 28, LT-03224 Vilnius, Lithuania

Jonas Matijošius and Artūras Kilikevičius
Institute of Mechanical Science, Vilnius Gediminas Technical University, J. Basanaviciaus St. 28, LT-03224 Vilnius, Lithuania

Jacek Caban and Andrzej Marczuk
Department of Agricultural, Forestry and Transport Machines, University of Life Sciences in Lublin, 28 Głęboka Street, 20-612 Lublin, Poland

Zhiying Cui and Fengwu Bai
Key Laboratory of Solar Thermal Energy and Photovoltaic System, Chinese Academy of Sciences, Beijing 100190, China
Institute of Electrical Engineering, Chinese Academy of Sciences, Beijing 100190, China
University of Chinese Academy of Sciences, Beijing 100049, China
Beijing Engineering Research Center of Solar Thermal Power, Beijing 100190, China

Fuqiang Wang
Harbin Institute of Technology, Weihai 264209, Shandong, China

Eduardo González-Mora and Ma. Dolores Durán García
Ingeniería en Sistemas Energéticos Sustentables, Facultad de Ingeniería, Universidad Autónoma del Estado de México, Cerro de Coatepec S/N, Toluca 50100, Mexico

Linrui Ma
Key Laboratory of Solar Thermal Energy and Photovoltaic System, Chinese Academy of Sciences, Beijing 100190, China
Institute of Electrical Engineering, Chinese Academy of Sciences, Beijing 100190, China
Institute of Electrical Engineering, University of Chinese Academy of Sciences, Beijing 100049, China
Beijing Engineering Research Center of Solar Thermal Power, Beijing 100190, China

Li Xu
Key Laboratory of Solar Thermal Energy and Photovoltaic System, Chinese Academy of Sciences, Beijing 100190, China
Institute of Electrical Engineering, Chinese Academy of Sciences, Beijing 100190, China
Beijing Engineering Research Center of Solar Thermal Power, Beijing 100190, China

Noboru Yamada
Department of Energy and Environment Science, Nagaoka University of Technology, 1603-1 Kamitomioka-machi, Nagaoka 940-2188, Japan

Masakazu Nakatani
Department of Energy and Environment Science, Nagaoka University of Technology, 1603-1 Kamitomioka-machi, Nagaoka 940-2188, Japan
Sun Marion Co., Ltd., 1603-1 Kamitomioka-machi, Nagaoka 940-2188, Japan

Fangyuan Yao, Dongqiang Lei and Pan Yao
Key Laboratory of Solar Thermal Energy and Photovoltaic System, Chinese Academy of Sciences, Beijing 100190, China
Institute of Electrical Engineering, Chinese Academy of Sciences, Beijing 100190, China
University of Chinese Academy of Sciences, Beijing 100049, China
Beijing Engineering Research Center of Solar Thermal Power, Beijing 100190, China

Ke Yu, Yingying Han, Quanxi Fang and Qiao Hu
Royal Tech CSP Limited, Changzhou 213163, Jiangsu, China

Dongchang You, Qiang Yu and Feihu Sun
Key Laboratory of Solar Thermal Energy and Photovoltaic System, Chinese Academy of Sciences, Beijing 100190, China
Institute of Electrical Engineering, Chinese Academy of Sciences, Beijing 100190, China
School of Electronic, Electrical and Communication Engineering, University of Chinese Academy of Sciences, Beijing 100049, China
Beijing Engineering Research Center of Solar Thermal Power, Beijing 100190, China

Zhifeng Wang
Key Laboratory of Solar Thermal Energy and Photovoltaic System, Chinese Academy of Sciences, Beijing 100190, China
Institute of Electrical Engineering, Chinese Academy of Sciences, Beijing 100190, China
School of Electronic, Electrical and Communication Engineering, University of Chinese Academy of Sciences, Beijing 100049, China
Beijing Engineering Research Center of Solar Thermal Power, Beijing 100190, China
Joint Laboratory, Institute of Electrical Engineering, Chinese Academy of Sciences and Guangdong Five Star Solar Energy Co., Ltd., Dongguan 523051, China

Index

www.ingramcontent.com/pod-product-compliance
Lightning Source LLC
LaVergne TN
LVHW081014210926
848485LV00004B/863
* 9 7 8 1 6 4 7 4 0 4 3 5 2 *